SCIENCE UNFETTERED

Series in Continental Thought

Science Unfettered

A Philosophical Study in
Sociohistorical Ontology

J. E. McGuire
Barbara Tuchańska

Ohio University Press
ATHENS

Ohio University Press, Athens, Ohio 45701

© 2000 by James E. McGuire and Barbara Tuchańska

09 08 07 06 05 04 03 02 01 00   5 4 3 2 1

Library of Congress Cataloging-in-Publication Data
McGuire, J. E.
    Science unfettered : a philosophical study in sociohistorical ontology /
James E. McGuire, Barbara Tuchańska
        p. cm. — (Series in Continental thought ; 28)
    Includes bibliographical references and index.
    ISBN 0-8214-1350-3 (alk. paper) — ISBN 0-8214-1351-1 (pbk. : alk.
paper)
        1. Science—Philosophy. 2. Science—Social aspects. I. Tuchańska,
Barbara. II. Title. III. Series.

    Q175 .M418 2000
    501—dc21

                                                    00-042765

Only by going too far will you go anywhere at all.
—Francis Bacon after T. S. Eliot

To the indomitable Irish and the unvanquished Poles.

CONTENTS

ACKNOWLEDGMENTS

THIS BOOK WAS CREATED within a dialogue, not only between the two of us, but also with our friends, who never stopped encouraging us, our antagonists, who never stopped criticizing our ideas, and our students, who never stopped asking questions. We are truly grateful to them all. We want to thank the Department of History and Philosophy of Science at the University of Pittsburgh and the University of Łódź for financial support toward the publication of our book.

INTRODUCTION

FOR REALISM, A DISTINCTION between what is so in itself and what is merely so for us is a fundamental truth beyond dispute. This view supposes that we are "worldless" subjects, bound within the subject/object opposition, spectators whose access to an objective world is through the mediation of inner states and representations, isolated cognizers remotely looking out upon the passing show. However, the distinction between what is so in itself and what is so for us is drawn from within our understanding of being and not, as the realist supposes, on the basis of an independent world from which humans could be entirely absent. In *Conquest of Abundance,* Feyerabend states the implications aptly: "being in the world we not only imitate and constitute events, we also reconstitute them while imitating them and thus change what are supposed to be stable objects of our attention. This complex interaction between what is and the (individual and social) activities leading up to what is said to be makes it impossible to separate 'reality' and our opinions in the way demanded by the realist" (p. 128).

We make a difference: there is no way things are independently of our practices and conceptual choices. What is known in acts of understanding is neither a pure subject nor a pure object, but a subject-in-knowing-an-object. Things come to be thus and so as correlatives of our practice; what is brought forth was not there previously, but its coming into existence is not entirely our free making. If we do not make up the world, neither do we encounter extramental objects in themselves and then subjectively project values and meanings upon them.

We are always already in specific situations that set our available possibilities and frame our moods and modes of being; we are always already involved with others and alongside nonhuman entities within the world; in short, our being is being-together-in-the-world to which we belong essentially. This means that our primordial access to the world is through concernful dealings with things, dealings that are direct involvements of manipulating and using them and not just bare perceptions of their presence and their concurrent states. We and our world are complementary and equiprimordial.

Against this background, derived critically from Heidegger and Gadamer, we argue for a communal approach to human cognition. To this end, we elaborate a fundamental sociohistorical ontology that articulates the underlying unity structuring our being-together-in-the-world and our constitutive modes of understanding. The ontological structures of our being are interrelatedness and participation. All entities, human and nonhuman alike, are what they are in relation to other entities: inseparably intertwined in these relations are the doings that we are. Crucial features of our fundamental ontology are sociality, historicity, self-making, and the appropriation of tradition. They allow us to characterize the many interrelated variants of the ontic-ontological circle and to illuminate scientific research as a form of practice, as an objectifying cognition constituting itself within the dynamics of history.

Today, as never before, science is the subject of three distinct disciplines: philosophy, history, and sociology of science.[1] Of course it was studied earlier. Since the seventeenth century, at least, philosophers have paid attention to scientific knowledge. But it was not until the middle of the nineteenth century that disciplines emerged for studying science from an intellectual, professional, and institutional standpoint. In the present century these disciplines have increased in scope following the immense growth in scientific knowledge. Knowledge of science has spawned ideas and approaches that are composed and recomposed in kaleidoscopic fashion. Routinely, philosophers, historians, and sociologists of science find recently established truths questioned or rejected. It seems as if these disciplines harbor a cornucopia of perspectives, problems, and solutions that defy ready synthesis. If unification seems impossible, this is not a simple matter of excessive diversity. Anyone who reads carefully in these disciplines finds gaps, missing arguments, internal incoherences, unclarified ideas, unnoticed assumptions, and modifications often running in opposite directions. On a less sanguine note, it may well be that analytic philosophy, history, and sociology of science have now exhausted their initial possibilities and need profound modification if they are not to turn into empty formal exercises or projects of blind data gathering. This judgment, in our opinion, is especially plausible in reference to analytic philosophy of science: as it is now practiced in the

Anglo-American tradition, its *philosophical* character has largely been eclipsed.[2] The (re)establishment of that character is necessary if science is to be understood as an element of modern life.

Not surprisingly attempts to fill gaps, to work out misunderstandings, to advance arguments, and to settle controversies constitute a Sisyphean endeavor that we refrain from undertaking in this study. Indeed, the scope of this book prevents us from giving a detailed picture of all three disciplines, and we restrict ourselves to some salient points in chapter 1.[3] Their achievements are of course happily relied upon when they are found relevant to our project, which is to understand — from a philosophical standpoint — the social and historical nature of science, more precisely, its sociability and historicity. Certainly, these aspects of science have been recognized (a step not easily taken by analytic philosophers of science), and largely studied both as empirical phenomena and philosophically by the phenomenological and hermeneutic tradition. We locate our project within a philosophical reflection on science and take seriously its philosophical character and its reference to science as a sociocultural and historical phenomenon.

This task requires overcoming the limits of analytic philosophy of science with respect to its conceptions of the scientist, scientific cognition, and the objects of science. It requires going beyond the subject/object dichotomy that underlies scientific cognition and most modern philosophy. Certainly the subject/object opposition has been problematized. The names of Kant, Hegel, Nietzsche, Dewey, Husserl, Heidegger, or Gadamer come readily to mind. Their aim, however, was purely philosophical: to reveal elements underlying both the subject and the object of cognition predominately from the point of view of an individual cognizer. However, apart from the work of Hegel, Heidegger, Cassirer, Collingwood, and Gadamer, few attempts have been made to situate human cognition ontologically within the social, the cultural, and the historical. Our study aims to address these dimensions of cognition, especially in regard to scientific cognition.

Let us consider some preliminary questions. What does it mean to problematize the subject/object opposition, and why is it important? Can this be done while preserving the key elements that are essential for establishing objective scientific knowledge? To problematize the subject/object opposition is not to abandon it. It is to reject the tacit assumption that the subject and the object are already fixed, as the cognizing subject and the cognized object, prior to entering the very relation of cognition itself. In this endeavor, neither the objective nor the subjective should be presupposed as given. "Both subject and object are derivative and secondary, in that both precipitate out of the more primordial unity of being at home in the world."[4]

To problematize the subject/object opposition is not to embark on a metaphysical search for the ultimate, ahistorical foundations of cognition. On the contrary, it entails undertaking an ontological inquiry that transcends the realms of the sociocultural and the natural as objectively understood. Put otherwise, from the problematizing standpoint, both terms of the relationship are considered, not as given, but as constituting themselves in the course of cognition itself. Consequently, it is not nature, or mental processes, or the method (of science) that constitutes the subject of (scientific) cognition, but the very activity of cognizing. This implies an answer to the question why it is important to problematize the opposition. Problematization allows us to reveal and to discuss the dynamics of cognition that constitute its subject and object, as well as the sociohistorical dimensions of cognitive activity. The account of cognition we intend to develop does not destroy scientific cognition and objective scientific knowledge. On the contrary, it aims at revealing them as possible (and real) within the subject/object nexus.

In order to problematize the subject/object opposition, we critically mobilize the resources of German philosophy. In particular, we refer to Heidegger and Gadamer, who provide the ontological and hermeneutic basis from which we launch and develop our position. From that standpoint three interrelated elements can be problematized: the subject/object opposition; the contention that both a subject and an object of cognition can be seen, as it were, as objects; and the notion that there is a pre-given basis of human cognition. Certainly, the myth of the given has been exploded from within the analytic approach to science together with the idea that there is a nonconceptual element in cognition. There is, even so, no concerted effort to problematize the subject/object opposition itself, despite the work of American pragmatism and the efforts of German philosophy. Once we establish the basis of our hermeneutic ontology, we will show how it embraces core aspects of human cognition and relates to objectified forms of cognition central to the scientific enterprise. To do this demands a careful analysis of variously interrelated features of the ontic-ontological circle. The argument for this larger picture begins in chapter 3.

Our reference to German philosophy locates our project within the field of a hermeneutic philosophy of the natural sciences, or—as we prefer to see it—in its very close vicinity. We refer to that tradition in German philosophy that sets science in critical opposition to philosophy while simultaneously undertaking an understanding of science as an element of life and culture. It "does not consider scientific research as an end in itself but, rather, thematizes the conditions and limits of science within the whole of human life."[5] This view is opposed to the positivist idea that scientific cognition provides the fundamental or only way of discovering how entities really are. Moreover, it criticizes the tendency to impose a scientific way of thinking, with its characteristic results, on our understanding of what there is. After all, as Coffa aptly observes: "Science solves the problem of reality by telling us what there is; but science does not tell us WHAT it tells us when it tells us what there is."[6] Clearly there is a job for philosophy to do, and it cannot do that effectively if it simply models itself on the cognitive stance of science as such.

It is crucial that our philosophical reflection on science should not be misunderstood. We are not offering a philosophical account of specific sciences or types of science: for instance, our work is not intended as a contribution to genre such as the philosophy *of* biology, or *of* physics, or even *of* the natural sciences. Nor does our work fall within the purview of the sociology of scientific knowledge. The reader will not find, for example, a microscopic account devoted to the experimental setting of a particular laboratory as in Rheinberger's recent study or of a particular scientific community as in Galison's account of microphysics.[7] Work of this sort is essential if our grasp of types of science is to develop. But essential, too, is a general grasp of science as a special form of human practice that realizes itself within the changing sociocultural and historical settings of human existence. This task requires a discourse and a philosophical landscape that differs significantly from those appropriate to philosophical investigations into the methods, procedures, and truth-claims of science or of particular scientific disciplines. It is our aim to provide that discourse and to illuminate that landscape.

A crucial issue for us is the task of defining our project against the background of the hermeneutic philosophy of the natural sciences.

There are two reasons why this is particularly difficult. First, in company with central currents of continental philosophy, a hermeneutic philosophy of the natural sciences "from the start sees science as an institution in a cultural, historical, and hermeneutical setting."[8] This is a perspective that we share. Second, the classical thinkers, to whom hermeneutic philosophers of the natural sciences refer, offer different views that in some respects conflict. If someone refers—as Heelan does—to Husserl, Merleau-Ponty, Heidegger, and Gadamer, the hermeneutic phenomenology of the natural sciences that results is substantially different from the pragmatic hermeneutic phenomenology of Crease, based on Dewey, Husserl, and Heidegger, or from Kockelmans's view based on Heidegger and Gadamer.[9] Nevertheless, there is a basic way in which our project needs to be characterized against the background of the hermeneutic philosophy of science: our aim is not to contribute to it through elaborating and developing its view of science as a field of interpretive practices. If we adopt Heelan's view that a philosopher of science in the phenomenological or hermeneutic tradition "would do research into constitutional problems, human embodied subjectivity, and world (life world) as reality—problems that do not enter into the purview of analytic philosophy of science,"[10] we can say that we not only undertake an analysis of the subject (human embodied subjectivity) and the object (life world), but, most importantly, we undertake an analysis of their underlying ontological features and structures. Unlike hermeneutic phenomenology of science, we do not take for granted that there is a phenomenon of scientific cognition that requires a new—hermeneutic—interpretation. What we attempt, on the contrary, is an ontological inquiry into the human way of being from which scientific cognition stems and to which it belongs. In other words, what we attempt is a fundamental ontology of science inspired by Heidegger's *Being and Time:* in particular, by his idea of the priority of being over beings, and by his way of philosophizing—that is, by his way of articulating an ontological analysis of being. Heidegger's way of philosophizing is existentialist, phenomenological (in a rudimentary way), historicist, and hermeneutic. The last two stances provide crucial perspectives for our analysis of science. It is an ontological analysis in the sense that we do not presuppose phenomena such as values, life worlds, culture, history, or society as they are and as they affect the constitution of scientific knowledge; on the contrary, we want to discuss them ontologically, as basic components of the human way of being.

Our "concern with science" is not simply "to interpret its historical conditions within human society";[11] it is, rather, to ask what science *is* within the human way of being. Therefore, Heidegger's ontology and Gadamer's hermeneutics provide a perspective that enables us—after criticism and substantial modification developed in chapters 2, 3, and 5—to articulate in chapters 4 and 6 a sociohistorical ontology of science understood as a social and historical phenomenon.

In general, our approach to the thought of the classical German philosophers represents an attitude described by Gilson: "Even those who did not want to learn from them what to think have felt that they could at least learn from them how to think."[12] Indeed, our approach to Heidegger's ontology fits Bernstein's observation: "we might think with Heidegger *against* Heidegger." [13] In other words, we will remain faithful to the spirit of German philosophy, but not necessarily to the letter of its doctrines.

Although we follow the view of German philosophy that reflecting on science philosophically is not parasitic on science itself, we part company with German philosophy when it claims that philosophy's contribution to human understanding is superior to that of science, and we combine its critical and culturalist approach to science with insights into the workings of science achieved in Anglo-American philosophical and sociological thought.

Of course, the inevitable challenge to our project is this: do we not risk undercutting the cognitive status of science and thus endanger its claim to deliver secure, reliable, and warranted knowledge? And will this not destroy, or considerably weaken, the firm ground necessary to distinguish science from pseudoscience? Our response is simple. It is science that defends itself against those who would challenge its claim to produce unique, reliable, and secure knowledge; it is science that articulates its canons of evidence and its justificatory procedures; and it is science that distinguishes itself from pseudoscience. Insofar as philosophers enter the fray in defense of science, they need to look carefully at the evidential patterns and justificatory strategies of the sciences themselves, a task confirmation theorists have largely failed to do. To enter into such matters is not the aim of our study. It does not articulate yet another philosophy of science, as standardly understood in the Anglo-American tradition, whose sole aim is to lay out in abstract and formal fashion the conceptual structure of scientific explanation, theory, or method. On the contrary, we question the adequacy of the

view of science this very outlook presupposes as well as the philosophical sufficiency of its techniques. So our concern is different; its
task is to provide an ontological basis for understanding science as a
modern sociocultural phenomenon. It is no part of this concern to
claim that our approach to cognition answers justificatory questions regarding the warrant of scientific knowledge given that these are standardly posed in philosophy of science within the subject/object nexus,
the very nexus we seek to problematize but not reject.

In constructing an ontology of science, we follow Heidegger's distinction between the ontic and ontological approach to what there is.
Ontic studies remain within the subject/object dichotomy and are
concerned with entities and facts about them or with events and processes which are understood objectively. Ontological inquiry problematizes the subject/object dichotomy itself and is concerned
primarily with the being of entities, and in this sense it directs its critical stance to a more basic level. Our aim is to articulate a fundamental
ontological analysis of the features, structures, and conditions of the
subjective and the objective, as well as the entire spectrum of relations
between them, which stretches from being with each other to acts of
cognizing the object by the subject.[14] These ontological features,
structures, and conditions are basic, constitutive, and necessary for
any form of human activity, including scientific research. For Heidegger, they are located in the existential structures of an individual human being; for us, in contrast, they are located in the structures of our
communal being understood as practice. And although we agree that
inescapable finitude belongs to the conditions of our being, we think
that individual existence is also conditioned by human plurality and
interrelatedness, as well as by the historicity of human practice. It is
these features that make our individual life socially and historically
situated. Importantly, ontological features, structures, and conditions
of being are not given, as such, within the empirical realm; they underlie it and must be revealed through hermeneutic analysis. This is to say
that they do not exist by themselves awaiting description by pure metaphysics; they are embodied in historically local and sociocultural phenomena. Therefore, in studying them we must interrelate ontological
features of being as a whole with particular ontic phenomena in a
hermeneutic and dynamic way, within an ontic-ontological circle that
possesses historical dimensions and expresses itself in various connected forms. This can be made clear by formulating a striking paral-

lel with Kant's dictum concerning the interdependence of the categories and the forms of intuition: the ontology of sociohistorical being without history is empty, and history without ontology is blind.[15] Our ontological approach to sociohistorical being concentrates on its historicity and reveals social self-constitution as the ontological structure of historicity.

To follow Heidegger's ontic-ontological distinction means, in particular, that we accept three main and orienting ideas of the hermeneutic philosophy of science: the priority of meaning over technique, the primacy of the practical over the theoretical, and the priority of situation over abstract formalization. We do not accept these as an initial framework within which we study science, but in order to undertake a study of those ontological features and structures that underlie, respectively, interpretive practices in science, "engagement with the world which is prior to the subject and object separation," and the fact that "scientific knowledge can never completely transcend these culturally and historically determined involvements."[16]

Mobilizing an ontological perspective, together with the idea of self-constitution, allows us to conceive science as unfettered. Although, from the ontic perspective, science must be understood as structured by different factors—for example, methodological values and rules, protocols of social interaction, constraints imposed by what is objective, and historical conditions within human society; from the historical-ontological perspective scientific research appears as an element of our global self-making. Indeed, since nothing is absolutely prior or given to human self-making, science is a process of self-constitution: in its very development it constitutes itself as science. Construed from the ontological standpoint, science is not defined by an essence, but by its own self-constituting practice.

Our book is a contribution to an understanding of science as knowledge or cognition. It is also a contribution to a hermeneutic sociohistorical ontology. The first aim is addressed by elaborating ontological categories that pertain to the structures and conditions of being; the second by developing the historical and social categories, understood both ontically and ontologically, central to the philosophical views of Heidegger and Gadamer. In our view, the activities of science cannot be addressed philosophically (ontologically) in the absence of historical categories, such as historicity or situatedness, or social categories that pertain to human relations and practice as the human way of being. On the other hand, if scientific activities are addressed empirically (onti-cally) without thematizing their sociability, historicity, and cognitive na-ture philosophically, they may be objectively described and explained but not understood hermeneutically according to the role they have within modern life.

The distinction between objectivist description and explanation and hermeneutic understanding is crucial to our approach. The hermeneutic character of our ontology indicates that it is an under-standing of ourselves as elaborated by ourselves, influenced as we are by the historically limited structures of our being; in particular, this understanding is influenced by existing (philosophical) concepts and prejudices formative of our thinking. We refer to this interrelation be-tween (our) epistemological views and the ontological conditions of our situation as a hermeneutic-ontological circle, which will be clarified later.

To characterize our position in more detail, let us separate four ways of doing philosophy: (1) scientific philosophy, which is descriptive and explanatory in a way that mimics the features that it attributes to science; (2) political philosophy that undertakes critical and normative projects, instances of which are the positions of Marx, Nietzsche, or the Frankfurt School; (3) metaphysical philosophy, which attempts to show "the absurdity of regarding a relative world as a self-sufficient re-ality" and opens up an interpretation of the world as conditioned by the Absolute;[17] and (4) philosophy construed as the search for an under-standing of ourselves, of our being and the reality within which we re-side.[18] As we have already indicated, we adopt this last perspective but accompanied it with the pluralist and relativist attitude of the jester, which Kolakowski contrasts with that of the priest.

The antagonism between the priest and the jester results from be-ing "for or against a search for support in absolutes."[19] In epistemol-ogy, the attitude of the *priest* is revealed in a search for "an immutable principle of cognition that is inaccessible to criticism and that consti-tutes an infallible prop for our thinking"; that is, in a search for "the monistic concept of cognition," and for "intellectual dominion over reality with the help of a set of laws, both the highest and most elemen-tary, which explain everything."[20] Even in science itself, as well as in any scientistic philosophy, the attitude of the priest may be found. It is instantiated in the view that science, as an objective, logical, and me-thodic enterprise, gives us access to reality as it is in-itself while escaping the deceptions and distortions of psychological, cultural, or social influences. Accordingly, scientific knowledge is seen as the only form of knowledge, or minimally as the chief candidate for knowledge, in its best possible form. On this view, the role of philosophy of science is to perform foundational and normative tasks.

The jester, on the other hand, "exposes as doubtful what seems most unshakable, reveals the contradictions in what appears obvious and inconvertible, derides common sense and reads sense into the absurd." The jester recognizes that "what is sacred today was paradoxical yesterday, and absolutes on the equator are often blasphemies at the poles."[21] This sort of critical, anti-absolutist philosophy—as elaborated by Nietzsche, Heidegger, Gadamer, Cassirer, Foucault, or Ricoeur, and exhibited by the work of Arendt, Bernstein, Dreyfus, Elias, Fackenheim, Kolakowski, Margolis, Rorty, and others—is motivated "by distrust of a stabilized system," by "an active imagination defined by the opposition it must overcome."[22] The jester's approach to science "denounces science" and claims that a scientific image of the world "does not represent an organization actually inherent in the world itself."[23] The jester also shows that the naturalist philosopher's search for the foundation of human uniqueness in nature is not really different from the search undertaken by the religious philosopher: both annihilate themselves as individuals because they try to define themselves and their behavior by reference to something they are not.

In most forms of contemporary philosophy the adoption of the anti-absolutist, pluralist and relativist, attitude follows from the recognition of human finitude that precludes any search for definite foundations.[24] Significant elements of the jester's attitude can be found in various philosophical and sociological studies of science, from the views of Continental philosophy, Feyerabend's tirades, Kuhn's historical constructions, Rouse's analyses, and Polanyi's self-reflective view of scientific activity, to forms of social constructivism in the work of Fleck, Knorr-Cetina, Latour, Woolgar, Mulkay, and others. From their viewpoint, scientific knowledge is but one form of knowledge and not necessarily the best; it is a social product, influenced by social context and historical change. Moreover, for them, science does not reduce to knowledge and is largely a social enterprise which does not justify itself, and is not good simply because it is. It may be studied as any other sociocultural phenomenon that emerges from human activity. Furthermore, it may be genetically related to other, more fundamental or simply earlier, spheres of culture. Indeed, the concern of Continental philosophy has been largely to situate the sciences with respect to other social phenomena.[25]

Apart from the guardians of the absolute, who essentialize being, nature, logic, language, self, etc., and who picture all differences in terms of abstract dichotomies, and the anti-absolutists, who refuse to essentialize anything, there are philosophers of the middle way, who look for reconciliation and agonize over details that do not fit into a systematic whole. Also there are those who believe in the possibility of abolishing existing dichotomies and first principles in order to create philosophy without assumptions, to initiate an absolutely zero point from which to launch all thinking. That the project of philosophy without assumptions is impossible is a lesson painfully learned from at least two cases: Husserl's phenomenology and logical empiricism. It is not so obvious that the middle way option is also a bad bet. Those who adhere to this option cannot be sure that the different vocabularies they strive to unite are in any way harmonizable, that the positions they want to combine do not harbor invisible yet contradictory presuppositions, and that the different aspects of things they intend to consider are in fact visible from their perspective. Anyhow, the middle way option seems to us to hold little promise. A disseminated form of this option is present in the burgeoning literature of the new science studies. These studies tend to remain "on the surface" of discourse and subject matter and link together different vocabularies in the name of a new lexical gestalt, even though those vocabularies may presuppose divergent ontologies.

To accept the anti-absolutist attitude does not necessarily result in a nihilistic demolishing of all truths, rules, and values. It may embrace an understanding and appropriating of tradition in order to build the relativists' own position, their own "truths," even though they are critical of those truths that claim to be absolute. They do not proclaim that their conceptions embrace absolute or ultimate truth; they see them as arising contingently as socially conditioned and historically limited ideas. Despite this, they may certainly claim that their conceptions illuminate an understanding of ourselves and the world we inhabit. As Wachterhauser emphasizes, some hermeneutic thinkers occupy such a position; they "would argue that there are no knowledge-claims that can claim to be true *sub specie aeternitatis,* but they claim only to be 'true' in a 'pragmatic' sense of being the best solution at the present time to a problem that has been generated out of a set of historically mediated understandings, interests, and practices."[26] So every piece of knowledge is but "a way of seeing things from the standpoint of a historically mediated set of concerns and preunderstandings."[27] Moreover, if—as we argue later—philosophy is produced within the dynamics of dialogue, a relativist, who would like to remain a perfectly consistent skeptic, cannot take part in its production, because the price of his scrupulous consistency is either silence or madness. The reasonable position is *inconsistent skepticism* and *inconsistent universalism.* The inconsistency stems from two sources: an attitude that characterizes, according to Kolakowski, European culture in general— namely, one's uncertainty about one's own standards[28]—and the fact that "*the world of values is not logically dualistic,* as opposed to the world of theoretical thought": two values may exclude each other without ceasing to be values, whereas truths cannot contradict each other and both remain true.[29]

We declare ourselves in favor of such inconsistent skepticism and universalism. Moreover, we believe that these are assumptions any relativism must presuppose to be consistent. Notable examples in current literature are Rorty's ethnocentrism that adopts the standpoint of twentieth-century Western liberal intellectuals and the Eurocentrism of Kolakowski.[30] In regard to this issue we are inclined toward Kolakowski's position.

Thus, in pursuing an understanding of science we remain antidogmatic and open to influence as well as to change. It does not follow that we are willfully eclectic and welcome every possible philosophical and ethical position: at the heart of our endeavor there lies a clear choice of values and not simply axiological (that is, ethical) indifference. We are self-uncertain, but not paralyzed by awareness that a choice between values in conflict is unavoidable. Moreover, we believe that our position is universal in the sense that it is suitable for (self-) propagation, though not, of course, by any and every means. We believe that voting for inconsistent skepticism and universalism and accepting the burden of problematizing fundamental oppositions are necessary for undertaking a philosophical understanding of science as part and parcel of our life. And we critically appropriate Heidegger's and Gadamer's belief that their standpoint is able to create tools of self-criticism. "The real power of hermeneutical consciousness is our ability to see what is questionable."[31] Hermeneutic tools reflect on their basis of understanding, and from this stems their vitality and ability to embrace change.[32]

In particular, from a hermeneutic standpoint the philosophical perspective of *a participant* in understanding can readily be inferred. To understand science from this perspective does not entail that it be perceived from within, in the manner of an active scientist-philosopher, such as Fleck or Polanyi. Contrary to Feyerabend, it is not necessary to be a scientist to understand science as a "form of life."[33] This claim is yet another exposition of insiderism, long ago criticized by Merton as being a version of social solipsism. According to insiderism, only members of a group or a collectivity (ethnic minority, women, homosexuals, etc.) have a monopoly on genuine knowledge about their group.[34] A weak version of insiderism, which states that one cannot have participants' knowledge (about science) without participating (in science), is tautological. Here we have in mind a possibility that is opened for an outsider. This is the perspective of a participant in the form of life of Western societies that is, to a great extent, constituted by modern science, its types of knowledge, and the practical effects of scientific research. Clearly, philosophers can attempt an understanding of science as a crucial element of the form of life in which they participate. Philosophy—as Gadamer emphasizes—can give "an account for our life as shaped by science," and it alone can fulfill this task because science itself is irresponsible "in the sense of its incapacity and its lack of any perceived need to give an account of what it itself means within the totality of human existence, or especially in its application to nature and society."[35]

Adopting such a position in reference to science has important consequences. First, in studying science, philosophy can contribute to "a better self-understanding of what we are as knowing agents," and this better awareness can "help us to overcome the illusions of disengagement and atomic individuality that are constantly being generated by a civilization founded on mobility and instrumental reason."[36] Second, the departure point for a philosophical understanding of science is not this or that position as it already exists inside the philosophy of science. It must be a philosophy attached to a form of life that is the philosopher's own form of life. Third, the philosopher's understanding of science will undoubtedly be different from the scientist's. This does not in any way preclude a dialogue between them. Fourth, scientific cognition, as conceived from the perspective of modern forms of life, does not reduce to linguistic and logical operations, nor to mental acts or physical manipulation. These features are interrelated, and—as we shall argue—theoretical holism needs to be replaced by practical holism. Scientific research is not an activity of an isolated subject of cognition; it realizes itself through interactions among scientific practitioners, in particular, through the ongoing dialogue of scientists.

CHAPTER 1

Ontic and Ontological
Perspectives on Science

NOWADAYS THE VISION of the scientist as a pure seeker after truth is largely viewed as a myth. Even a cursory glance at developments in the philosophy, the sociology, and the history of science reveals a tendency either to weaken or eliminate the traditional epistemic picture of science. True, there are philosophers who remain reluctant to accept the view that styles of reasoning prized in science have significant reference to social forces. Yet there are few who view scientific rationality in absolute terms. The tendency to weaken the epistemic picture of science is part of a larger movement in contemporary thinking to demythologize Reason. Reason is now seen in concrete, historically indexed terms that refer to the particular circumstances within which it operates.[1]

In philosophy of science the task of de-idealizing science has begun. The belief that it can establish ultimate, necessary, and certain Truth is skeptically received; indeed, to claim that scientific knowledge is ahistorical has either been rejected or seriously qualified. Moreover, the claim that science is methodologically driven by individual cognitive subjects is no longer a central epistemic stance.

Sociologists, always more eager than philosophers to view science as located empirically in particular social contexts, approach it as a continuous part with the rest of culture. Seeing it in this light makes us "more ready to use it for what it is, to value its insights and wisdom within rather than without the political and cultural process."[2] This perspective leads in a positive sense to the disenchantment of science. Science is viewed as an enterprise pursued in an arena in which knowledge is socially constructed. From this standpoint, it gains the flesh of social relations and interactions; of conversational exchanges and controversies; of networks, agonistic fields, and competition; of quests for credibility and scientific authority; as well as for prestige and recognition.

Finally, historians have begun to destabilize science. Individual scientists are seen within historically local contexts, which are routinely approached through historicized categories. Evident is a deeper appreciation of how the contingencies of history affect science together with the historical indexed character of the period in which it develops.

Undoubtedly, these efforts have produced a more realistic picture of science that better serves the need to understand it as part of our modern condition. It is not, however, a unified picture, and it leaves aside some important questions that need to be posed. It is essential to our position to argue that ontic perspectives on science constructed in philosophy, sociology, and history need to be supplemented by an ontological picture if science is to be seen as an essential part of human life as a whole.

TWO PHILOSOPHICAL APPROACHES TO SCIENTIFIC COGNITION

There are two main pictures of cognition in philosophy: the conception of an individual subject cognizing an independent object, and cognition understood within the framework of hermeneutics, a view we accept.

The first picture has a simple structure. It presupposes that subject and object stand in opposition and that within this framework epistemological issues can be raised and considered. Furthermore, the cognitive subject is viewed as standing in a cognitive relation to an independent object of cognition. Traditionally, philosophy attempts to bridge ontic "gaps" by occupying a chosen side, viewing it as unproblematic and inquiring into the other side. In such cases the adequacy and appropriateness of epistemic representations are assumed and no attempt is made to problematize the opposition itself. Within this dyadic picture the subject and the object are open to a variety of interpretations.

According to *subjectivists,* the subject is the key element in establishing claims to know, the object is reduced to the subject, and cognition, as the relation that connects them, is explained away or absorbed into the intrinsic (mental) processes of the cognitive subject. Phenomenalism is a classic example: it deals with the dualism of experience and the world by constructing the world out of individualized experience. In philosophy of science, logical empiricists initially took the subjectivist stance; but they then turned toward naturalism and assumed a realist stance when they accepted the physicalist interpretation of protocol sentences. Most analytic philosophers now take an epistemically realist stance. Although the structure of the subjective is variously articulated, opposition between mental states and representations and outer objects that are schematized or represented is typically presupposed. In this light, the statement "The cat is on the mat" (uttered on direct confrontation with the animal) is a paradigm instance of a cognizer who stands in direct cognitive relation to an occurrent object.

A different, nonsubjectivist standpoint is evident in *methodological rationalism*, a view developed by Popper, Lakatos, Laudan, Musgrave, Watkins, and Worrall. They accept a strict separation of the subjective from the natural and reduce the subject of cognition to the objective, rational methods of scientific cognition. For them, the scientist is not so much a lover of wisdom and truth as he is a "cognitive machine" driven by the rational imperative of searching for truth (or for empirical adequacy, or explanation, or predictive success). In this picture, the issue of the subject and the object fades and goes to the periphery, and cognition itself becomes the "object" of attention. This posture is also evident in the cognitive sciences with their modeling and computational approaches to knowledge, thought, and language.[3]

Finally, naturalists, such as Quine, and behaviorists like Dennett, reduce the cognizing subject to what is considered to be natural. When a naturalist stance is taken, direct causal connections are posited between pre-given objects and their "copies" in the mind. The most draconian reductionist strategy in naturalism is eliminative: it simply reconceptualizes its subject matter replacing without argument references to non-natural entities with those to natural entities. Still draconian are attempts to reduce the mental to the neurological, the character of which is taken to be uncontroversially natural.[4] This view simply assumes that accounts of the mental or the cognitive couched in epistemic or psychological terms are translatable into neurological terms without remainder or loss of meaning. Epistemic naturalists transform the cognitive subject into an objectified system to be studied in the self-same manner as the objects of the physical sciences or according to an evolutionary picture if they accept an adaptational story of how human biopsychology emerged from units of evolutionary change producing en route specific modular structures inside the brain.[5]

Objectifying strategies generally presuppose that the cognitive subject is self-transparent,[6] that it can cognize itself entirely either as a natural object, according to the methods of science or the latest information that science provides, or in terms of its own internal mental processes. Naturalists provide an ontic picture of human cognition and understand their own enterprise as the construction of theories that "serve to explain the phenomenon of science itself in roughly the way that scientific theories explain other natural phenomena";[7] or, more loosely, as the elaboration of hypotheses that may be articulated and rationally criticized. In both cases their aim is to search for a nomological picture of science, a universal description, explanation, or rationalization of scientific activity in its varying contexts. As Nickles states: "it is certainly an idol of the tribe of philosophers to stress unity and generality."[8] Make no mistake, however; to the extent science treats human beings in an objective way, it throws considerable light on our physical, neurological, biological, or evolutionary characteristics. This is fine. But the fact that scientific achievements are goods in themselves provides no automatic reason for straightforwardly modeling the entirety of human cognition on the techniques and assumptions of science.

Why is this so? An answer emerges if we grasp the consequences of transforming the human subject of cognition into a natural object. This move assumes that our conscious self-awareness, our sense of ourselves as cultural, social, and historical beings, is an illusion. This is apparent from the perspective of evolutionary theory, according to which the immediate elements of human experience are simply by-products of evolutionary adaptations. It is adaptation that has made the neurological and biopsychological structures of our makeup what they in fact are. On this view, science, religion, language, history, culture, and philosophy are late arrivals on the scale of evolutionary time; they are "spandrels," contingent yet dependent by-products of the evolutionary process itself. The issue is whether such an interpretation is justified and whether it tells us all we need to know. Are not history and culture autonomous forces? Is their autonomy not compromised if they are viewed merely as dependent products of the processes of evolution? Can we seriously attempt to reduce social and cultural domains to the hard sciences, even if for some writers reducing the mental to the neurological seems to be a genuine option? Can we speak "neurologese" about culture and society as well as in reference to ourselves?

For us, culture, history, and human consciousness are forces constituting a fundamental reality in their own right. This means, of course, that in our view reductionism is neither a threat nor a possibility. Why?

First, it is intuitively improbable that reductive steps will take us directly from the categories of philosophy, psychology, and history through neurobiological and evolutionary theories to quantum field theory understood as the fundamental science of the way things are. What nonbiased reason is there for believing such draconian reduction is possible, even in principle, where "in principle" can only mean what is "theoretically" possible for a sufficiently large computer or the divine mind? Indeed, to take such a stance is to presuppose the efficacy of reductionism.

Second, even modest reductive approaches raise serious problems. Suppose the claim is simply that sociocultural phenomena depend upon, but are not reducible to, the underlying reality of the things science tells us there are. Here thoroughgoing reductionism is not an issue, since the aim of this strategy is to take each case on its own terms. But the question becomes: according to what categories is the dependency of one level on another to be understood and how are connections between them to be articulated?[9] What decides whether it is causal or noncausal, deterministic or nondeterministic? If deterministic, are sociocultural phenomena determined completely by the natural base or are they only determined in the "last instance"? Are they merely spandrels, ghosts, and shadows of ongoing natural processes, or are they relatively autonomous? And is there reciprocity between domains? Clearly, there are as many senses of dependency or supervenience as there are interpreters who problematize the issue!

To these questions Pinker delivers an equivocal answer. On the one hand, cultural and historical processes exist, and, moreover, history's "rules" of change are dissimilar to those of natural selection.[10] On the other, these processes cannot be understood unless they are understood through the biopsychological substrate from which they arise. Thus, social and psychological phenomena and the physical world are linked by evolutionary biology, and culture is tied tightly to evolutionary forms of adaptive behavior. All questions concerning how the "mind can throw off fascinating but biologically functionless activities"—that is, throw off nonadaptive cultural activities such as art and music—must be directed to specific modules that make up the "hardware" of the brain. Pinker's basic gambit is to "reverse engineer" the evolved brain so as to capture its design function given the now-contested story of an ancestral environment in which our ancestors hunted and gathered and faced survival problems. He posits a tight fit (unconvincingly argued for) between a modular conceptualization of cognitive processes in the individual brain and a strong adaptationalist interpretation of evolutionary theory. Not surprisingly Pinker concludes that successful adaptive strategies designed for domain specific cognitive tasks have been genetically conserved. These are, of course, the strategies that solved problems in the past.[11]

It is notorious, however, the extent to which Pinker's evolutionary psychology largely ignores current work on the archaeological, fossil, and palaeoenvironmental records that refer to an ancestral environment and provide evidence for the kinds of problems our ancestors may actually have faced in the struggle to survive. Only if these records are seriously exploited can we hope to answer the *historical* question whether we are dealing with spandrels or with engineered mental solutions to adaptive problems.[12] Moreover, it is surely one thing to address the question of how brain mechanisms encode certain patterns of colors, shapes, and sounds; it is quite another to ask why a piece of art, music, or literature is good or bad. The first question seeks a causal story in reference to postulated mechanisms and looks at the "what" and the "why." The second takes an interpretative stance, presupposing an entirely different framework of intelligibility. It looks to the "how," "when," and "where" of things, to the contingencies of cultural response, to the fact that the inevitable seldom happens and the unexpected always.

If we reduce our understanding of human thought and action to the perspective of objectifying models, we obscure the ontological conditions that are presupposed by the very possibility of their articulation. Once our thought is closed within these models it is abstracted away from the ontic-ontological circle within which our life, everyday

experience, as well as scientific cognition proceed. That circle characterizes, in particular, dynamic interactions between our objectifying
experience of the world and the ontological features of our situated
existence that constitute the condition of making ontic claims about
the world. What we are placed within is a holistic situation with respect
to which objectified cognitive models are but a part. The point can be
made by contrasting our position with Sperber's scientistic view of culture. For him, our cultural life must be viewed as a product of cognitive processes understood in terms of a modular model of how the
mind works, and communication must be explained by causal chains
of representations originating in cognitive inputs and outputs. Since,
in his view, only material things (substances) exist within a causal
nexus, cultural items can have no ontic status or import in their own
right.[13] But, for us, cultural and sociohistorical items are primordial
features of human life. So in considering them, a hermeneutic perspective on human understanding is required, one that preserves the
holistic features of our way of being. The perspective must be hermeneutic in the sense that we cannot accept reductionism or objectifying
models of cognition if we hope to deal adequately with our situatedness (which includes our cognitive orientation to everyday experience). Here a hermeneutic philosophy of the natural sciences comes
to the fore, since in studying scientific cognition it does not undertake
the objectification of the subject or the object of cognition, but problematizes that opposition itself. From our perspective, therefore,
Sperber's position is simply an ontic picture embedded in the sociohistorical ontology of our situated existence and is itself in need of
interpretation.

In general, a hermeneutic philosophy of the natural sciences rein-
troduces "history and culture into the philosophy of the natural sciences" and helps rid us of the need for both "postulating a final,
hidden truth" and for "calling truth a matter of arbitrary convention."[14]
It treats science as "a *form of human culture*" constituted by inquiry and
the search for meaning.[15] Within the hermeneutic philosophy of the
natural sciences there are two main approaches and areas of study.

The first area is a "weak" hermeneutics of communication that
attempts to interpret science (scientific knowledge and research) in
terms of interpretative practices or discourse or narrative situations. It
can study "the literary, graphics, and mathematical materials (textual
material in the strict sense)."[16] Its basic presuppositions are that

scientific knowledge "involves a disclosure (saying) of something to somebody," that "it deals with meanings,"[17] and that scientific inquiry is itself hermeneutic because "it is a search for a theoretical meaning to be fulfilled in experience."[18] Concentrating on meanings, the "weak" hermeneutics of science understands them as shared social entities embodied in language and fulfilled in public experience,[19] and it attempts to study "the circumstances of continuity and change in the historical transmission of scientific meanings," discontinuities of meaning, and the coexistence of new and old meanings.[20] Within "weak" hermeneutics one can also distinguish an attempt to see within the natural sciences, in the very dialogue of scientists, hermeneutic practice itself because scientists reflexively understand what they are doing.[21]

The second area is a "strong" ontological hermeneutics that undertakes a phenomenological-hermeneutic analysis of perception and scientific practices, such as laboratory experimentation and measurement.[22] These activities are viewed as meaningful situations, performances, or as historically and culturally situated hermeneutic processes. The aim of a "strong" hermeneutics of science means, for instance, understanding "how quantitative empirical methods function in science to give meaning to empirical contents, in particular, how measurement equipment plays a double role creating both theoretical and cultural meanings."[23] The "strong" hermeneutic philosophy of the natural sciences also undertakes a study "of that region which natural science discloses"—that is, of scientific phenomena or entities as present in the life-world.[24]

Hermeneutics thematizes the insight that fundamental to our way of being is the fact that modes of understanding ourselves and modes of understanding our everyday experience are always already an irreducible part of what we are. In any occurrent cognitive act we use operative concepts that function immediately and cannot at the same time be thematized. Moreover, in expressing cognitive commitment we presuppose concepts implicitly interwoven in networks of implication and implicature. This means that in acts of thinking and knowing interconnected cognitive networks are constantly reiterated as we move from cognitive situation to situation. These intrinsic traits of cognition imply our partial, but never total, reflexivity. And they raise the possibility that certain cognitive features that drive our thinking and knowing might never be completely before us. This motivates the contention that an unthematized field of thought—within which and through which we think—contributes to the conceptual structuring of intentional objects of thought.

We doubt whether the subject/object schema has adequate resources for throwing light on these implicit elements in human cognition. The schema idealizes a human knower as free from conceptual ambiguity and unaffected by perceptual error. From this standpoint, the key and sole elements in a cognitive situation are explicitly present to a knower. In contrast, from the hermeneutic perspective, human cognition suffers from thrownness, from the fact that it is always finite and situated within a local horizon. Seen in this light, cognition can never be perfect and completely transparent, "not because of imperfections in our methods but because of the finiteness of our being."[25] The cognitive subject is never completely self-transparent and fully intelligible to itself.

We come now to positive reasons for abandoning naturalist reductionism. Clearly, determinism and naturalism are unable to grasp the phenomenon of human conscious, cultural and historical experience, our awareness of being "a strange empirico-transcendental doublet"[26] and having a life suspended between being a subject that constitutes and being an object that is constituted, between the clarity of the *cogito* and the opacity of the unthought manifesting itself in the unconscious, in the other, in oneself, in the biologically determined, and in dreams, errors, madness, and illusions. If our sense of ourselves is shaped through conscious, cultural, and historical life, if we experience ourselves as actively choosing between alternatives, the reductive strategies of determinism and naturalism are not helpful. The consciousness we possess, the cultural, linguistic, and social frameworks that allow us to interact, the historical forces that impinge upon us require a different approach. What is needed is a hermeneutic ontology of science that places science in relation to these human realities, not one that attempts to reduce them away in the name of science. We therefore turn to Gadamerian hermeneutics and Heideggerian ontology. Their views provide resources for a sociohistorical ontology that will adequately address human experience, our modes of understanding, as a product of the human cultural enterprise. Of course, modes of understanding do not lie on the reflective surface of human self-awareness. They must be reclaimed and illuminated by an interpretive process that preserves a holistic picture of human reality and deals with scientific cognition. These related themes are treated in detail in chapters 5, 6 (especially under the section "Historicity of Science"), and 7.

Hermeneutic ontology does not presuppose nature as the absolute and independent foundation for what there is. Within its framework the core of our conception of reality is constituted by culture and history dialectically interconnected with self-transforming human beings. Since human choice-making implies difference, the who that each individual becomes is local, plural, and diverse, not common and universal. In our view, there is no one notion of human flourishing, no one form of life that makes us uniquely human, no one mould in which we are fashioned, and no simple convergence to a single human identity anchored in either a transcendental self or in the structures of human biology. In short, from our perspective both the content and the idea of human "nature" are open to historicity.

At the risk of simplifying our ontological view, we can say that acts of choice-making in the face of alternatives we can weigh and evaluate put us apart from the causal nexus of the natural order. Ontologically speaking, what is distinctive about human beings is the fact that they *have a world* and are not merely embedded in a physical environment in the manner of other living things. Indeed, to have a world and to have a language are inextricably intertwined; they together express the human capacity to rise freely above the environment, to take a "distanced orientation" toward it through the free and variable use of language and other symbolic resources.[27]

From the perspective of hermeneutic and historicist ontology, language and cognitive acts are not evolutionary spandrels. Nor are they simply instruments for ordering and classifying the world. They condition the possibility of knowing it and are the medium through which things become objects for us as we interact and come to grips with them. This perspective allows us to concentrate on how the knower-known complex obtains at all, rather than on the traditional Cartesian project of taking the subjective knower as given and pondering how objective knowledge is possible. The point can be articulated in this way. We are situated in a world and engaged in coping with things. We grasp our world as agents, and our disengaged descriptions and representations of it are realizable only because they are based on the ways in which we deal with things. Indeed, the very idea that things can be represented (whether linguistically or nonlinguistically) by a disengaged subject, presupposes an understanding of ourselves as historically and collectively placed in the midst of things. The real is not what is represented by the cognitive power of an abstract individual subject, and understanding does not reduce to representations that are identifiable independently of what they point to and are about. To ask if represented entities exist externally is meaningful only within the subject-object relation, and its priority evaporates once we see that our collective being is linked by language, history, practice, and the presence of things which are not our doing.

From the standpoint of our activity, entities relate to us but are not primordially grasped by our representations: the latter result from an epistemic stance that depends upon our situated modes of dealing with things.[28] Thus, the assumption (common to both empiricism and representationalism) that immediate perception alone delivers fundamental knowledge must be disavowed. The world is revealed first and foremost as a world we care about, one in which we dwell, and not first and foremost as something simply noticed from a spectator point of view—for example, "that thing over there looks red."

Our hermeneutic ontology provides an instrument for performing an essential task, that of analyzing the sociohistorical features of the larger whole within which all forms of human cognition are placed. No form of human experience or action takes place apart from symbolic mediation. Ontically conceived, when speakers typically engage in cooperative conversational exchanges, they make utterances that convey certain meanings, explicitly intending that others take them to have that content. Seen in this light, cognition involves making inferences from what is said to what is meant and reveals an inferential structure underlying interactions between speakers or groups.[29] Well and good. But clearly there is much that is presupposed and nonexplicit in such simple exchanges: conventions of language, forestructures of understanding, norms of communicative behavior and cultural meaning, the cultural formations of the society in which the speakers live, not to mention their implicit orientation toward the way things are. Indeed, the idea of intentional pickups in explicit conversational exchanges presupposes that we have more control over intentional signaling than is in fact contextually warranted. In short, even highly explicit cognitive exchanges presuppose an ontological view of how we are situated culturally and socially. None of these elements is revealed, or successfully problematized, within an ontic model. What is needed is a wider perspective that allows them to come to the surface of the overall situation, an interpretive task that hermeneutic ontology is eminently suited to perform.

From the hermeneutic perspective, an epistemic situation is not simply identical with objectifying cognition; to make that identification is a particular interpretation of that situation, and its further claim to exclusive objectivity must be qualified. Accordingly, there is no ultimate foundation for the belief that objectifying cognition has representational access to what is pre-given, because what it cannot make present here and now is its own situatedness.[30] In other words, hermeneutics rejects the possibility of privileged access to the fundamentals of cognition and knowledge, either to Cartesian subjectivity or to the Kantian transcendental conditions of possible cognition.[31]

Conversational exchanges are instances of cognitive action, whose picture we will develop. A cursory glance at the relevant literature indicates that the very notion of knowledge has shifted over a number of fronts. On the whole, it is no longer uniquely viewed in propositional terms. For many, knowledge is a process locally embodied in diverse strategies and competencies, which means it is no longer viewed as a unitary product of abstract thinking, but as a pluralistic, historically indexed phenomenon tied to particular purposes and goals. Moreover, it is now germane to speak of knowledges collectively generated and sustained within specific forms of human activity. In our view, forms of knowledge and their modes of acquisition are best construed in processual terms and seen as emerging from ampliative processes pressing continually beyond their boundaries.[32] Moreover, no one formal structure is characteristic of all types of knowledge as is assumed by computational modeling in cognitive science. We reject the tendency to treat knowledge as a substantial, thing-like commodity with the fixity of a "fact," a view cousin to the claim that it is embodied in special types of mental states. For us knowledge is inseparably directed within action with respect to both its origin and ongoing character; the materiality of praxis underlies the ideality of ideas, and certain of our cognitive abilities, such as grasping the qualities of a good wine, are not conceptual at all. Of course, to be a cognitive agent is to engage in cognitive processes, activities such as thinking, reflecting, or observing, that are elements of our participation in processual interactions with others, such as asserting, inquiring, questioning, conjecturing, discovering, and communicating.

Consider, for example, what is involved in asserting that something is the case. The standard analytic view sees assertions as entities, distinct from reality, language, acts of asserting, and the context within which something is said to be the case. On the ontic view that we accept, asserting is a performance, an act performed within language and directed telically toward another. It is epistemically derivative in so far as it presupposes a preinterpretive context of values, meanings, foreunderstandings, and conventions. To concentrate solely on "what" is asserted, the statement, is to obliterate the wider epistemic situation necessary for understanding the act of asserting. Thus, according to the view of language we accept, assertion is not simply what is asserted, the statement made, but crucially the deed performed and expressed in making it.

For the same reasons, we also reject the "Platonic" view of concepts. In this picture, concepts are treated as abstractions divorced from the inferential patterns that support them. This creates an unbridgeable gap between their semantics and the material contexts in which they are employed. On the other hand, we conceive concepts ontically in terms of the inferential roles they play in structuring various forms of human action. This provides background for grasping the historicity of conceptual configurations, a view that is central to our position. Rather than construing concepts as templates that stamp out modes of possible experience, we view them as specific patterns of inference carried by human action that stretch diachronically across the boundaries of the past, present, and future. So construed, conceptual interrelations lack the incommensurate boundaries endemic to the Platonic view, and an intuitive framework is provided for hermeneutically fusing past and present horizons, an interpretive element central to the sociohistorical ontology we want to articulate.[33]

But what does our conception of sociohistorical ontology entail? Nothing less than a thorough reconceptualization of the interrelations of the social, the historical, and the cognitive at the level of fundamental ontology. From this standpoint, little is to be gained by cobbling together a reconciliation of the received views on science and those of social constructivism in the manner of Kitcher. In Kitcher's case, the result juxtaposes various ontic models of cognition so that an integrated picture of how individual cognizers and epistemic communities interrelate is not achieved. Moreover, Kitcher resolutely sides with epistemic and explanatory individualism, despite his talk of joint cognitive efforts undertaken by epistemic communities. The idea that science is driven by "external" or "social" facts is a phenomenon he merely links to individual actors by cognitive strategies already established in terms of formal decision-theoretic models.[34] So, instead of reconciliation we have an unequal division of labor.

Indeed, his vision of rationality as a means-end activity for deciding what is rational in terms of how well selected procedures satisfy projected goals has roots in cognitive science. Needless to say, his aim is to show that his vision of what is legitimate and plausible in terms of "epistemic" rationality (the idea that there are enduring goals with respect to which scientific change can be evaluated) is better than epistemic relativism and social constructivism. Of course, he never denies that humans are subjected to social, historical, and cultural forces. Nevertheless, in Kitcher's view the noncognitive surface can be separated from the cognitive core and articulated in idealizations that explain the cognitive success of science. This strategy is designed to ensure, in fact, that the social and cultural present no obstacle to understanding scientific cognition epistemically. Indeed, for Kitcher the epistemic core necessary for defending cognitive progress in science is realism, psychological naturalism, and the correspondence theory of truth.[35] As Machamer points out, however, his case is far from proven, since it fails to rule out philosophical views of cognition epistemically more community oriented, such as coherentism and pragmatism.[36] Decidedly, cognitive individuals take center stage in Kitcher's world, and the social is nothing more than a backdrop for the playing out of scientific cognition. But the social has a more potent role in cognition than that. Indeed, there is no sharp antithesis, we shall argue, between the cognitive and the non-cognitive, and no principled way of distinguishing the cognitive from the social. This does not make them identical, or reduce cognition and thinking to forms of the social, but it does point to their inextricable and overlapping connections. Let us now turn to some ontic views belonging to the sociology of scientific knowledge that are relevant in various ways to our sociohistorical ontology.

THE SOCIAL NATURE OF SCIENCE

In globalist philosophy of science, such as Kuhn's, or in the Strong Programme, the social is addressed mainly as a context within which individual cognitive acts proceed.[37] No wonder, then, that social context is characterized as consisting of shared goals, rules, techniques, meanings, and knowledge. In other words, although the making of science is situated within cultural sites, its description is still based on the traditional epistemological conception of cognition as an individual activity.[38]

Locating scientists within social context and studying the development of scientific constructs (theories, problems, conventions, paradigms, etc.) provides an insufficient basis for understanding science and scientific cognition as a social enterprise, even if it is presupposed that they are products and properties of scientific communities. On this account, the concept of the social is always secondary for the simple reason that the social does not enter into the making of cognition. This point must be faced by any socialized or cultural epistemology: in order to "socialize" cognition, the social cannot simply be added as a context to the cognitive. Epistemology of this type cannot show, as Kitcher advocates, "how individuals reason" and "how their efforts, in admittedly artificial social context, combine to yield distribution of cognitive effort";[39] instead, a concept of cognition as a social enterprise must be developed.[40] This does not mean, as Kuhn retrospectively recognized and criticized, that research groups can be treated as real subjects of cognition. "We badly need to learn ways of understanding and describing groups that do not rely upon the concepts and terms we apply unproblematically to individuals."[41]

A way of understanding science that avoids the danger of personification is offered by constructivist sociology of science. From this perspective, knowledge "is constructed not by individuals, but by an interactive dialogic community."[42] In order to avoid needless personification, the emphasis must be placed—as Longino notes—on the dialogic or, more broadly, the interactive nature of cognitive communities. The actual content of the activity of knowledge-construction lies within the network of social interactions of individuals who participate in such communities.[43] The important consequence of such an approach is rejecting the view that scientists are determined subjects in favor of the view that they are participants. Constructivists ask how science operates as a social institution, and their answer is interaction-centered. To erase any doubt, we emphasize that in concentrating on interactions, constructivists do not have in mind, as is the case with philosophers,[44] interactions between scientists and the world. Their concern is with direct interactions among scientists, or indirect ones that address "what other scientists have said about the world."[45]

Science is social, not simply because it involves teamwork or because its elements, including the uses of technical equipment, have a social character. It is social in the sense that "individual action is necessarily oriented to a language community."[46] In other words, scientists direct their activities toward "the agonistic field: the sum total of the operations and arguments of other scientists."[47] Addressing the social nature of science in detail, constructivists interpret it in terms of discourse or controversies, or in broader terms of social practices, especially laboratory practice.[48] From the first perspective, science is "a constructive, socially situated, and socially contingent discursive enterprise";[49] and scientific controversies are "a strategic anchoring point for the study of consensus formation, that is, of the mechanisms by which knowledge claims come to be accepted as true."[50] Scientists are regarded "as actively constructing and reconstructing the meanings of their own actions," and from this standpoint their interpretative practice of defining and redefining meanings is the subject of study.[51] From the second perspective, scientific practices, as enacted in their actual sites (frequently scientific laboratories), should be studied "in order to examine how objects of knowledge are constituted,"[52] and to show how these practices "give rise to 'logical' arguments, the implementation of 'proofs', and the operation of so-called 'thought processes.'"[53] Both approaches concentrate not on knowledge but also on the process of its construction, and they extend the concept of social construction to the objects of knowledge. Science is a dialogue in a strict, not just a metaphorical, sense;[54] scientists' operations take the form of "*discursive interaction* directed at and sustained by the arguments of others."[55]

The interaction-centered approach marginalizes individual activity and sees no need to explain it directly and causally. The constructivist approach is characterized by "a tendency to give priority to the question HOW scientists go about talking and doing science over the question WHY they act as they do."[56] This approach stems from the stance of interpretative sociology, which is represented, for instance, by symbolic interactionism and ethnomethodology:

> The symbolic interactionist cares less about why the members of a group invoke certain meaning-frames than about how they negotiate and monitor a definition of the situation. The ethnomethodologist seeks not to explain, but to learn how we proceed when we convince ourselves to have something explained in everyday life.[57]

Since constructivists focus on the complex of shifting practices in which scientists participate, they search neither for universal regularities and causal determinations of action nor for contextual, social determination by existing social constructs, such as shared knowledge and conventions. Scientific practice, for them, is open, spontaneous, or even improvisational, and to some extent self-determining.[58] Unlike philosophers,[59] constructivists need not refer to shared contents, values, or goals to grasp the social nature of science. Philosophers fail for the most part to realize that appealing to consensus and to what is shared is far from the best way of grasping the social nature of science; the constructivists' path is much more illuminating sociologically. They assume that social integration "is based not upon what is shared, but upon what is *transmitted* between agents."[60] The existence of shared, intersubjective knowledge is not in itself explanatory. On the contrary, it begs for explanation. In order to explain its existence, Knorr-Cetina refers to the dynamics of conversation, its relational nature and its elements of mutual influence, and shows how ideas are formed and transformed within conversation. What is equally important within conversation is that people "are continuously shaping and reshaping themselves in relation to each other."[61]

When the activity of scientists in the laboratory is studied without prior epistemological assumptions, such as the claim that science is the practice of representing or of discovering facts, scientists may be seen as engaged in fabrication, a process that is constructive rather than descriptive.[62] Constructivists believe that social practice constitutes the natural world and that this can be demonstrated even in the case of (scientific) discoveries.[63] It is representation that "gives rise to the object" and not the other way round; objects are constructed as being out there.[64] They argue that "out-there-ness" is an ontological category elaborated in the course of human cognition—in particular, in scientific cognition.[65] Recognizing that the concept of out-there-ness is not a transcendental category prior to human cognitive practice, Latour and Woolgar "feel it necessary to alternate between two asymmetric explanations for the solidity of reality—constructivism and realism," since the two meanings of the term *fact*, as something given and something created, are never present together.[66] They hold that in the laboratory the deconstruction of reality, its conversion from facts out-there into statements and the construction of reality, are merely two sides of scientific practice. In other words, for Latour and Woolgar, an adequate description of scientific practice holds, paradoxically, that facts are simultaneously fabricated by scientists and yet not fabricated by anyone.[67]

For constructivists, science is not closed within laboratories or other scientific institutions. Scientists are involved perforce in trans-scientific fields or networks of interaction and discourse in which administrators, grant agencies, publishers, etc. also participate.[68] In virtue of their involvement within transscientific fields, scientific results "are continuously transfigured and disfigured as they circulate in transscientific fields": they turn into literature studied by other scientists, into political arguments, and into industrial tools.[69] In order to describe relations between science and society, Latour talks about technoscience and actor-networks understood in terms of systems of social relations.[70] They are "networks of practices and instruments, of documents and translations" connecting microlevels and macrolevels within social practice.[71] Linking people and things, laboratory practices, and the wider social field of technical and political negotiations, networks are the basis for the transfer (translation) of phenomena from one social location to another. Pasteur, for instance, translated the anthrax disease from its initial farm environment to his laboratory and subsequently moved it back from the laboratory to the field.[72]

To sum up, the idea that scientific research has intrinsic social characteristics is nowadays almost a cliché, but it confronts two attitudes, the implications of which move in strictly opposite directions. The epistemological attitude does not take the cliché seriously; and although it recognizes the social nature of scientific cognition, its advocates immediately proceed to disregard it. The sociological attitude takes the cliché so seriously that it proceeds to deconstruct the very notion of scientific cognition. Social constructivists, social historians, and even certain philosophers tend to reduce scientific cognition to social practices. Often they pay the price of losing entirely the epistemological content of the concept of scientific cognition, when they claim that epistemological concepts and dichotomies should not be presupposed by a thoroughly socialized view of science.[73] In scientific laboratories distinctions "between the cognitive and the social, the technical and the career-relevant, the scientific and the non-scientific are constantly blurred and redrawn."[74] Clearly, sociologists replace acts of representing with acts of constructing, the process of discovery with that of invention, and truth with credibility. Even the standards of ra-

tional acceptability are perceived as "*practical*, based upon the perceived needs of ongoing research activities" and "*situated* within
localized social networks."[75] The moral that stems from attempts at
"deconstruction" says that there is nothing in scientific activity that is
epistemologically self-evident: neither the concept of the object of
cognition nor that of the subject—nor the concept of cognition itself.
On this view, the social is certainly not an inessential cover that has to
be removed from science in order to reveal its true and purely epistemological content.

Two objections can be raised against the sociologization of scientific cognition by constructivists. First, it is either an illusion or a
groundless opinion to believe that real science is in laboratories,
whereas what remains of scientific activity is only scientists tinkering
about in order to justify the results; "there is no a priori basis for in-
sisting that laboratory shoptalk is a more authentic part of science
than Nobel speeches, proposals for funding to congressional committees, and popularized accounts and pedagogies."[76] Second, since
constructivists must admit that an already existing reality out there (in
a life-world) is deconstructed, melted, and fused into statements in
the laboratory, their initial picture of laboratory construction is significantly weakened and becomes in a sense noncontroversial, in so
far as they admit there is a reality independent of scientific construction. This clearly indicates that the relation between scientists and
their statements, on the one hand, and reality, on the other, cannot
be understood without going significantly beyond sociological studies of scientific practice. The scientific world may be constituted in
scientific discourse, but the world, its objectivity and its out-there-
ness, together with concepts that refer to them, have been present in
human practice long before laboratory science came into being.
What we should do, therefore, is reconstruct the epistemological
content of the concept of science on the basis of the initial idea that
science is a form of practice. However, this task requires much more
than simply developing or even correcting the sociologists' picture of
science. It needs ontology and, first of all, it needs to overcome the
subject/object opposition.

Although Heelan and Latour occupy different positions, both
insist that it is necessary to study concurrently human embodied
subjectivity and world as reality[77] or "understand simultaneously
how nature and society are immanent—in the work of mediation—

and transcendent—after the work of purification."[78] According to Latour nature and society are immanent in the sense that they are constructed, changeable, and determined (by science); they are
transcendent in the sense that they are objective, given, and determine (science). Moreover, the natural cannot be separated from the
social; they are mixed together, though modern cognition has a
strong tendency to separate and purify them. Indeed, for Latour, the
social and the natural, far from being fixed and preexistent, are co-
produced along a symmetrical axis by the action of the self-same
nexus of human practices. One of the chief purifying operations for
Latour is the segmentation of scientific social networks "into three distinct sets: facts, power and discourse." Such a segmentation is clearly
a distortion, because they "are neither objective nor social, nor are they
effects of discourse, even though they are real, and collective, and discursive."[79] On the other hand, with respect to the actor-actant-network, an
entity is both natural and social in proportions that are shaped by its diachronic travels within the network and its interactions with other entities. Latour's own study on the Pasteurization of France shows that

> Pasteur's microbes are neither timeless entities discovered by
> Pasteur, nor the effect of political domination imposed on the
> laboratory by the social structure of the Second Empire, nor are
> they a careful mixture of 'purely' social elements and 'strictly'
> natural forces. They are a new social link that redefines at once
> what nature is made of and what society is made of.[80]

Pasteur's microbes belong to the world that is at once natural and
social: once studied by Pasteur they became empowered to change
both aspects of the world. We are trained (by our modern constitution) to divide this world into nature, political power, technology, science, and the discourse of the media; in its real nature, however, it is
not divided, it is hybridal: "The ozone hole is too social and too narrated to be truly natural; the strategy of industrial firms and heads of
state is too full of chemical reactions to be reduced to power and interest; the discourse of the ecosphere is too real and too social to boil
down to meaning effects."[81] In particular, the content of science, its
interior, and the context of science are no longer separated, such that
the context is able to serve as explanation for the content of science or
of scientific knowledge. The boundaries between science and technology, government, medicine, agriculture, or daily life are constantly

shifting. Latour sees our world as a global and hybridal network. It is clear enough why he insists on calling his conception anthropological. It is neither sociology of knowledge nor of science, nor an exercise in philosophy of science; it is not even a conception referring to relations between science and society or between society and nature, since they are no longer interrelated: they are primordially one. His perspective is anti-essentialist: when we abandon modern ways of thinking "we do not land on an essence, but on a process, on a movement, a passage. . . . We start from a continuous and hazardous existence— continuous because it is hazardous—and not from an essence; we start from a presenting, and not from permanence."[82]

Unfortunately, Latour does not undertake a construction of the ontological foundations of his symmetric anthropology. Consequently, it is far from clear how we can simultaneously embrace and articulate premodern ideas of the "non-separability of things and signs" with the modern "final separation between objective nature and free society."[83] Certainly, Latour is right to suggest that if we want to articulate how science is constituted, we have to accept this phenomenon as he articulates it in its entirety. We believe, in contrast to Latour, that the reconstruction of the concept of the cognitive nature of scientific research requires a fundamental ontology that must embrace a social and historicist position. In such ontology, unlike in ontic social studies of science, science is not reduced to social context. So reduced, it becomes a system of conventions or consensus, or of interactions, or of practices and units of power, and so forth.[84] Consequently, ontological issues, such as the nature and genesis of science, remain hidden behind surface descriptions of the workings of science. At the end of the day, the central philosophical questions not only go unaddressed, they are in fact not even recognized.

A position akin to Latour's is Pickering's. However, his view of scientific practice is no better placed ontologically than is Latour's. Pickering rightly sees experimental practice as an open-ended dialectic of resistance and accommodation, a real time process productive of unprecedented events. Clearly, it is a path-dependent work of cultural extension and transformation within which human and material agency are emergently reconfigured by the performative mediation of machines. Consequently, in the processes of real-time, practice, human agency and material agency get "mangled."[85] Pickering certainly puts flesh on Latour's actor-network picture. Nevertheless, he tends to plot his account of what emerges experimentally along an epistemic axis. What is missing is a large-scale account of the ontology of the practice such as we establish here. Only from an ontological perspective (we will argue) can the objectivity, temporality, and historicity of human action and material agency achieve meaningful articulation. This standpoint reveals both the social and the natural not as causes but as an indeterminate product of the dynamics of underlying practice.

In abandoning the view that cognitive activity is simply an activity of a given individual working in a given social context, we leave behind the foundationalist hope of discovering universal cognitive mechanisms, procedures, or conditions of cognition. If cognitive communities are local and historically limited, their activity has an unavoidable historical aspect. Let us turn, therefore, to issues of scientific cognition from the historical perspective.

THE HISTORICAL NATURE OF SCIENCE

History of science has significantly reoriented its approach: it has shifted concentration from the intellectual structures of science to its sociocultural context. It has reconceived the basic units of analysis, revised relations between intellectual structures and their contexts, and adopted new attitudes toward the developmental patterns of science.

Most earlier writing concentrates on great names and views historical change in terms of subject-oriented categories and as transitions from one culminating period to another. It reifies and decontextualizes ideas and treats them as if they were autonomous historical agents. At bottom, this stance is nothing more than a diachronic version of Platonism: it seeks omnipresent principles in the warp and woof of the historical process. Certainly, the intellectualist approach recognizes science as a significant cognitive enterprise. It fails, however, to provide necessary perspective on the subtle ways in which science is integrated into varying social and cultural contexts.

One important ontic device for linking the cognitive and the sociohistorical is the notion of styles of thinking. In his study of scientific styles, Crombie says: "The Western scientific movement has been concerned with man's relations with nature and his fellow beings as perceiver and knower and agent. It can be identified most precisely as an approach to nature effectively competent not only to solve problems, but also to determine what counts as a solution."[86] For Crombie, bodies of scientific knowledge are organized in terms of various historical dimensions that characterize styles of thinking. Styles of thinking have historical beginnings and ends, possess cultural trajectories, and carry forward positive bodies of knowledge. Crombie shows in detail that science, throughout its developmental history, has been organized into a variety of styles of inquiry, demonstration, and explanation. From the historical standpoint, these differences rest mainly on choice of subject matter, on how nature is conceived, and on the ways in which experience is used. Emerging historically, forms of inquiry, oriented to these desiderata, have generated questions to be investigated and have set criteria for what counts as providing successful answers.

Viewed in this manner, scientific knowledge advances by identifying solvable problems that become configurated by particular styles of thinking. Scientific styles have also organized the investigative procedures of the sciences into distinctive disciplinary complexes. Crombie identifies six main styles that have emerged in the history of Western science. He does not address the issue of whether the notion of styles carries explanatory power above invoking the relevant historical, social, and cognitive activities of science, and its efficacy as an explanatory device remains an open question.

Crombie's position has made an impact on the philosophical community. For example, Hacking approaches Crombie's views philosophically while recognizing the nuances of social and historical change that Crombie presents. In Hacking's view, styles of reasoning (his term) establish the *forms* of knowledge; they delineate the methods and discursive formations by which we reason. Moreover, they establish the possibility of truth and falsehood within a discourse — namely, the prior capacity (in virtue of reasoning in a specific way) that a particular discourse possesses for making reference to truth conditions. From this perspective, being objective and rational is just to reason in a determinate way within a recognized style of reasoning that invokes certain values and presupposes certain cognitive commitments. Hacking notes, for example, that prior to statistical reasoning, there was no such thing as a true or false statistical proposition "because there was no procedure of reasoning about the relevant ideas."[87]

Crombie's and Hacking's positions have affinities with Kuhn's recent view that theories are holistic, lexical clusters whose interrelated elements make truth claims possible relative to a particular semantic framework. Indeed, Kuhn himself explicitly recognizes the connection between Hacking's work and his own.[88] These views—we will argue open a space for understanding the historicity of human thought. This is true especially in regard to those changes in historical context that inform shifts to different styles of thinking or to different semantic frameworks.

Interestingly, these views relate to Foucault's nuanced, historical conception of "games of truth" relative to types of discourse embedded within a distinctive discursive formation. A discursive formation is a "field of dispersion" in which different conceptual configurations find expression, even those that are contradictory. Its unity lies in its governing rules and in the regularities of its practices.[89] The totality of such formations, in a particular society, culture, or civilization, Foucault calls the *archive*. The archive refers to "systems that establish statements *(enonces)* as events (with their own conditions and domain of appearance) and as things (with their own possibility and field of use)"; it contains bodies of knowledge and their individual discourses—that is, science, literature, history, and economics.[90] For Foucault, discourses have historical life independently of individuals and link together the patterns of conceiving, saying, and doing that are present in human action.[91] From this standpoint, practices constitute the ground for actions in virtue of their judicative and veridicative dimensions: they establish and apply norms, controls, and exclusions and thereby make the discourse of truth possible.[92]

In his archaeological writings, Foucault treats historical documents as monuments, as positivities to be studied according to their position, relation, and configuration within historical networks.[93] The historical task is to uncover rules that jointly characterize the objects to which the statements refer, the cognitive status and authority they possess, the kinds of concepts they employ, and the cognitive strategies in which they are embedded.[94] Proceeding this way, archaeology brackets questions of causation, transmission, and development; these are the province of genealogical strategies that examine the existence conditions under which particular discourses emerge and enter connections within sociocultural space. If archaeology seeks the anonymous rules that govern discursive practices, genealogy reveals their status as temporary integrations within the changing nexus of power relations among historical actors.[95] Both perspectives concentrate on "historical ontologies" and show that the ahistorical, the self-evident, and the essential are deeply rooted in the contingencies of history.

If history of ideas essentializes the contingent and brackets the incidental, the social approach in history of science recognizes the concrete role they play within the vicissitudes of history. The social history of science asks how social, cultural, and cognitive elements relate historically and how they bind together into temporally integrated historical complexes. A possible answer is supplied by the layer-cake model developed, for example, in Marxist theorizing about ideas and praxis. Its explanatory outlook is substantialist and structuralist. Social structures, strata, and institutions are viewed as constituting a hierarchical "reality" that embraces human action in both its individual and collective forms. An opposing model sees historical elements in terms of integrated dynamic wholes, the "parts" of which continually interact and form ever-changing dependencies, with no one element dominating. Some historians, such as Biagioli, speak of the sociocognitive and virtually identify the social with the cognitive.[96] For others, such as Merton or Shapin, social, cultural, and cognitive elements function together as interdependent phenomena in historical space. For instance in Merton's structuralist-functionalist model, emerging English science of the seventeenth century is understood in terms of the shifting interests and values of a society in transformation, the motives of which can be excavated by the techniques of sociohistorical analysis.

Merton considers science as dependent historically on the social and the ethical, but he separates these elements from the cognitive.[97] He also stresses its connections with warfare and its interrelations with the technological and the economic contexts of English society. Merton's sociohistory clearly illustrates structuralist thinking. His discourse is filled with terms such as "strata," "patterns," formations of "legitimation," cultural and social "frameworks," spheres and institutions of "interdependence," and so forth. This is nothing less than sociohistorical externalism: its aim is to analyze individuals and collectives within a sociocultural context. He does not ask whether scientific cognition is integrated within social formations, but asks in what ways are science's cognitive aspirations legitimized in social terms. Merton sees science as if it were a sociohistorical "black box" suspended within networks of social relations. Certainly, he speaks of human practices, of shifting directions in human interest, and of ways in which these relate to the general value systems of culture; but his primary objective seeks to show how these phenomena become polarized and structured by existing social strata. Little consideration is given to how socially embedded activities make scientific knowledge what it is or to how they maintain science as a form of life. Nevertheless, Merton's conception of a creative alliance between English Puritanism and the development of natural science and technology in England has much influenced historians of seventeenth-century science, notably Shapiro and Webster.[98]

A more audacious approach to the social and the cognitive is offered by Shapin and Schaffer. They consider relations between science and its historical context as a mutually interactive complex. Their study of cultural interactions between Boyle and Hobbes assumes that science is a social activity performed within institutionalized frameworks guided by social norms. Given this sociohistorical standpoint, they see a clash between Boyle's individualist views concerning the rights of scientific authority within the larger polity and Hobbes's perspective on what is necessary for maintaining civic authority within the state. For them, the rise of modern science and the rise of the new social order in England are manifestations of one and the same process: a solution to the problem of order. Science is an integrated pattern of activity, a "form of life," and scientific disputes are "different patterns of doing things and of organizing men to practical ends."[99] According to their view, scientific knowledge is produced and sustained by situated practical activity, and its claims to authority are managed by codes embedded in the moral economy of society as a whole. They refrain from studying the interior of science; their aim is to weave a complex political, social, and cultural story. From their standpoint, "an intimate and important relationship [obtains] between the form of life of experimental natural science and the political forms of liberal and pluralistic societies."[100]

Well and good. They fail, however, to appreciate fully the symmetry implicit in their historical perspective on individuals, society, and nature. Indeed, they privilege society over nature for explanatory purposes; for them, it is the social that is out there, not the natural. This introduces an unnecessary asymmetry into their analysis, an artifact of their tacit assumption of the society/nature dichotomy, despite their avowed aim of problematizing it! At one level, they take the social to be the primordial basis for explanation. At another, their approach tracks the path of "things" as they emerge historically into forms of stabilization. True, it is no explicit part of their position to assume that "things" are always already "concretized" on either side of the society/ nature pole. Nevertheless, they fail to capitalize on the potential of their analysis and remain locked into a sociologically driven ontic picture of the history they articulate. This impasse cannot be satisfactorily resolved unless that picture is placed squarely within the framework of a sociohistorical ontology.

Social history of science clearly replaces the subject-oriented categories of the history of ideas with various nonsubject-centered categories. One useful notion is the concept of an "intellectual field," articulated by Bourdieu and developed by the intellectual historian Ringer. An intellectual field is a network of relationships composed of individuals or groups each variously related to larger cultural fields. Within such fields, agents compete for the right to define or to co-define "what shall count as intellectually established and culturally legitimate."[101] Mobilizing the concept of an intellectual field, the historian concentrates on shared intellectual habits and cultural and collective meanings rather than on individualized meaning and intentional action. To characterize the roles an idea plays is to relate it to other elements within a field; and to ascertain its significance demands an understanding of its interrelations with these other configurated elements. Intellectual fields are dynamic realities; their orientation shifts from one to another orthodoxy. What is orthodox today may be heterodox tomorrow, and the entrenched roles of orthodoxies help to "shape the heterodox reversals they call into being."[102]

Ringer, following Bourdieu, conceives culture as comprised of the visible elements of intellectual fields; but it also includes hidden and often unarticulated assumptions embedded in institutions, practices, and social relations. In this way, intellectual stances that would separate cognitive dispositions from institutions, practices, and social relations are undercut, enabling historical analysis to disclose empirically what lies hidden and tacit in the realm of the cultural preconscious.[103] For Bourdieu and Ringer, what is important here is the historical nature of habitus. "The *habitus*—embodied history, internalized as a second nature and so forgotten as history—is the active presence of the whole past of which it is the product."[104]

Seen in this light, the cultural preconscious is simply the "forgetting of history which history itself produces" in constituting the habitus of an age.[105] This is fine. But the idea that a habitus underlies cognitive fields raises questions of how parts relate to wholes, of how patterned cultural dispositions are reproduced in each individual and transmitted from individual to individual and among groups. The difficulty lies in the need to posit a stabilizing structure that can fix the active configurations of intellectual fields in the absence of a central role for individualized intentional action. To avoid this difficulty, we mobilize Foucault's conception of human actors cognizing and communicating in fields of action in which they are continually reconfigurated by relations of power-knowledge. According to the picture we will develop, parts and wholes are conceived in dynamic interrelation, and individuals are viewed as acting in significant situations, not only in virtue of their individualized intentions but in terms of the normative expectations of others.

Undoubtedly, the categories chosen to construct a piece of history push the analysis to one side or the other of the dialectical relationship between continuity and discontinuity. This dichotomy is clearly related to cognitive issues seen from the historical perspective. Without difference there is continuity but no grasp of change; without continuity there is difference but change remains unconstituted. Overemphasis on tradition reduces continuity to sameness; but overemphasis on difference renders historical change meaningless. Even at the practical level of historical understanding, what gets emphasized is dictated by the goals the historian espouses.[106] After all, whether or not history is considered significant for understanding the present clearly depends on the stance taken on this issue.

Certainly, historians seek to write history that is diachronically fluid and synchronically mobile and at the same time maintains resources for overcoming the paradoxical elements in the continuity/discontinuity dichotomy. Not made evident, usually, is the extent to which these ontic categories merely ride on the surface of historical writing. Indeed, the ways they play out in a piece of writing is clearly dictated by the historian's grasp of the historicity of the period under study. The manner in which existence is grasped historically profoundly conditions the ways change and stasis are understood in reference to historically indexed processes and events. This in turn affects the manner in which past conditions actively at work in the present are understood. To portray the full significance of this, these ontic pictures must be integrated into a deeper ontological perspective.

These observations prompt another: how do we understand interrelations of historical contingency, becoming, and historicity? If science is viewed within these categories, how are its claims to cognitive universality and truth to be understood historically? Such questions are scarcely raised by historians of science. And if they are, they are not integrated into a perspective that pertains to the ontological structure of history. History of science abounds with ontic pictures that weave historical events and processes into networks for integrating individuals and societies into historical formations—such as symbolic worlds, styles of thought, intellectual fields, or discursive formations. Such pictures, naturalistic and culturalist alike, are based on the subject/object opposition; they posit, either implicitly or explicitly, a characteristic spatio-temporal (diachronic) structure with reference to which the interplay of entities is understood. Mostly ontic histories either objectify space and time, treating them as a pre-given framework within which persistence and change occur, or they simply fail to thematize them at all. Clearly, the temporally indexed configurations of historical change cannot be reflected within the homogeneous space-time structures posited by the physical sciences.

History is often conceived as existing independently of the historian who studies it, either as a series of states of certain privileged entities (usually understood in terms of substance-ontologies) or as an objectified, self-sustained process that differs entirely from those entities. These standpoints leave the ontological categories of becoming and historicity unanalyzed precisely because they ignore how historicity and becoming condition the ways of being of individuals and human communities. This prompts a question. Which ontic entities—substances, processes, or events—are compatible with the sociohistorical ontology we want to develop? From the standpoint of ontology there is, of course, no necessity to warrant one ontic picture over another. This tendency to either/or dichotomizing is characteristic of ontic thinking. Nevertheless, some ontic pictures, we maintain, are *primi inter pares*.

HISTORY AND DYNAMIC ONTOLOGIES

All ontological views stretch between the timeless permanence of Parmenidean Being and the pure flux of Heraclitus. Substance-ontologies, best illustrated by the Aristotelian conception, presuppose things as the ultimate bearers of predication and forms or preexisting actualities

that define things as what they are. What comes to be something comes
to be that from an underlying substratum that makes it what it is.
Although a thing appears perceptually in contrary ways, ultimately
what it is in-itself is what it is essentially at all times. In actualizing its
potential to become what it is, a thing possesses an end and passes
through stages that fall short of its actualized being. What always
becomes and perishes, never strictly *is*, and being and becoming from
this standpoint are irreconcilable. That Aristotle restricts "what it is" to
the category of substance is evident in the *Metaphysics*, where he says
that "'what it is' unqualifiedly applies to substance, whereas it only
qualifiedly applies to other kinds of things."[107] He holds (or came to
hold) that to ask of something "what it is" is to say what it is to be for that
kind of thing, to specify the kind of being that is essential to it.[108]
Hence, nonsubstances—for example, being such-and-such a color—
differ from substances whose "essential being is not just the accidental
being of something else."[109] Moreover, change is an imperfection. And
since history is a species of change, from this perspective it is a process
of becoming within which things are always coming to be only to pass
away. In this light, history exemplifies the imperfection of imperma-
nence; it is a temporal process that never fully is, since it always passes
away. This implies a substantialist view of change and history since
change implies things that change, just as qualities imply things that
are qualified. On this view, historical features are merely accidental
details superadded to the changeless *archaei* that make things what they
are. Given the substantialist foundation of his ontology, Aristotle is able
to conceive everything as substance-in-process. Each thing is involved
in dynamism, forever in transition as it drives through the interplay of
potency, activity, motion, and change.[110]

    Event-ontologies and process-ontologies are alternatives to sub-
stance ontologies. If substances are static entities enduring self-samely
through change, events and processes seem to embody change itself.
Indeed, events are often viewed by historians as the "stuff" of history.
Foucault provides a case in point: he views history in terms of events
and attempts to capitalize on the way they differ from processes and
the states of things. For him events directly concatenate historical
change and do not take things as their ultimate bearers after the fashion of properties and states. Viewed in this light, their primary mode
of being is to happen so that a given event "takes effect, becomes effect, always on the level of materiality."[111] History is a theater of

configurated events, unassociated with the regularities of determining causal processes. For, unlike Hegel and Marx, Foucault does not see history as a plethora of events necessarily organized by the predetermining unity of the historical process. From his perspective, it is the unending ways in which events are contingently dispersed across the flux of history that take center stage. They are "presences" that make a contingent difference to the situations in which they participate.

But if events are contingently concatenated moments, simply disseminated breaks in historical movement, the dynamic of history turns on difference. This means that history is change and that the drive of flux is what constitutes history's way of being. From this standpoint, everything that comes to pass seems unconnected to any unifying process of historical becoming. How is this so? Simply because Foucault replaces substance-ontology, and the univocal categories of being, with the dispersed ontology of the event. From this standpoint, he seeks to "permit difference to escape the domination of identity" and "the law of the same." His view that difference recurs endlessly within historical complexes connects at this point with his view that event-ontology is fundamental. For him, the ontology of historical change lies outside the univocal tropes of being as understood by the metaphysical tradition. On the contrary. "Being is that which is always said of difference; it is the *Recurrence* of difference."[112] This expresses Foucault's conception that being is becoming. To Platonic essentialism, which sees identities as the repetition of sameness, he opposes the dynamism of events in their multiple seriality. From this perspective, the ontological trajectory of history is delineated by the contingent play of events, and its temporal patterns are set apart from the Platonic picture of the causal-teleological nexus. Events alone—multiple, dispersed, unexpected and polymorphic—define the contours of historical change.

From our perspective Foucault's position remains on the surface and is an ontic view. Admittedly, events portray the movement, shape, and texture of history, given their disparate character. They are singularly unable, however, to structure an ontology of historical becoming; they fail to connect those features of the historical complexes in which they participate in a truly integrated way. Certainly, Foucault's event-ontology and his conception of endlessly recurring difference fit well together. But they fall short of connecting those self-same features trans-temporally. Clearly, his picture lacks the "glue" necessary for the unification of historical becoming. Indeed, Ricoeur notes that the theory of the event as difference produces a negative ontology of history: "the notion of difference does not do justice to what seems to be positive in the persistence of the past in the present."[113] In affirming an event-ontology, and rejecting other ontologies, Foucault leaves himself without adequate resources for unifying historical complexes through time. Events, no matter how polymorphous, disparate and multiple, cannot fill the ontological gap.

In our view, processes provide an apt ontic picture for underwriting historical becoming. To anticipate, unlike events, which simply occur and recur, dynamic masses—processes and activities—are multiply occurrent in time and space.[114] Moreover, they bear features of both the particular and the general. These features are germane to understanding the ontological structure of historical becoming.

Nietzsche combines perspectives that are important for our analysis. He mounts a philosophical attack on substance-ontology, affirms the ontic significance of processual phenomena, and outlines the structure of historical becoming in a genealogical critique of historical origins. Indeed, his philosophical attack on substance connects with his genealogical critique of historical origins. Both reject the idea that invariant structures ground change. As a search for the basis of what we say, think, and do, genealogy exposes critically the historical origin of ontology. At its core is the philosophical view that thought's proper function is criticism, diagnosis, and demythologizing. In this role genealogy is the wedge that opens a "space" of plausibility through which ontologies may be put into historical perspective.

This is fine. But what reasons does Nietzsche offer for rejecting the claim to ultimacy of substance-ontology? He argues that Indo-European languages are structured so that their speakers are compelled to talk of objects as unities persisting identically through change and distinct from their effects and properties.[115] But given the contingent character of lexical features, we cannot infer that their ways of constraining how we conceive the world necessarily reflect its reality. Famously, Nietzsche holds that substance is a projected shadow of the cognizing grammatical subject: "the concept substance is a consequence of the concept subject: not the reverse."[116] For Nietzsche this standpoint creates a two-tier ontology. The idea that every deed presupposes a doer adds a substratum to every act of thinking. But it is the doing that is everything, and "the doer" is a fiction added to the act of doing.[117] Adding a doer to every deed, far from representing a fact, is instead the impact of meta-physical commitment. Even "The 'subject' is not something given" in direct awareness, but is an interpretation, "something added and in-vented and projected behind what there is."[118]

Indeed, language seduces us into an overinterpretation of phe-nomena. To say "lightning flashes" tricks us into thinking that there is a substratum (lightning) beneath the activity of flashing itself. But this is a fiction. Inventing these metaphysical fictions, however, is deeply rooted in our cognitive habits. For Nietzsche it has a double root. First, we construe ourselves as causes, as intending agents, as doers of deeds. Accordingly action is explained by a two-tiered, dualistic ontol-ogy: mental states (intentions and desires) are posited as inner causes of outer behavior, the former the basis that distinguishes action from other events.[119] Second, this schema is then projected upon the hap-penings of outer events, and causes are posited as underlying all events or things to order and shape experience.

If substances do not necessarily underlie patterns of spatio-temporal change, so likewise patterns of change that eventuate in the present and move to fulfillment in the future have no need of underlying, pri-mal origins. Patterns of change, affecting persons and things alike, are patterns of contingency. Genealogy problematizes change and reveals history working in the construction of the self and the various historical contents it expresses. However, things cannot be understood simply as a process of change. Relative stasis must be recognized, the fact that things persist through stretches of time. From the Nietzschean stand-point, historical becoming is a process of attainment, of achieving rela-tive stasis within change, where attainment is understood dynamically and not as a permanent stage that emerges within becoming. Darwin's theory colors Nietzsche's view of historical contingency. Darwinianism illuminates the open-ended and undirected nature of change and be-coming. Seen in this light, societies, cultures, languages, and human practices are processes of becoming; each begins and ends within tem-poral extension, and each is a plurality occupying various historical, temporal, and spatial boundaries. Clearly Nietzsche reverses the prior-ity of permanence and change; instead of a metaphysics of substance he posits a historized version of processual features theorized by process-ontology.

Let us look more closely at these features. Processes have their own intrinsic dynamic and manifest distinctive developmental patterns in space and time. Most importantly, there are autonomous processes, which do not reduce to the properties and activity of things and do not presuppose them as their bearers and subjects.[120] Unlike particulars, things that are in one spatial location at a time, processes, like masses and activities, are repeatable and occur in various locations at the same time; their patterns of persistence are multiply occurrent in time and space. For example, the referents of mass terms, such as wine and water, and verbal phrases referring to activities, such as "snowing now," display traits characteristic of both particular and general entities. "Masses and activities are concrete entities like particulars, since they have concrete physical properties; on the other hand, they are multiply occurrent like general entities—you and I, we may have the same stuff in our glasses and when we drink from them we engage in the same activity."[121] Such entities are as much now as then and as much here as there. Further examples come easily to mind: hurricanes, tornadoes, a performance of Beethoven's Ninth, shifting dunes of sand, lightning, electromagnetic action, thunder, and so on. Historical processes share an important feature (among others) with dynamic masses: they too are multiply recurrent in time and space.

Process-ontology also provides a coherent notion of persistence in time that does not conceive a thing's endurance at one time as identical to its existing at another. A process is a unitary item, not in virtue of the continuance of a set of essential properties, but in virtue of its history which dynamically integrates, consolidates, and constitutes it in terms of the temporal structuring of its successively patterned actions. Process-ontology need not assume identifiable particulars to establish objective identification and reidentification. As Rescher points out: "identity rests on identifiability, and identification is something interactional, that is, it involves being identifiable as a single unit by an interagent. And such identification is always and unavoidably processual."[122] That is, the integrity and stability of things consist in the ongoing unity of process, as things integrate and consolidate themselves by acting in characteristic ways in relation to one another. There is no need to posit the underlying presence of a bare substratum to unify and coordinate their activities and doings. As Nietzsche indicates, the duplication characteristic of a two-tiered ontology is unnecessary, the unity of things just is the unity of process.[123]

Event-ontologies and process-ontologies carry their own difficulties. As we have indicated, events, given their disparateness, cannot structure historical becoming, since there is no necessity that they depend upon the continuants within which they participate. This particular difficulty is remedied by the fact that process-ontologies aptly underwrite historical becoming. Processes provide a framework for understanding the coherence of historical becoming. Even if event-ontologists are right in claiming that the material of history presents itself in the form of dispersed events, history is more than sheer becoming, the wheel of recurrence, the endless substitution of the future, present, and past. If there is a balance between sameness and difference, so also there is one between change and stability.[124] Process-ontologies do provide a basis for narratives that structure the sheer recurrence of events dispersed in the flux of time. They cannot, however, reveal the character of the historical in the flow of time as distinct from that flow itself, or reveal it as it characterizes particular historical processes: processes provide a framework for the historicity of our being but do not address it as a philosophical issue. To go beyond the ontic view of history as simply a set of historical processes, understood in an objectivist way as temporal sequences of events moving through past, present, and future and governed by patterns of change, we need to de-objectify the concept of historical process. This requires a narrative and hermeneutic account of history and, moreover, it requires a fundamental ontological analysis of historicity understood as a condition of the mutual relations that obtain between us, events, and processes.

SCIENCE AS A COGNITIVE SOCIOHISTORICAL PHENOMENON: THE NEED FOR A FUNDAMENTAL HERMENEUTIC ONTOLOGY

We believe that Heidegger's philosophical standpoint, if developed in a nonexistentialist way, provides a strategy that frees the ontological perspective from the essentializing attempts of substance-ontology, from the insufficiences of Nietzsche's and Foucault's perspectives, and from the one-sidedness of Heidegger's own ontology. So developed, it points beyond objectifying considerations of what there is, which presuppose as their fundamental condition the "mutual autonomy of subject and object."[125] Indeed, overcoming this autonomy is a necessary condition for an understanding of science as a social and historical form of human practice that contributes to the constitution of both the natural and the social. Moreover, an ontological perspective allows the task Guignon poses in reference to hermeneutics to be realized. It is from the hermeneutic perspective that

> traditional epistemological questions about truth, rationality, jus-
> tification, and relativity should be reformulated and addressed
> only after we have worked out a general account of our situation
> as being-in-the-world. Old notions of correct representation and
> Cartesian certainty, along with the inner-outer and subject-object
> distinctions, will probably prove unhelpful except when discuss-
> ing certain regional activities.[126]

We want to mobilize Heidegger's way of philosophizing to elaborate a fundamental ontology differing importantly from his own. In undertaking this task, a clear exposition of Heidegger's thought must be established since his writings are difficult and open to serious mis-interpretations and trivializations. This caution applies to Gadamer as well.

Heidegger's ontology is historicist, hermeneutic, and existential-ist. Its merit lies in rejecting the substantialist-objectivist perspective, in the shift from substances, entities, or objects, to *being*, in short, from it to *is*.[127] Traditional ontology starts from entities, presupposes that they are, and equates their being with actuality, with their presence; so considered, they occur presenting themselves as objects equipped with essential features.[128] We will return to this issue in chapter 3. Al-though he rejects substantialism, Heidegger does not simply embrace process-ontology. Rather, he constructs a being-ontology that consid-ers existence (a human way of being) in a rudimentary phenomeno-logical fashion—that is, as something that we experience and as constituting the *sense* of our being and the being of entities. Construct-ing a being-ontology he moves from philosophizing within the sub-ject/object dichotomy to problematizing the dichotomy itself.[129] He also breaks with other metaphysical dualities: mind/body, the intelli-gible/the sensory, the real/the apparent, the eternal/the temporal, being/becoming, and the permanent/the changing. His conception stands, therefore, in stark contrast to the tradition of Descartes and Husserl, who give priority to knowing (to immediate perception) over being.[130] Heidegger's aim is to grasp being as underlying entities and to uncover the conditions of cognition and, in general, of our being. For him these conditions do not lie in eternal, immutable Being or in the transcendental subject; they are structures of our being. Although being underlies all entities whatsoever, it is particular and historical, and it reveals itself in various successive disclosures—that is, in the movement of history.[131]

Believing that being is the basic theme for ontology rather than Being, and presupposing a phenomenological requirement that philosophizing should begin with phenomena as we experience them,[132] Heidegger starts his analysis from everydayness; but, unlike Husserl, he does not bracket everydayness. Instead, he undertakes the analysis of human being as it realizes itself in everyday life and in this sense his ontology is existentialist.[133] In reference to human beings Heidegger uses the term *Dasein* to emphasize that each of us is a way of being that can be distinguished ontically and ontologically; a way of being for which its own being is an issue, so it is always my-own-being, "in the possibility of existence."[134] Being is a practical issue for us because we lead a life, and we have it to be, whether or not we consider living in one way or another.[135] We are who we are, and our being reveals itself to us as taking a stand on who we are in relation to the individuals we have already become and in virtue of the yet-to-be-realized possibilities of our being. Hence, Dasein is neither an inner-worldly thing located in space and time (even if distinguished as a thinking substance, or a rational being, or an "I") nor a worldless subject. For Heidegger, all metaphysical attempts of ancient philosophy, Christian theology, Cartesianism, or modern science to determine the essence, or the (invariant) nature, of human beings have failed "to give an unequivocal and ontologically adequate answer to the question of the *kind of being* of this being that we ourselves are."[136] In such objectifying pictures, our being is forgotten and we are relativized to something nonhuman, such as Reason, divine transcendence, or biological life. To overcome this metaphysical reduction is, for Heidegger, to analyze, in a phenomenological fashion, the very existentiality of our existence, and to recover our authentic selfhood through the restoration of our existential features: finitude, mortality, and humanity.[137] Human facticity and historicity can neither be explained away by reference to a metaphysical ground nor shown to be constituted by transcendental subjectivity. They are the most fundamental structures of all being and, consequently, of knowing as well. Clearly, Heidegger is not interested in studying human beings empirically in different cultures or historical periods. He wants to "lay out for us general, cross-cultural, transhistorical structures of our self-interpreting way of being."[138] In his own terms, Heidegger is not interested in ontic *(ontisch)* but in ontological *(ontologisch)* inquiry.[139]

The existentiality of our being means that we are in-the-world not as spatially contained things, but in the sense of being-in-the-midst-of-doing-something or of dwelling-in-the-midst-of-things. This separates us from other entities: they are encounterable in the world, but Dasein leads its life and has a world. Facticity of (our) being means it happens in-the-world and consists of different ways of being.[140] The common meaning of these different ways of our being is *taking care of,* or *concern,* in its proper or deficient mode (as in the acts of omitting or neglecting something). We are always toward-something, we are in a process of transcending being-here toward what is-there. Every particular way of our being, including our entire activity, is a form of *understanding being,* our own being and that of other beings, and it aims at its disclosure.[141] Heidegger turns the concept of understanding into an ontological category: it is "the original form of the realization of Dasein."[142] We understand things when we can handle them, when we are up to them, when we can do something with them.[143] Our most primordial way of being is our concernful dealing with entities in everyday life.[144] This activity discloses and appropriates entities as ready-to-hand tools. "The seemingly mystical 'understanding of being' is simply the know-how and can-do that comes from living familiarity with others among things in one or another cultural world."[145]

Also cognition, as it proceeds within the subject/object opposition in the form of philosophical speculation or scientific research, is a way of human being, but not by any means the most fundamental.[146] Thus, Heidegger comprehends the relationship between a subject and an object of cognition in ontological, existentialist terms.[147] Cognition (representing) is based ontically upon our being-in-the-world and conditioned ontologically by the structures of being, especially by the temporality of Dasein.[148]

Human existence is a primordial way of being from the ontological perspective, but it is not the only one. There are entities other than Dasein and dependent on it, so that their way of being is derivative. Basically, entities that differ from us are tools, pragmata, used by us in everyday life. Their way of being is readiness-to-hand or handiness.[149] Together with equipment, we encounter in our acts of taking care of things the public world and the environment, the surrounding world of nature, that is at hand along with equipment.[150] A third fundamental way of being is objective presence, presence-at-hand, which characterizes entities we encounter within the world as opposing us as subjects of cognition. They are objects, substances, things of Nature, which "is a boundary case of the being of possible innerworldly beings."[151] Entities present-at-hand are the occurrent objects of scientific or metaphysical cognition, deliberation that is based on the subject/object opposition. All these ways of being do not form a hierarchy of genera and species of being as such; being is always the being of an entity, but it is never itself an entity. All forms of being are interconnected into a whole.

Heidegger's ontology is historicist. Historical understanding "permits a distanciation from an immersion in immediacy."[152] It also allows us to see history and uncover historicity as its ontological structure, thus rendering it a possible object of historiography. In his analysis of historicity, Heidegger shows that what is ontologically prior to the historical character of things and events is human being, constituted by anxiety and historizing—that is, stretching along from birth to death.[153] Human temporality means that we exist simultaneously along three temporal dimensions: at home, in our immersion within the present, as thrown into being; already in time, as we take our past along with us; and ahead of ourselves, when we project ourselves into the future. These are the *ekstases* of temporality.[154] Being-toward-death is at once the foundation of Dasein's connectedness, its lifelong unity and its only possible way of transcending finitude.

Conceiving understanding in an ontological way, Heidegger replaces the search for ultimate foundations with the unfolding of the hermeneutic-ontological circle. As we are before we begin to understand the sense of being, so likewise we understand our world before we begin to cognize it. As being gives rise to understanding and conditions it, so preunderstanding gives rise to thought and conditions it. The *hermeneutic* character of Heidegger's ontology means that it aims at the interpretation and understanding of Dasein and its different ways of being, and not at their objective description and explanation.[155] It means also that ontology must understand itself as an event of Dasein's being. What leads Heidegger to hermeneutics is the recognition that (1) when we ask about the sense of our being, we discover it constituted by our understanding of being; (2) when we understand, we "recover an understanding in which we already stand";[156] and (3) when we try to find the source of our preunderstanding, of our prejudices, we arrive at tradition, results of earlier human thinking. Thus, there are no ultimate beginnings, "we are always in the midst of an ongoing process."[157]

The hermeneutic dimension of Heidegger's ontology has been elaborated by Gadamer. Gadamer's hermeneutics is *ontological* and *historicist*. He goes beyond earlier views of hermeneutics as the art of understanding a text or as a methodology of the human sciences.[158] For him hermeneutics is philosophy per se, "a universal ontology of experience and language," or even broader, a philosophy of culture.[159] It seeks an ontological analysis of understanding without, of course, presupposing the subject/object distinction. Only then, Gadamer stresses, is it able to do "justice to the historicity of understanding."[160]

Although Gadamer bases himself on Heidegger in constructing his ontological hermeneutics, he substitutes for Heidegger's ontic/ontological opposition his own distinction between ontic and historical (hermeneutic) discourses and rejects Heidegger's view, present in particular in his later works, that Being is an ultimate source of language. In Gadamer's hermeneutics "the question of language supersedes that of ontology," and language becomes a medium of understanding.[161] For Gadamer, understanding is neither a psychological act nor a relation between the knower and what is known. Existing factually as interpretation, it "constitutes the original structure of 'Being-in-the-world'"[162] and "pervades all human relations to the world."[163] Certainly we exist in a hermeneutic universe because the source of its unity lies in historical tradition and in the natural order of life; but also because this unity is secured by the way we experience one another, historical traditions, and "the natural givenness of our existence and of our world."[164]

The conditions of understanding uncovered by Gadamer's hermeneutics are neither methodological nor existentialist. They are onto-historical and are located in tradition: indeed understanding is "participating in an event of tradition," by way either of embracing or criticizing it. Although in Gadamer's view, understanding remains an ontological concept, in the sense that it constitutes the original structure of our being in the world, it is not just Dasein's way of being; it is a dynamic relationship between our participation in tradition and tradition's shaping us.[165] Consequently, Gadamer is able to consider not only those ontological conditions of understanding which constitute our existentiality, but also those which are supra-individual structures:

language and history, as they manifest themselves as tradition. Gadamer stresses the significance of the past and the "effective" power of tradition.[166] Understanding is an event of our being and simultaneously an achievement of language, since our being in the world is itself linguistic: we exist through language.[167] Language is prior to the individual and independent of any individual; it is "the place where subject and object, thought and world, meet—or, more precisely, where they are at home together prior to being split asunder by conscious reflection."[168] Because of language humans have a world.

Having a world means that we are intentionally directed toward it, providing a basis for disclosing it, for positing the epistemic opposition of the subject and object of cognition, and for engaging in representational thinking. Here again Gadamer modifies Heidegger's view. The condition of the uniquely human phenomenon of having a world is not the existential constitution of Dasein, but language that is a possession of a community and that determines the way we experience the world. The relationship between a world and a language is reciprocal; they are ontological conditions for each other. "Not only is the world world only insofar as it comes into language, but language, too, has its real being only in the fact that the world is presented in it." Every world is verbally constituted and thus "always open to every possible insight and hence to every expansion of its own world picture, and is accordingly available to others." Consequently, every world is exposed to historical change.[169]

Hermeneutic understanding is the only form of inquiry capable of being applied to ourselves and to human history and as such offers a nonobjectivist view of them; "as our situation cannot be completely objectified because we are in it, so our own being cannot be perfectly known because we are."[170] Hermeneutics offers us nonobjectifying self-knowledge because it is (1) critical of the subject/object dichotomy and of all forms of objectification we undertake; and (2) self-reflective and so able to recognize its own historicity and its direct involvement in the entirety of human experience. However, hermeneutics possesses no miraculous method for self-cognizing that allows us to grasp ourselves as we are-in-ourselves. Since we are finite, historical beings we are always on the way to self-knowledge that can never be completed. Hermeneutic (self-)understanding, the process of exploring meanings, is an endless undertaking "in the twofold sense of being without an aim in which it terminates, and thus in being an

ongoing process which continually renews itself in repetition."[171]
Moreover, we can never reflect ourselves out of language.[172] Our consciousness cannot be entirely self-transparent precisely because it is
linguistic: acts of thematizing ourselves cannot at once thematize language as a medium and as a means of thematizing. On the other
hand, when we thematize language, when it becomes an object of
cognition, we are not "focally aware" either of ourselves as thematizing, speaking, and communicating or of the conditions of objectification. "Every revelation is also a disguise: every consciousness
of an object is made possible by a nonobjectivizing consciousness."[173]
The hermeneutic effort directed toward human self-understanding is
plainly endless. The same interminability applies to all forms of understanding: "the discovery of the true meaning of a text or a work of
art is never finished; it is in fact an infinite process."[174] Understanding
is always contingent and limited by its horizon.

   Horizon is "something into which we move and that moves with
us"; it is "the range of vision that includes everything that can be seen
from a particular vantage point."[175] Its fundamental role is not negative and restrictive, but positive and constructive for three main reasons. First, its existence is an ontological condition for the dialectic of
familiarity and strangeness and, therefore, for the very possibility of
understanding and interpreting, which aims at something different.
Second, it works as a frame of reference for cognition: it allows us to
recognize the relative importance of different subject matters that appear inside the horizon. Three, precisely because every horizon is local and finite it can be acquired, or altered; in particular, it can be
expanded to a higher universality. An absolute, nonlocal and ultimate
horizon cannot be changed (expanded) without ceasing to be the absolute horizon. Every horizon is finite (temporal), because it is neither
fixed and ultimately determined by a historical tradition nor protected from future changes.

   Understanding that expands the interpreter's own horizon leads
to the *fusion of horizons* that constitutes "the one great horizon that
moves from within and that, beyond the frontiers of the present, embraces the historical depths of our self-consciousness" into "heritage
and tradition."[176] The fusion of horizons, like understanding itself, is
a historical phenomenon.

   Gadamer's hermeneutics offers a profound analysis of the historicity of understanding. The contingency and finitude of understanding

are inseparable from our involvement in history; one cannot reflect oneself "out of the historical involvement of his hermeneutical situation."[177] History is not just our environment, mediated by language and manifesting itself as tradition, it is an ontological ground of our being.[178] "In fact history does not belong to us; we belong to it."[179] Belonging to history, we understand our historicity long before understanding becomes self-conscious. An act of understanding is itself "a historically effected event."[180] Being-affected-by-history is a fundamental state of any hermeneutic situation and the consciousness of being-affected-by-history "is primary consciousness of the hermeneutic situation."[181] Hermeneutics belongs to the sphere of historically effected consciousness. Gadamer uses this concept "to mean at once the consciousness effected in the course of history and determined by history, and the very consciousness of being thus effected and determined."[182]

History, and the traditions through which it manifests itself, does not, however, exist independently of us as an external, supra-individual reality determining us and our acts of understanding. It is the interaction between us and tradition that "constitutes both the reality of history and the reality of historical understanding."[183] Because of the involvement in history, understanding has an unavoidably circular structure: "the understanding of the text remains permanently determined by the anticipatory movement of fore-understanding."[184] This hermeneutic circle of understanding is not a logical or methodological circle, but the ontological structure of understanding (interpretation) that underlies the subject/object distinction, in terms of which logic and method operate.[185] The ontological basis of the hermeneutic circle is "the interplay of the movement of tradition and the movement of the interpreter."[186]

This means that any cognitive act, from a hermeneutic interpretation of a text to argumentation in an objectivist way, is already within the hermeneutic circle, whether or not this is recognized. The cognizer is always informed by tradition: prejudices and foremeanings "are not at his free disposal";[187] they are "constitutive of the historicity of our being."[188] Thus, "we belong to a tradition before it belongs to us: tradition, through its sedimentations, has a power which is constantly determining what we are in the process of becoming. We are *always already* 'thrown' into a tradition."[189] Consequently, specific acts of interpreting, though new and different, are not merely willful or unrelated to other acts: they are guided by shared assumptions and thus saved from arbitrariness and subjectivism. That we can discriminate between different interpretations, given the open-endedness of the human situation, is ensured by the fact that all valid practices of interpreting and all genuine acts of innovation are grounded in the continuity of tradition. In this sense human understanding is tradition-impregnated.

Seen in this light, tradition mediates between us and the past and fills a temporal distance between us and any text or phenomenon we interpret.[190] Temporal distance is an obstacle to understanding, but "once we have freed ourselves from the supposition that the meaning of a text resides in reconstructing the sense it had for its authors or their contemporaries, we are in a position to experience historical distance as something positive."[191] Tradition is also a medium through which the historicity of our own being is revealed, because it is connected not only with the past but also with the present and the future: prejudgments or prejudices handed down to us are open to transformations, and they change because we produce the tradition ourselves, "inasmuch as we understand, participate in the evolution of tradition, and hence further determine it ourselves."[192] Moreover, our activity is necessary for the very existence of tradition that "needs to be affirmed, embraced, cultivated."[193] The interrelation between us and tradition is the reason why understanding can never achieve completion. "We are always understanding and interpreting in light of our anticipatory prejudgments and prejudices, which are themselves changing in the course of history."[194]

The operation of the hermeneutic circle is visible in any interpretation as a tension between the strangeness of an interpreted text and its familiarity, as a play between "being a historically intended, distanciated object and belonging to a tradition."[195] The conditions of understanding are our critical response to what is familiar and our openness to what appears as strange—that is, our ability to listen to its message and its claim to truth.[196] Because of our involvement in the hermeneutic circle, the understanding of the past cannot aim at nomological knowledge of past phenomena as instances of universal laws. "Its ideal is rather to understand the phenomenon itself in its unique and historical concreteness."[197] Gadamer opposes understanding to all forms of objectifying explanation, although he does not deny the possibility of explaining historical events.[198] Such an explanation must ignore, however, the historical way of being of its subject matter that is not an object-in-itself.[199] Since it exists in history together with its interpreter, and as they interact in the course of hermeneutic understanding, the historical entity is not independent of our interpretation; it is always a correlative of "the present and its interests" and ultimately of our need to know "the sense of what is feasible, what is possible, what is correct, here and now."[200] Thus, to think historically is to enter into a relation with the past, into an interplay between the preconceptions of an interpreter and the meaning embodied in what is interpreted.[201]

Gadamerian hermeneutics "wants to be a post-metaphysical philosophy, a *prima philosophia* without metaphysics" that leaves behind the relativism/absolutism dichotomy.[202] The concept of horizon allows Gadamer to mobilize arguments against this opposition. The threat of relativism appears only from the perspective of those who cherish the ideal of objective knowledge.[203] They believe that it is possible to transcend any limitation intrinsic in a cognitive horizon and achieve a point with no limitations whatsoever. For Gadamer, this would be transcending into nowhere; whereas, in fact, we can only transcend our horizon into a new, fused horizon.[204] The impossibility of reaching "a nowhere point of view" means that there are no eternal truths and no objectively better ways of understanding. "Truth is the disclosure of being that is given with the historicity of Dasein."[205] As many interpreters emphasize, *prima facie,* this implies relativism.[206] In fact, however, for Gadamer the recognition that understanding is horizonal and related to tradition does not imply nihilism.[207] Historical relativity is not a limitation of truth.[208] Indeed, every type of knowledge has its own claim to correctness and cognitive value, as well as its own ground for justifying this claim. In the case of scientific knowledge, the claim to truth is justified by the objectifying nature of scientific cognition. Hermeneutics's claim to correctness stems from the fact that the process of understanding, which takes place in language and dialogue, "is simply *the concretion of the meaning itself.*"[209] However, what we accept as cognitively valid, within any discourse, has validity that overcomes this particular discourse because it is rooted in the hermeneutic universe of the social practice of (our) linguistic and historical community. Although truth-claims are relativized to culture-bound discourses, "which cannot in principle command universal assent," there "remains the meta-hermeneutical pre-condition," which binds speakers together in an effort to achieve "mutual recognition and openness."[210] Moreover, in understanding, we are related to a tradition "out of which 'the things' can speak to us," but we are also related to these things themselves, as they come to language through tradition.[211]

The expression used in the title of this section—namely, "hermeneutic ontology"—illustrates the general direction of our criticism and revision of Heidegger's and Gadamer's positions undertaken in the next chapter. In their thought both ontology and hermeneutics are present: Heideggerian ontology is hermeneutic and Gadamerian hermeneutics has an ontological objective; but neither conception constitutes a truly hermeneutic ontology that is at the same time an ontological hermeneutics. Heidegger uses hermeneutics to explicate the forestructures of understanding, and to express the nonobjectivizing approach of his ontology; but the ontological structures he reveals are existentialist rather than hermeneutic.[212] Gadamer uses ontological analysis to discuss the historicity of understanding, but he fails to go beyond language and reveal its ontological foundation.

CHAPTER 2

Beyond Heidegger's Existentialism and Gadamer's Linguisticism

OUR APPROPRIATION OF HEIDEGGER and Gadamer is critical in reference to four points in their work: first, the purely existentialist way of understanding the ontological constitution of human being in Heidegger's ontology, which neglects the question of the ontological basis of the sociocultural; second, Heidegger's reductive strategy for grounding the historicity of the world in the temporality of Dasein; third, Gadamer's deontologization of understanding that allows him to posit language as the ultimate medium of understanding; and fourth, their understanding of scientific cognition and the idea of the alienation of the natural sciences.

THE ABSENCE OF THE SOCIAL IN HEIDEGGER'S ONTOLOGY AND GADAMER'S REMEDY

Heidegger's ontology is intended as a fundamental ontology. It seems, however, that it is not really a fundamental ontology, in his own and Husserl's sense, but a regional ontology that analyzes existence in its everydayness. Having rejected the epistemological bias of Husserlian philosophy, the idea that the fundamental way in which we deal with the external is cognition (experience), Heidegger takes it for granted that everyday life is primary to all other human ways of being.[1] So for him the opposition between the cognizing subject and what is present-at-hand (a scenario typical of scientific and metaphysical cognition) is one form of the dichotomy between a doer and what is acted upon. He accepts as given, and in a sense as absolute, the opposition between

Dasein's way of being and the way of being of equipment. This duality is not, literally speaking, a particular form of the subject/object dichotomy; it underlies it.[2] It leads to the opposition between the human (Dasein) and the nonhuman (everything else), or between understanding and merely being-understood. Underwriting the subject/object dichotomy, this opposition provides a foundation for scientific cognition that is "objectifying thematization" and allows Heidegger to show how science emerges from a more primordial way of being.[3] What is particularly confusing, however, is that the human/nonhuman opposition is by no means identical for Heidegger with the opposition between the sociocultural and the natural; for, in fact, both Dasein and equipment are clearly sociocultural.

Although Heidegger's ontology establishes the ontological fundamentality of the social nature of human life, it leaves unclear the ontological status and sources of the sociocultural. This line of criticism, advanced by Buber, Fackenheim, Habermas, Jonas, Levinas, Löwith, Sartre, Theunissen, and Tugendhat,[4] is considered unfair by many students of Heidegger who stress the social nature of Dasein.[5] Buber and Levinas criticize Heidegger for his inability to include in his ontology the relation between I and the Other as Thou, such as love or friendship.[6] Other authors note that ontologically Dasein remains monadic and its being-with is being in the world shared by myself and others.[7]

It is of course true that in Heidegger's analysis human existence is social not only ontically—because we encounter other people factically—but ontologically: being-with is an ontological structure of human existence.[8] That is, the social is a constituent of Dasein, just as Dasein and its world are internally related, inasmuch as Dasein's being is being-in-the-world. The Heideggerian concepts of human being and human life are more thoroughly "socialized," therefore, than in traditional philosophical anthropology, although no more so than in other post-Kantian German conceptions of humanity. One might expect, therefore, that for Heidegger Dasein's understanding is inescapably entangled within social interactions and institutions and embedded in language, habits, and rules, as well as in other cultural structures; and second, that Dasein's interactions with Others are ontologically different in character from its concernful dealings with useful things. Moreover, one might expect Heidegger to treat the social aspect of Dasein's existence as equiprimordial with its practical aspect, with taking care of things, since being-with-Others is equiprimordial with being-in-the-world. Such expectations are not, however, fulfilled. "We vainly search his writings for an account, or at least an acknowledgement, of the rich variety of ways in which one human existence can enter the existence of another: conflict ranging from quarrel to war, physical violence, political and economical subordination, barring access to information, etc."[9] Habermas sees the primary reason for this failure in the fact that Heidegger adopts the framework of Husserlian phenomenology.[10]

Thus, contrary to what Guignon claims, Heidegger fails to show that "To be Dasein is essentially to be a nexus of the socially constituted relations of a culture."[11] At least, he fails to show this in the case of Dasein's authentic being.[12] We also disagree with Brandom, who believes that Heidegger's idea of being-with "can be understood according to the Hegelian model of the synthesis of social substance by mutual recognition."[13] This reading imposes on Heidegger's analytic of Dasein the Hegelian dialectic of the relations of recognition in the master-slave narrative and does not derive it from the assumptions of Heidegger's own position. The fundamental definition of Dasein is constructed by Heidegger, as Guignon states, "in terms of a teleological and hermeneutical self-relation."[14] There is no *mutuality* in the relations of Dasein toward Others; one's own being remains monadic and monological; it remains, in fact, being-there-too, being in the same way that only "implies the Other's being."[15] Heidegger himself stresses that it cannot be "conceived as the occurring together of several Subjects,"[16] but he does not otherwise clarify how it should be understood. Okrent rightly notes:

> To say that Dasein is essentially 'being with' is to say that every individual Dasein, in order to be Dasein, must as a logical necessity inhabit a shared world; and that insofar as it acts in a way appropriate for its community, it has an implicit practical understanding of the other members of the community as purposive agents who also use things as they ought to be used.[17]

Interactions between humans do not constitute the ontological structure of our being. The ultimate ontological foundation of human being-in-the-world is its temporality, its stretching out between birth and death. "The existential-ontological constitution of Dasein's totality is grounded in temporality."[18] In our opinion, Heidegger's phenomenological analysis does not reveal the source of Dasein's sociality, and his existentialist conclusions seem deficient in reference to both Dasein and its world. Heidegger seems to believe that Dasein's relations with others reduce to the fact that Dasein can refer to them and to the destiny of a community. Moreover, all of this is, at best, of secondary importance in comparison to Dasein's dealing with pragmata, as if being-with-Others were only a context for our being-in-the-world. One exists as Dasein, not in virtue of any mutual relations with others, but in virtue of one's unidirectional relation to useful things, objects, Others, and foremost, to one's own death. The only reflexive relation attributable to Dasein is its self-understanding.

Given his phenomenological perspective, Heidegger not only cannot grasp the reciprocality of human relations, but also he cannot grasp their *directedness*. This is a crucial point. In *The History of the Concept of Time,* Others belong to Dasein's world as producers, owners, or users of equipment encountered "along with environmental things."[19] The idea of encountering Others environmentally leads to the conclusion that they are within the world, like tools; whereas Dasein has its world. Nor does this idea preserve the obvious intuition that the way of being of Others is different from the way of being of useful things and objects. Heidegger emphasizes, therefore, that Others "really do not have and never have the world's kind of being."[20] He seems to miss, however, that on the basis of this reading, social relations between Dasein and Others are always indirect, mediated by things, so the way of being of other people is substantially different from my way of being: they exist "as what they do"; I exist as "who I am." Moreover, since we meet other people in the world, Heidegger must admit that in the context of such encounters neither our mode of being-with nor their mode of being is Dasein. Indeed, he acknowledges this: Other's way of being is co-Dasein; my mode of being is self-Dasein. This result could not, of course, satisfy him entirely.

Looking for a more satisfactory view, Heidegger decided that using the term *being-with* at the outset would help to solve the problem.[21] Consequently, in *Being and Time* the concept of being-with is fundamental: Being-with is equiprimordial with being-in-the-world, but Heidegger still claims that Others are encountered inner-worldly "as what they are" and that "they *are* what they do"; even though he emphasizes that they are like Dasein and that Dasein does not refer to them in the same way as it does to things at hand or to objects.[22] Others now become objects of more direct attention than care, namely of *solicitude* or concern in which "the other is initially disclosed."[23] Solicitude has two opposing forms: inauthentic, in which "the other can become one who is dependent and dominated even if this domination is a tacit one and remains hidden from him," and authentic, which "helps the other to become transparent to himself *in* his care and *free for* it."[24] It seems, however, that authentic, liberating solicitude for somebody else is a paradoxical relation because "while seemingly uniting people, it refers them to their radical unrelatedness."[25] Moreover, the only way that two persons can become authentically bound together is through devoting themselves "to the same affair in common."[26]

To sum up: existentially Dasein is always alone in its monadic and monologic being, although it may feel solidarity with the Other to the extent its liberating solicitude lets the Other be authentically—that is, to face his or her own anxious self-responsibility.[27] An authentic alliance, togetherness that is a direct and mutual relationship between Dasein and Others, who themselves are also Dasein, is not present in Heidegger's ontology. In opposition to being-in-the-world, Dasein's being-with-Others lacks an internal ontological structure, and being-with remains the existential situation of Dasein itself. However, if we are not connected by direct relationships, we do not and cannot form a community.[28] In fact, in Heidegger's ontology there is no possibility of the existence of a community of any sort.[29] His ontology is able to provide for only two primordial ways of being: that of Dasein and those of entities different than Dasein (handiness and objective presence), but communities—we will argue later—cannot have any of these ways of being. In Heideggerian ontology, communities, together with their internal social relations, institutions, systems of meanings and values, in short, with their public worlds, do not have their own, ontologically defined, way of being.[30] If Dasein and Others do not form a community, the world in which Dasein dwells is not a social-historical world.[31] This judgment is also considered unjust by writers who refer to the general inclination of Heidegger's philosophy, to his criticism of Husserl's solipsism, and to a few direct remarks, such as the following:

> the world is always already primarily given as the common world. . . . the first thing that is given is the common world— the Anyone—, the world in which Dasein is absorbed such that it has not yet come to itself . . . This common world, which is

there primarily and into which every maturing Dasein first
grows, as the public world governs every interpretation of the
world and of Dasein.[32]

They emphasize that in Heidegger's ontology the world in its very
nature is sociocultural and historical. Dreyfus states that "a plurality of
Daseins . . . *must* uncover a single shared world because background
familiarity and ways of being Dasein are not a matter of private experiences but are acquired from society."[33] Mulhall points out that the
readiness-to-hand "of objects for a particular Dasein is not (and could
not conceivably be) understood as their readiness-to-hand for that
Dasein alone." This is true. But, contrary to what Mulhall claims, this
alone does not imply that Dasein dwells in a genuinely social world.
Nor does it support the conclusion that "Dasein's inherently worldly
Being is essentially social."[34]

Undoubtedly, in so far as he draws an ontic picture of our everyday living, Heidegger is aware of the social, communal nature of the
world;[35] but nowhere does he undertake an ontological analysis of its
sociality. Instead, he appears to treat the borderline between inauthentic and authentic as if it duplicates the boundary line between
public and private; and, moreover, he "focuses upon the pejorative aspects of social reality as repressive and constrictive of the human disclosure of Being."[36] One could say, however, that the social exists for
Heidegger because it is the source of the intelligibility of the world,
and of its worldliness, its being a structure within which being happens. The indispensability of a sociocultural framework for Dasein's
understanding of the world is beyond doubt, even though Heidegger's statements give the impression that useful things are "immediately and non-discursively accessible to Dasein."[37] It is certainly the
case that Heidegger's view is different from that of Husserl or Sartre:
the relation between Dasein and the web of meanings is a dialectical
and mutual interrelation, and Dasein is not in itself a source of meanings.[38] Nevertheless, the social is not analyzed by Heidegger as that
which constitutes worldliness.

When he inquires into the ontological structure of the world, he
asks: "In what way must the world be for Dasein to be able to exist as
being-in-the-world?"[39] Since being-in-the-world is composed of dealings with equipment, this question contains a problem: In what way
must the world be, if Dasein is to be able to deal with things as

meaningful tools? It is obvious that in answering this question Heidegger cannot offer a metaphysical theory based on the concepts of substance and essence. His inquiry proceeds as an analysis of readiness-to-hand as the way of being of entities and the worldliness of Dasein's world.[40] He begins by stating that reference and assignment make an entity into a tool. This idea leads to a question: "In what sense is reference the ontological 'presupposition' of what is at hand, and as this ontological foundation, to what extent is it at the same time constitutive of worldliness in general?"[41]

His further analysis of the phenomena of reference and assignment is, however, far from clear and convincing. He turns to a discussion of significance.[42] He sees Dasein in its familiarity with significance as "the ontic condition of the possibility of the disclosure of beings encountered in the mode of being of relevance (handiness) in a world that can thus make themselves known in their in-itself"; but he also says that the significance itself "contains the ontological condition of the possibility that Dasein, understanding and interpreting, can disclose something akin to 'significations' which in turn found the possible being of words and language."[43] That Dasein's familiarity with significance is a condition of discovering things as ready for use is obvious, since significance is a source of their handiness. However, it is an ontic, not an ontological, condition, because it is a phenomenon that takes place within the world. Heidegger's later statement that an ontological condition lurks within significance itself is either a misleading way of saying that significance itself is an ontological condition that makes possible the disclosure of things as significations or a hint that this ontological condition is hidden within significance. However, Heidegger does not continue his analysis, and worldliness remains the ontic condition of significance.[44]

This move of attributing significance to Dasein—not really surprising within an existentialist ontology—raises serious problems. First, any real difference between significance itself and Dasein's ability to signify and to understand significance disappears; both are existential states of Dasein. Second, the world and worldliness as its ontological condition lack an ontological foundation that differs from Dasein's existential states and structures. Heidegger believes, it seems, that the question of how the world must be, if Dasein is to be able to deal with entities as meaningful tools, may be answered without presupposing two ideas: first, that the world has objectively, or rather intersubjectively, a meaningful nature; and second, that there exist language, as the medium of meanings, and tradition, as their source. For Heidegger neither of these assumptions is necessary, since he straightforwardly reduces ontologically both significance and language to the existential state of Dasein.

Heidegger's construction is not convincing from an external point of view; it contains a sort of "existentialist solipsism": the world is the world of Dasein, even though it is always the world that "I share with the others."[45] Without a source of meaning independent of Dasein, and an encompassing web of social relations and interactions, there is no guarantee that the world of ready-to-hand things is common to different human beings. There is nothing in Dasein's existential structures, or in its circumspective concern, or—for that matter—in individual pieces of equipment, that coerces the "socialization" of the world. Unlike Dreyfus, we think that nothing guarantees that meanings understood by Dasein are "public possibilities provided by society."[46]

To sum up: in Heidegger's ontology there is no autonomous place for the sociocultural. It appears neither as an ontological basis of the existential structures of Dasein nor as a ground of the meaningful nature of Dasein's world, even though the way of being of both is sociocultural. In both cases Heidegger discloses the existential structures of Dasein as the ultimate ontological level that allows us to understand Dasein in its being-with and the world in its worldliness. The basic problem, therefore, in reference to Heidegger's ontology is whether this dissolution of sociocultural facticity does not go too far. In other words, can Heidegger claim that Dasein's existence is necessarily being-with-Others-within-the-common-world without attributing to Others, or to social relations, or to human communities an existence whose structure is not in fact reducible to the existential structures of Dasein?

Our view is that the ontological structures of the historical, sociocultural life-world cannot be reduced either to the existential structures of Dasein or to the structures of the overwhelming, almost Hegelian, Being of the later Heidegger.[47] The existentialist understanding of Dasein must be grounded in the sociocultural; or rather, Dasein and the sociocultural should be considered as mutually dependent, as we will argue in the next chapter. What needs to be emphasized here is that being-with-Others cannot be successfully conceptualized as a state or an activity of Dasein, as its turning-toward-Others; it must be construed as a movement back and forth between the communal and the individual. Moreover, since Heidegger reduces all relations between ourselves and the external to our coping with things, he cannot include interactions between us and others into our activity. It seems, therefore, that one of the sources of his inability to grasp sociality is the fact that his way of reading Dasein's activity is based on the Greek concept of *poiesis,* but not on the concept of *praxis.* Heidegger reappropriates, at best, only certain components of the latter concept.[48] Poiesis is an individual activity, characterized by its means, goals, and a specific use of the product; whereas praxis is thoroughly social and cannot be defined by goals, means, or capacities.[49] As we shall argue in the next chapter, the conception of science as a social enterprise requires the full, but critical, appropriation of the concept of praxis.

The perspective of praxis allows one to realize that there is another simplification in Heidegger's concept of Dasein's activity. Circumspect dealing with equipment constitutes a *technical* aspect of our being-in-the-world; however, there is also the *mythical* aspect of our being that cannot be reduced to acts of manipulating tools. Practical concern can reveal only the practical meaning of entities that are within-the-world—namely, their serviceability and readiness for use. This does not exhaust, needless to say, their entire web of meaning even within the scope of everyday life. Pannenberg is right to ask: "Is it not true that the question concerning the possibilities of human existence is after all always referred for its clarification to the questions about the world, about society, and beyond both of these, about God?"[50]

It is myth and not poetry, as Heidegger believed in his later days,[51] that constitutes a primordial form of our making ourselves at home in the world. In any case, it is not practical concern that makes the world a familiar and intimate home for Dasein; this is always threatened by the possibility of equipment's malfunctioning, absence, or collapse, and is thus always undermined by the possibility of discovering the otherness, alienness, and even the hostility of entities different than Dasein. If—as Gadamer believes—the experience of both familiarity and strangeness is ontologically more basic than understanding and interpretation, it is myth that turns the world into an intelligible and friendly whole in which we dwell.[52] Myth "is the primordial event of human linguisticality, giving Dasein its first orientation within the

world."[53] It supplies us with a kind of global structure in which everything has its own meaning and value not reducible to handiness.

From the viewpoint of our criticism of Heidegger, Gadamer's
perspective seems more promising. Just as Heidegger rejected Husserl's early epistemology, with its view of the subject as a spectator
"who has no place in the world," and conceived the subject as an entity
dwelling within the world, so Gadamer overcomes the Heideggerian
view of understanding as separated from "communication with another
person."[54] Gadamer's hermeneutics is based on ontological premises
that are as anti-metaphysical as anything in Heidegger's ontology. In
Gadamer's view, there is no ahistorical substratum that underlies understanding,[55] but he goes beyond existentialist ontology and replaces
the Heideggerian concept of being-in-the-world with the idea of our
being-in-culture understood as constituted by language, tradition,
and history. Neither of these cultural phenomena is reducible to the
existentialist structures of Dasein. History, embodied in tradition and
accessible through an understanding of tradition, is for Gadamer what
mediates between the existential structures of Dasein, its finitude and
thrownness, and Dasein's ontic activities, such as interpretation. While
for Heidegger facticity is simply an ontological feature of the phenomenon of being-thrown into being, for Gadamer it has a historical dimension: it is realized in virtue of human belonging to tradition. The
human world is primordially—according to Gadamer—a social
world, a world of speech communities.[56] Understanding and agreement are for them "the culminating form," and nothing is excluded
from this form, neither "the specialization and increasingly esoteric
operations of the modern sciences nor material labor and its form of
organization, nor the political institutions of domination and governance which bind the society together."[57]

HISTORICITY OF HISTORY AND TEMPORALITY OF DASEIN

When Heidegger analyzes typical meanings connected with the term
history, he has the historical nature of entities in mind and the general
opposition between history and nature, to which thinkers in the
humanities refer.[58] According to this tradition, historicity is essentially
human, and nature becomes historical by entering into the human
world. Undoubtedly, as Dauenhauer notes, Heidegger is right in arguing that one "does not exist as an extrahistorical subject or ego which

at some moment becomes historical by becoming intertwined with circumstances and events."[59]

That Dasein does not enter history means, for Heidegger, that it lacks ontological conditioning by history, and this leads him to conclude that it is Dasein itself that is the ontological foundation of history. The ontological roots of historicity, which is the way of being of the historical and of history in general, are located in the temporality of Dasein.[60] The temporality of Dasein itself, and the historicity of beings, are constituted by Dasein's authentic historizing—that is, by its happening in a historical way; or, in fact, by Dasein's temporalizing that is its being-toward-death.[61] Heidegger's understanding of history and historicity is, therefore, clearly existentialist and, consequently, reductionist.

The reductionist nature of Heidegger's existentialist interpretation of history is particularly problematic in reference to the historicity of entities different from us. Their historicity is *world-historicity*, which means that it is always derivative and restricted to the sphere of our inauthentic existence, to our factical thrownness; as such, it has no ontological roots other than the authentic historicity of Dasein. From the perspective of existentialist ontology, world-history, the history of entities existing within-the-world is only ontically objective—that is, external and given to humans. It is not genuine from the ontological viewpoint; "the ontological enigma of the movement of historizing in general"[62] dissolves itself—in Heidegger's analysis—into the question of the ontological content of Dasein's authentic historizing. Heidegger does not offer an ontology of human (social) history; rather, he annuls the very possibility of an ontological analysis that could differentiate such history from the existentialist dissolution of history itself. Nor does Heidegger provide an ontological ground for social history when he considers being-with-Others. From this perspective, Dasein's historizing is "a co-historizing and is determinative for it as *destiny.*" Destiny is "the historizing of the community, of a people."[63] The reference to fate, destiny, and community does not, however, provide any nonexistentialist foundation for historicity. "Everything indicates that Heidegger here confines himself to suggesting the idea of a homology between communal destiny and individual fate."[64] Dasein's fate, communal destiny, and history remain firmly grounded in the existential structures of human being. In sum: "Authentic being-toward-death, that is, the finitude of temporality, is the concealed ground of the historicity of Da-sein."[65]

That history remains rooted in the historicity and the finitude of Dasein comes as no surprise in Heidegger's ontology. It has, however, some serious consequences. Heidegger does not squarely consider history as a succession of generations, a process within which the handing down and inheriting of cultural heritage from predecessors occurs. The only way he takes notice of historical heritage is in its connection to Dasein. This is revealed when Dasein chooses a hero to follow and repeats the hero's life, or the unrealized possibilities inherent in it, as "a possibility of existence that has been handed down."[66] Repetition "occasions a reopening of the past by translating that which has been into possibilities to be chosen time and again."[67] In virtue of repetition the past is not simply a collection of isolated facts; it acquires meaning and thereby becomes appropriated by Dasein. However, there is no mutual interaction between Dasein and the past; repetition as fate remains monadic.[68] It is Dasein's resoluteness that grounds the repetition and hands down the past. In taking a stand on its life, Dasein is resolute in the sense that although its life is directed toward the future, confidence is derived from a return to the past, to those possibilities that were inherent in it but not realized.[69] Moreover, the role of historical traditions is often seen by Heidegger from a negative perspective; they "obstruct understanding and constrain existence."[70]

Consequently, for Heidegger historiography is neither a study of "the set of unique things or events which have happened just once and are irrepeatable," nor "some universal scheme impervious to specific things and events"; its aim is "to reveal to the presently historizing Dasein the possibilities already met with and lived through by previous Daseins."[71] Both history and historiography have a key role in serving us: history as the heritage, which Dasein "can hand down to itself as its fate in the course of its own futural taking up of its own possibilities,"[72] and historiography as the analysis that discloses and displays the heritage that is projected before Dasein.

Heidegger's existentialist interpretation of history raises serious doubts. His historicist analyses "remain undeveloped, and this leads to serious problems in the understanding of the relationship between history and social existence, on the one hand, and ontological inquiry, on the other."[73] Constrained by the principles of his existentialism, Heidegger can do justice neither to the mutuality of our relations with others, as the ontological foundation of our individual historicity, nor account for the positive, constitutive role of a dialectic between us and past generations, as the ontological foundation of the sociability and temporality of our world. It seems impossible to explicate one's condition as a temporal being apart from one's placement in a historical context that makes reference to a dialectic relationship between an individual and a society. These doubts prompt the suggestion that Heidegger's analysis of human historicity is not free from metaphysical overtones, especially from the presumption of universality.

Even if the ontic experience of death "as a phenomenon of life" is a universal human experience, Heidegger's existential (ontological) interpretation of death need not be so, notwithstanding his belief that it "is prior to any biology and ontology of life."[74] It discloses death as Dasein's "possibility of no-longer-being-able-to-be-there," which is non-relational and cannot be overcome, since Dasein is always already thrown into it. Heidegger's existentialist analysis of death also reveals that the experience of death lived through by Dasein, who is authentically, is the awareness of its "ownmost potentiality-for-Being, which is non-relational and not to be outstripped."[75] In other words, it is the awareness of being doomed to unrealizable self-making in ultimate solitude. This is not, however, the experience of people who legitimize death within symbolic universes[76] different from Heidegger's; people who believe, for instance, in reincarnation or eternal redemption and—what is more important—live in accordance with their beliefs. In their forms of existence there is no place for the acceptance of "the destiny of human being as final and irreducible," and at their disposal there are techniques that allow them "to annul or transcend the human condition."[77] It also may not be an experience of an archaic man who is free "to annul his own history through periodic abolition of time and collective regeneration" and who begins existence anew each spring repeating the recovery of "the possibility of definitively transcending time and living in eternity."[78] The experience of being-toward-death, understood in the Heideggerian way, is not a universal but a historical experience; it is an experience possible for people who live in the post-Nietzschean world. So what the anxiety of authentic Dasein discloses is "not a transhistorical limit of human being, but merely an historical limit of an age for which 'God is dead.'"[79] To consider being-toward-death as an ontological situation of any human being is a metaphysical move that is itself based on the distinction between authentic and inauthentic being. It is the result of treating human being as an object of philosophical analysis that seeks to discover its essence through the separation of Dasein from its self-understanding in everyday life. Heidegger admits, it is true, that "the interpretations of death in primitive peoples, of their behavior toward death in magic and cult" throw light on the understanding of Dasein; but he goes on to claim that "the interpretation of this understanding already requires an existential analytic and a corresponding concept of death."[80] However, since the results of the existential (ontological) analysis of death are not compatible with part (at least) of the content of people's experience, the analysis violates the ideal of the ontic-ontological circle accepted by Heidegger. This circle "exists because the ontological analysis (which must claim universal validity) rests on an ontical foreknowledge (which is radically particular)"; and it is said to be nonvicious because "the ontical foreknowledge *itself* rises to universality."[81] The claim to universality of the ontic foreknowledge of people who live in a world in which God is dead is, however, purely metaphysical. It is established per fiat. Heidegger in no way demonstrates that his existential interpretation of human situatedness and finitude is or can be the only possible one. We return to this issue in chapter 5.

In general, therefore, the existentialist reduction of historicity does not seem convincing. Gadamer's approach to the problem of historicity, with its stress on the mutuality of our relation with the past, and on the dialectical relation between humans and human culture as the source of their humanity, offers more promise for understanding historicity. It is not, however, free from its own difficulties. Given that there is no world-history that connects our interpretation with the past before we embark on it, we must admit that the interpretation itself establishes the fusion of our horizon with that of the past. Interpretation is at once the condition of the possibility of fusion between two horizons that are unconnected prior to the act of interpretation and the fusion itself. This is exactly the ontological way of reading the concept of hermeneutic interpretation: hermeneutic interpretation is a means not only of cognizing the past and thereby enlarging our own horizon; it is also a means of constituting history as such. Does this mean, however, that acts of hermeneutic interpretation of the past exist by themselves, separated from the nonhermeneutic content of human activity? In the light of the role Gadamer attributes to language, a positive answer seems unavoidable. Language is the medium that conveys both the past, the present, and the tradition that connects them; we have contact with things and each other in language. "We live *in* language and never just *with* language."[82] The Gadamerian solution raises, accordingly, the question whether the ontological grounding of our understanding in language is acceptable and convincing.

LANGUAGE AS THE MEDIUM OF UNDERSTANDING

For Gadamer, understanding certainly occurs in the cultural "space" of the "in-between-individuals." However, this and the fact that he considers the historical, sociocultural world as a sphere in which we have our being does not mean that his philosophy automatically embraces a conception of direct interactions between people. For him social interactions as well as human relations toward the world are pervaded by understanding that exists within language.

In Gadamer's view, as well as in Heidegger's, especially in the latter's later philosophy, "our belonging to language is essentially a way of being bound to Being and world."[83] As Gadamer says: "Language is the fundamental mode of operation of our being-in-the-world and the all-embracing form of the constitution of the world."[84] The world and our activities are linguistic: "the event of Being occurs in and through language"; language is the house of Being, and it reveals Being; our experience and understanding of the world take place within language, and "reason has come to be seen as thoroughly embedded in language."[85] It is language, or rather our linguistic practice, that "gives us a world, not as an instrument which we deliberately use to carve a world, but as an activity in which language and world first appear."[86]

Attributing to language the ontological role of "the medium and the register of all understanding"[87] may lead to its reification and ahistorical absolutization. Therefore, in his meta-critical comments on *Truth and Method,* Gadamer tries to erase any traces of a substantialist view of language from his position: "Language is not the ultimate anonymous subject . . . in which all social-historical processes and actions are grounded, and which presents itself and the totality of its activities, its objectifications, to the gaze of the detached observer."[88]

He insists that the unquestioned fact of the linguistic nature of human activity and social practice does not justify a view of language as the substance of history that conveys its events and mediates between them. Language is not an ultimate ontological structure in which our understanding is anchored, nor is it an ahistorical mediation between the present and the past or between individuals and the world. Precisely, as in the case of history and tradition, language is a phenomenon

of human interaction, "the game of interpretation that we all are engaged in every day."[89] Moreover, language, and the world as a whole, are
fundamentally united in a nonobjective way: language "is capable of revealing this whole because its relation to the world is not objective."[90]

If language is a game in which we participate, then *conversation* and
our engagement in dialogue form the model for hermeneutics. Language spoken in dialogue "performs the communication of meaning
that, with respect to the written tradition, is the task of hermeneutics."[91]
Hermeneutic interpretation of a written text "consists in dialogue with
tradition."[92] Hence, it should proceed according to the inner dialectic
of a living conversation happening in the common sphere of meanings. In conversation "we take what the other says not as an expression
of himself but of the topic at hand, and we take this expression as addressed to us. Dialogue consists in mutual concern with a common
topic."[93] In placing language within conversation Gadamer emphasizes
its eventful and creative character.[94]

Although Gadamer adopts conversation as a model for hermeneutics, he does not analyze the interactional character of dialogue in
terms of social relations. He remains within the linguistic perspective
and stresses the importance of questions and answers in the process of
understanding. To ask a question is to bring its subject matter into the
open. "The openness of what is in question consists in the fact that the
answer is not settled. It must still be undetermined, awaiting a decisive
answer. The significance of questioning consists in revealing the questionability of what is questioned."[95] An interpreted text "puts a question to the interpreter," and to understand the text "means to
understand this question."[96] Gadamer emphasizes that recognizing the
inner dialectic of question and answer allows us to understand more
precisely the concept of historically affected consciousness. "Anticipat-
ing an answer itself presupposes that the questioner is part of the tradition and regards himself as addressed by it."[97]

Undoubtedly, the appeal to conversation illuminates an essential
feature of hermeneutic interpretation, as opposed to the traditional,
objectifying way of studying history. It recognizes the reciprocal, dialectical connection between a text and its interpreter. It also allows for
the realization that dialogue is a source of the intersubjectivity of understanding; dialogue does not require "the transcendent authority
formerly assumed by myth, the sacred, or religion."[98] Dialogue is by itself a critical forum "where arguments can be put forward to sustain

some of our views"; and though these arguments "are never free of rhetoric and human interests," their interaction produces limited, relative objectivity.[99] However, in focusing all attention on the linguistic aspect of understanding, and on the analogy between it and conversation, Gadamer is prevented from appreciating the social foundation of historical understanding. Moreover, his efforts at eliminating any possibility of a metaphysical understanding of language, understood as a medium in which interpretation proceeds, without establishing the ontological foundation of language, results in the deontologization of understanding in so far as he treats it as a purely linguistic phenomenon.

Gadamer, preconditioned by his intention to apply hermeneutic interpretation in philology, law, theology, and history,[100] pays little attention to hermeneutic experience in everyday life, although he does not deny it. Moreover, he decontextualizes hermeneutic interpretation in the sense that he abstracts it from the nonlinguistic elements of the disciplines he is studying.[101] This is a strategy that moves from analyzing real cases of interpretation, as performed by interpreters, with all their idiosyncrasies (no matter how strongly influenced by tradition) and material equipment, to conceiving understanding as a relation between two traditions that exist, in a sense, by themselves. Decontextualization may be one reason why Gadamer "too frequently speaks of 'tradition' writ large, as if the hermeneutical case would not have to be made differently every time with regard to different concepts of traditions, converging different cases such as 'traditions of inquiry' and 'cultural traditions.'"[102] The deontologization of understanding is manifested by the fact that Gadamer replaces Heidegger's ontological conception of fore-having, which underlies theoretical cognition, with pre-judice, "which seems for him to be an implicit belief or assumption."[103] Consequently, understanding is no longer a contentful feature of our being, as it is for Heidegger.

The decontextualization of understanding allows Gadamer to free an interpreted text from its context; a text becomes "detached from its author as well as from its intended reader, i.e., from all things psychological, it is virtually elevated into the sphere of meaning pure and simple."[104] Freed from its initial context, the text becomes open to new meanings. Accordingly, hermeneutics can study it primarily in terms of what is stated in the text, what the text tells us, and not in terms of what it expresses.

Even if such a decontextualized view of historical understanding is an adequate, though idealized, depiction of interpretation in the disciplines addressed by Gadamer, it is far from sufficient for the philosophy of science. Science does not reduce to forms of knowledge, scientific knowledge is not simply a text, and the boundary between dialogue and practice in science cannot be clearly drawn.

The conclusion of our criticism of the existentialist bent of Heidegger's ontology and his view of history, as well as Gadamer's idea of language as the nonultimate medium of understanding, can be stated in three objections. First, the ontological analysis of human being is defective without an ontological analysis of the sociocultural that prevents the reduction of the social nature of our being to our existential constitution. Second, the ontological analysis of history cannot be reduced to revealing its sources in the temporality of Dasein. Third, to protect the ontological reading of the concept of understanding, interpretation cannot be decontextualized; it must be considered as an integral part of the entirety of our practical activity. Although it is contrary to his position, the point can be put in Heideggerian terms: it is not only Others, who are what they do; we also are what we do. This emphasizes that there is more to our role as cognizers than the linguistic dimension. These three objections, however, by no means complete our criticism of Heidegger's and Gadamer's positions. Reservations also emerge in reference to their pictures of cognition and their views of the natural sciences.

SCIENCE AS OBJECTIFYING THEMATIZATION

The reductive, existentialist inclination of Heidegger's ontology in *Being and Time* has a significant negative impact on his concept of cognition. If the only ontological ground for cognition lies in the existential structures of Dasein, cognition remains an individual activity.[105] It is, in other words, a certain mode of human individual being, and its sense is to encounter something that is objectively present. It is true, admittedly, that for Heidegger language is historical and has a social basis; that meaning, or the totality of signification, is articulated primordially in discourse; and that the "intelligibility of being-in-the-world is expressed as discourse."[106] But he fails to draw implications for cognition from the assumption regarding the interpersonal nature of language. Consequently, his concept of cognition does not differ radically from what is found in traditional epistemology. Cognition remains an activity, though not a primary and undetermined one, of an individual human being, though not in isolation from other humans; it has, even if only in supposition, its object outside itself and represents it. It seems, all things considered, that Heidegger's ontology simply supplies the traditional concept of cognition with existentialist foundations.

It is true that from the hermeneutic perspective, objectifying, representational thinking is not the only type of cognition, nor indeed the most fundamental—that is, one that constitutes our primordial contact with reality. There also exists philosophical contemplation and cognition involved in everyday coping with things, which is itself prior to representational scientific thinking. Nevertheless, what Heidegger's ontology provides is an account of the ontological foundation and genesis of representational thinking, based on the a priori assumption that it is a relation between a subject and an object. This is a key issue: it concerns truth and knowledge, as understood by the traditional epistemic enterprise, to which we return in chapter 6. Here we want to emphasize that just as Heidegger's ontology does not provide a view of our being-with-Others, so his ontology of science is unable to overcome the traditional, epistemological and individualistic, approach to science.

For him, science is a way of being of Dasein, not of communities, in particular scientific communities. In reference to science, Heidegger executes a double substitution: he identifies it with scientific cognition and scientific cognition with the theoretical attitude toward entities. This twofold substitution is the reason why he does not undertake an ontological analysis of the social practice of science; and why, moreover, the ontologically constitutive structure of science is for him the theoretical attitude toward the world, which he contrasts with the attitude we have in everyday life. Although Heidegger sees science as based in human everyday practice, he does not consider it to be part of this practice; he excludes it "from the structures of meaning and significance which this practice displays."[107] For him, scientific cognition cannot be conceived as an instantiation of everyday concern in reference to entities within-the-world, because Dasein cannot discover entities as elements of nature in its everyday modes of being.[108] If this were not so, our scientific mode of being would not be able to question and transcend commonsensical, manipulative knowledge of the world.

The derivation of science from our more primordial activities is not analyzed by Heidegger (nor by Gadamer, for that matter) in a satisfactory way. Heidegger's ontological analysis of the transition from the everyday taking care of things to objectifying them is "disturbingly vague."[109] He merely alludes to the temporality of existence as a necessary condition for the theoretical attitude. He does not, however, show how existence, or its temporality, necessitates both the theoretical attitude and the transition from everyday coping with things to objectifying thematization. It seems that Heidegger's analysis is too narrow and one-sidedly existentialist to allow for an understanding of science and its historical emergence. It "results in a negatively defined science derived from the interpretations, interests and instrumentalities of everyday life."[110]

When considering science itself, Heidegger admits that "theoretical investigation is not without its own praxis."[111] This practice, however, does not constitute the essence of science; it remains for him essentially a matter of theoretical research. Rouse rightly stresses that Heidegger's hermeneutics, as well as the conceptions of Quine, Hesse, and Rorty, contains a theory-dominant view of science. For this reason, Heidegger perceives science as decontextualized cognition of isolated things.[112] He fails to grasp the centrality of scientific experimentation, the significance of the local, material settings of science, and the role of technical and practical know-how as revealed in the activities of scientists.[113] Moreover, Knorr-Cetina is right to state that his picture of science "as an abstract 'theoretical' study of isolated properties of objects is founded on nothing more than the *decontextualization* from which it was derived."[114] What is stressed in Heidegger's theory-dominant picture of science is the projective and historical nature of scientific research. In this respect, his picture bears "suggestive similarities with the work of Kuhn, Feyerabend, Polanyi, Hanson, Hesse and other post-positivist theories of science."[115]

In the light of Heidegger's emphasis on the projective character of scientific research, his idea that science discloses something that is a priori appears as the most puzzling element of his view of the objectifying nature of scientific cognition. If the natural sciences are locked into their forestructures—according to the argument of Føllesdal and Rorty—they are caught in the hermeneutic circle, and philosophers cannot consider them in terms of the recontextualization of objects, but only in terms of the recontextualization of beliefs.[116] This criticism is not, however, well aimed: it is not science and its forestructures that determine what can become an object of science.[117] This is the task of ontology, which leaps ahead "into a particular realm of being, discloses it for the first time in its constitutive being, and makes the acquired structures available to the positive sciences as lucid directives for inquiry."[118] Furthermore, there is an even more primordial inquiry—namely, existentialist ontology—that clarifies the very concept of being and discovers objective presence as the way of being of both scientific facts and nature.[119]

There is, however, an important danger connected with the primordiality of (existentialist) ontology. Determination of what counts as an object of science by existentialist (phenomenological) ontology can lead to overlooking the historicity and the facticity of the objects of science. It is true that for Heidegger "since scientific theories state deworlded relations between deworlded data, the fact that such theories are arrived at by worldly practices is in principle irrelevant."[120] There may be various ways, however, in which science deworlds facts, and, if this is the case, its practice is not irrelevant. Only inside scientific practice, inside historically changing modes of research, is one able to find conditions necessary for the scientific process of deworlding facts.

Also Gadamer's view of natural science, based on the idea that induction is the method of science, is traditional, monolithic, abstract, and essentially ahistorical.[121] The presence of method is for Gadamer an indication of an opposition between the natural or social disciplines that pursue methodical scientific research and that part of the humanities that applies hermeneutic understanding. The method of science, based on the subject/object opposition, produces alienation: it prevents their reunion, leads to the domination of the subject over the object and to attempts at changing the world rather than understanding it.[122]

An important consequence of Heidegger and Gadamer's view of the natural sciences is the idea that the sciences, in opposition to (hermeneutic) philosophy, are not self-reflective.[123] They are products of human activity, which they do not study and, therefore, do not describe or explain in their own terms. The existence of scientists, who establish calculations, experiments, or theories, is an ontological precondition of the natural sciences, but scientists as scientists cannot reflect upon it. Practice, which makes science possible, is "taken for granted and ignored by the science."[124] The nonreflective character of science is an unavoidable result of its objectivity, of "keeping the subject outside and separate from the object."[125] For Heidegger and Gadamer, sociological or psychological research that proceeds according to the methods of the natural sciences is also not (self)reflective; such research cannot give us an understanding of science or of our modes of consciousness. Any scientific methodical research takes its objects "as simply given," and the ontological relation between us and those objects becomes veiled, or, if it becomes a theme of scientific study, it is objectified, transferred from the ontological to the ontic sphere. In particular, science is unable to grasp any way of being of entities more primordial than objectified being—that is, being present-at-hand. This is why scientific knowledge is always an incomplete worldview. For example, physics deals with the path any entity takes as it falls freely on the assumption that it possesses a mass and, given appropriate initial and boundary conditions, that an equation governs the beginning and end of the entity's transit of that path. Physics has no concern for what the object is or for its history before and after the entity is described in this nonindividualized and abstract manner.

Adherents of the scientific worldview may question the claim that scientific knowledge is incomplete. For them, science itself is the foundation for denying the existence of any ways of being different from those studied by it. In other words, it gives reasons for rejecting any nonscientific philosophy, which claims the right to its own truth. Yet, such denials—no matter how decisively made—cannot prevent relations or phenomena, which science does not and cannot grasp, from existing, if they exist. This means that the only criterion, to which we may appeal when evaluating hermeneutic considerations of science, is whether they tell us something interesting about science or about ourselves. Moreover, these two objectives are not independent: understanding science is understanding an essential element of the form of life we nowadays lead.

There is another important consequence of the hermeneutic view of science. This is the rejection of metaphysical realism and the fallacy of privileged access to the world that is predicated on this view. The natural sciences are not the source of knowledge about the world as it is in itself, notwithstanding that this is the thrust of their realist claim. The hermeneutic approach also allows us to reject Kantian transcendental constructivism. The natural sciences do not give us universally and objectively (in the human sense) valid knowledge about the world of human experience. For Heidegger and Gadamer, scientific objectification of the human world means only that science tells us its own truth about this world as objectified in a scientific way. According to their view, the worlds of scientific disciplines are relative to their manner of inquiry and thus conditioned by them. In other words, scientific universes are existentially relative to their own horizons and to the human scientific cognitive attitude, which is different from, for instance, the attitude arising from everyday experience of the world. Moreover, the fact that these worlds are connected with language makes the concept of the world-in-itself problematic, because there is no "possible position outside the human, linguistic world," from which it could be discovered in its being-in-itself.[126]

On this point Heidegger and Gadamer's view of science is close to sociological constructivism. This juxtaposition allows us to recognize, without embracing constructivism, the necessity of going beyond the existentialist and linguistic purifications executed by Heidegger and Gadamer: without the concept of science as a subpractice, the idea of the constructive nature of scientific cognition cannot be explicated. An ontological reconstitution of the social nature of science is a primordial condition for analyzing its historicity. These three issues—the social nature of science, its historicity, and its cognitive nature—we discuss in the following chapters.

CHAPTER 3

Fundamental Ontology:
Communities and Practice

To give an ontological account of science we must overcome an explanatory strategy typical in philosophy and sociology of science. This is the strategy of metaphysical "reification," the practice of stabilizing certain realms as a basis for considering science. Those who mobilize this strategy locate science either within the social or the natural realms or between the social and the natural viewed as realities in themselves. When sociological relativists, like Collins and Yearley, say that to understand science we should be natural relativists and social realists who experience the social world "as the day-to-day foundation of reality,"[1] they articulate a socialized ontology of science. They stabilize Society in order to explain science as practice and its world, Nature, as a social construct. When naturalist and realist philosophers of science claim that science should be explained by considering the interaction of scientists with Nature, existing out-there independently of us, they articulate a naturalized (in the narrow sense of the term) ontology of science. They stabilize Nature as out-thereness in order to explain science as that which represents it. Finally, when writers like Knorr-Cetina[2] claim that we must assume the existence both of a recalcitrant material world and a social order and "invent countless combinations in order to mix Nature with Society,"[3] they articulate a naturalized (in the broad sense) ontology of science. They stabilize both Nature and Society in order to explain science as operating in the space between the social and the natural. Each of these views remains within the subject/object dichotomy, and each offers an ontic picture that essentializes nature or society, or both, through naturalization, sociologization, or "discursivization." In any case, none of these views "makes it possible to understand the modern world."[4] Moreover, none is judged ontologically successful in the light of its rival's efforts to destabilize and historicize Nature, or Society, or both. We believe that each of these positions lacks the resources necessary for understanding science as a social and historical enterprise. One possible way of avoiding reification is to elaborate an authentic conception of how human beings dynamically interrelate—especially of how they interrelate with themselves and with things in terms of activity, knowledge, and the exercise of power. Foucault puts forward the basis for such a view. He provides a framework that can illuminate mutually engaged practical activity in which power, knowledge, thought, and action are intimately interrelated.

INTERRELATIONS OF ACTION, KNOWLEDGE, AND POWER

Foucault articulates an ontic picture of collective action that interconnects human actors in fields of action. In this picture, human relations to things cannot be separated from social relations and from relations of individual self-understanding. Our ways of relating to things are mediated through mutual relations with others, and these in turn are implicated in relations with the self.[5] Relating to things through relations to others, and thus to ourselves, we act mutually to create and structure fields of action. In virtue of these interrelations, we so objectify ourselves as the subjects of action and cognition that we become, in fact, as subjects of thought-action.[6] Through this medium we grasp the uniqueness of ourselves in virtue of possibilities for action available in a given situation. Seen in this light, fields of action situate us and structure possibilities whether or not we engage in action, since every act affects a field's structure and thereby other possible and actual actions. Action endlessly overlaps action, and the effects of actions constantly superimpose upon the effects of actions. Certainly actions and their effects do not permanently "structurate" fields, but they decisively configurate and stabilize them through segments of time.[7]

From Foucault's standpoint, thought resides in every act of speaking, doing, behaving, or experiencing. It should be considered as the "very form of action—as action insofar as it implies the play of true and false, the acceptance or refusal of rules, the relation to oneself and others."[8] More precisely, thought resides in individuals' action so long as they are acting toward strategic ends recognized by others who also act within a field. Every thought, every utterance is a deed, and a principled distinction between thoughtful and practical action vanishes. In themselves, of course, action-events and their relations are meaningless; to carry the implications of intentional action they need to be inhabited by thought and organized into practices. The idea that no genuine action can eventuate without thought Foucault calls "the principle of the irreducibility of thought."[9]

This picture allows a reconceptualization, in part at least, of the traditional picture of action according to which action embraces both a doer and a doing. Standardly, the individual is epistemically central, and the doer and individual acts of intentionality carry the source or cause of action. So to explain others' actions we view them as caused by their intentional attitudes, mental states such as desires, beliefs, or goals. On the fields of action picture a nonindividualistic view of intentional action can be developed and mobilized. Often we act, not on individual intentions, but on the normative expectations of others independently of our own positive attitudes. Here reasons for actions are not located causally, but in reference to the satisfaction of norms and criteria of behavior that are publicly accessible. What is at issue is not acting for a reason but acting while having a reason in view. This will be addressed more fully below when we discuss the sociability of consciousness. Clearly, naturalized conceptions of intentionality are at odds with the ontic and ontological perspectives we are developing. On the ontic level our aim is an account of human action that integrates its linguistic, cognitive, and practical aspects. Here we mobilize Foucault's concepts of thought-action and power-knowledge. On the ontological level we view individual and collective action in terms of mankind's collective self-making and the participation of individuals in this project. Indeed, for us, the person and the community are both primitive actualities, the constitutive features and structures of which can only be investigated ontologically.

Broadly considered, on the fields of action perspective there are simply the situated thought-actions of individuals that endlessly interplay with those of others and which constitute the who that each individual becomes. So, individuals are constituted by their interrelatedness, and the sociality of their actions is immanent in such relations. Moreover, the traditional distinction between essence and existence is overcome, since everything is contingent and situated. The fact that actual and possible thought-actions are situated and contingent constitutes fields of action as spheres, which acquire "structuration" by the constant redistribution of relations of power-knowledge.[10] Indeed, discursive and practical acts are constantly interpreted and reinterpreted as they face new and situation-bound applications. In reality, relations of power and knowledge are coextensive and inseparable. But to get a hold of the conception of power-knowledge, power must be first considered as if it were separate from knowledge.

Foucault's conception of power has the following features: (1) "Power exists only when it is put into action";[11] (2) power is not the unique possession of individuals nor a disposition exercised on given occasions nor a relation of dominance in a hierarchy of social positions;[12] most importantly, (3) power's exercise "is not simply a relationship between partners, individual or collective; it is a way in which certain actions modify others";[13] (4) power is a dynamic reality; put succinctly, it is not merely prohibitive and repressive but productive and enabling; not exercised solely from above but distributed widely; not exclusively political or economic but dispersed in endless forms; and (5) every action is countered by an opposing action so that resistance is the necessary correlative to domination and freedom; there is no action that goes unchallenged or which remains unaltered. Power essentially involves the freedom to act and to resist so that freedom is the very condition of its existence: "at the very heart of the power relationship, and constantly provoking it, are the recalcitrance of will and the intransigence of freedom." It is this that makes power what it is and distinguishes it from forms of repression and violence. "Power is exercised only over free subjects, and only insofar as they are free."[14]

Foucault does not conceive power in individual terms but understands it in terms of an event-ontology according to which everything is copresent and interrelated. Power relations are interwoven with individuals as actors sited together at changing points within overlapping fields of action. To the extent it permeates disciplinary techniques directed toward bodily movements and underlies rules for forming statements of truth, power's exercise involves individuals as actors. It refers to their ability to affect themselves mutually, to alter fields of possibility available to themselves and others, and to reconstitute the reservoir of possible actions. Power is in its enactments, in its relations of alignment and counteralignment that configurate individuals and groups. Individuals as they concert together "are always elements of its articulation. In other words, individuals are the vehicles of power, not its points of application."[15]

For Foucault, knowledge-power, thought-action, and persons, viewed as dynamic processes, are deeply interconnected phenomena. The will to knowledge links thought-action and power-knowledge: thought and knowledge constitute a form of action that distributes power relations. This is not to deny that we refrain from overt action and that often we retreat into acts of contemplation, reverie, or memory. Foucault sees power-knowledge as a nexus: "power and knowledge directly imply one another; that there is no power relation without the correlative constitution of a field of knowledge, nor any knowledge that does not presuppose and constitute at the same time power relations."[16]

Relations between power and knowledge are not exterior and unidirectional; they comprise an inner connection of mutual productivity. However, although power-knowledge has an inner mutuality, power-relations have a primacy over relations of knowledge: "the latter would have nothing to integrate if there were no differential power relations."[17] What is important from this standpoint in reference to science is not the content of knowledge or changes in its theoretical form, but rather "a question of what *governs* statements, and the way in which they govern each other so as to constitute a set of propositions which are scientifically acceptable."[18] The subject who knows, the objects to be known, and knowledge itself are correlatives of power-knowledge relations and their historical transformation.[19] It is power that "produces reality; it produces domains of objects and rituals of truth,"[20] and truth and falsity are the "effects of power peculiar to the play of statements."[21] Each society possesses a political economy of truth that indicates which discourses enable truth-telling. These convey the mechanisms that distinguish true from false, embed recognized techniques for making claims to truth, empower those who say the truth, and prohibit certain actors from roles in specific games of truth. Discursive governance does not fall out of the blue. As Visker observes: "Power and knowledge are linked internally; they are linked by the referent."[22] This means that established forms of discursive governance are historically copresent with forms of power indigenous to particular historical periods and, as well, are generated by them.

Of course, discursive relations among statements reflect relations of power enacted in human action, but they are not reducible to them. Similarly, power relations as they permeate discourse are not identical to knowledge, nor indeed are they identical to the symbolic media of communication, although they continually pass through them.[23] Power relations by themselves are heterogeneous, formless, undifferentiated, and nonlocalized. They are the raw "stuff," the dynamic "thereness" that relations of knowledge integrate and stabilize. Forms of knowledge integrate, among other strategies of integration, the power relations they presuppose and actualize.[24] From this perspective, a piece of syllogistic reasoning or the mobilization of a scientific theory are examples of the integration of power-knowledge.

As Foucault rejects substance-ontology (metaphysics), and the Platonic-Cartesian conception of true knowledge as universal, ahistorical, individually acquired, and independent of the contingencies of human action, so he opposes the view that knowledge is an individualized state of mind, a propositional content reflecting (more or less accurately) a world ready-to-be-represented. He emphatically rejects the idea that "knowledge can exist only where power relations are suspended and that knowledge can develop only outside its injunctions."[25] Foucault's picture need not blur the distinction between the warrant for a piece of knowledge and the social conditions of acceptance of that knowledge. A warrant for knowledge is the normative question of how well or ill it stands to its evidence: its acceptance is the descriptive issue of what underwrites its acceptance in a group or society. We return to this issue in chapter 7.

Undoubtedly, Foucault elaborates a convincing ontic conception of the dynamic interrelations among human beings that involves thinking, knowledge, and power relations. His perspective is thoroughly dynamic and anti-essentialist. Every field of action, scientific research included, carries its own inner logic, values, and norms, and its own ways of carving up experience. There are no timeless truths identically present in all forms of human activity: moral and political, social and economic, scientific and artistic. There is no universal method, based on objective reason, that infallibly yields correct solutions to problems in all fields of inquiry. Cultural worlds do not evolve as objective wholes within which everything coheres, but are contingent artifacts of human making. His perspective is also thoroughly synergistic. Action resides in relations of power, power exists as

embodied in action, and knowledge is effective through power relations. Moreover, this perspective indicates that the categories of the
social and the cultural are neither primordial nor explanatorily primitive. On the contrary, they are part of a "surface ontology," of an ontic
picture that needs to be designated in terms of more basic ontological
relations among actors actively engaged in fields of action. Indeed,
this synergistic picture of engaged and thoughtful action embodies a
dynamic of becoming and historicity. Unfortunately, in Foucault's
conception notions of becoming and historicity are presupposed, and
his attention is focused entirely on the exposition of power-knowledge.
No matter how we view them, being, becoming, and historicity do not
reduce to the endless recurrence of difference, a central idea of Foucault's position.

Refusing "to dig below the surface," Foucault surveys the ground
before him which amounts to little more than showing the interrelatedness of knowing and acting as it works in empirical, historical societies. Undoubtedly, he lacks ontological categories. When he seeks to
clarify linkages between knowledge, power, thought, and action he
constantly invokes metaphors such as "mutually implying each other."
Of course, mauling over the sociohistorical data he studies, how these
interrelations work can be understood, but we do not understand
what and how they are. Instead of thematizing being-within-the-field-
of-action and interrelatedness, Foucault asks what sorts of entities
exist in particular fields of action, or—as in his later period—who individuals are and how they constitute themselves. Certainly, he iso-
lates some important ontic relationships, but their roots are never
exposed; one must go beyond metaphors and factual description in
order to explain connections between knowing and acting. Suitably
modified, a version of Heidegger's ontology of being allows us to
problematize what Foucault leaves unproblematized—namely, the
question of interrelatedness. From the Heideggerian standpoint, it
can be considered as a feature of the way of being of ourselves, our actions, and our thoughts, as well as of things that surround us.

We agree with Foucault's rejection of unidirectional, causalist
pictures of how things interrelate. But we disagree that it is sufficient
to replace causal explanation with an (equally ontic) picture of interrelations of power-knowledge and thought-action assumed to be self-
explanatory. Sharing his anti-essentialism, we agree that the foundations
of these interrelations cannot be found in the metaphysical picture of

permanence and change. As an anti-essentialist he might have followed Nietzsche and explained interrelatedness in terms of the
primordial interaction of wills to power driving for dominance and self-
fulfillment. This form of naturalism is, of course, inconsistent with
Foucault's avowed culturalist orientation. Or he might have invoked
Heideggerian fundamental ontology. Unfortunately, this entails an
acceptance of Heidegger's existentialism, an option incompatible
with Foucault's post-Nietzschean genealogy.

BEINGS AND THE WAYS OF BEING

Let us return to Heidegger's views on being. His approach holds
promise in the light of the collapse of various forms of essentialism. If
what is natural is not pre-given, as all antirealists and constructivists
claim; if the essence of society cannot be discovered in structures or
dependencies, as sociologists from symbolic interactionists and ethno-
methodologists to Giddens tell us; if essentialism cannot be justified
by reference to a supranatural Being; and if the transcendentalist or
phenomenological program cannot succeed, what profit is there in
maintaining the metaphysical belief that entities are what they are in
virtue of their essences? If indeed things do not possess essences prior
to their being, if they have features, structures, and relations within it
and not apart from it, what they are cannot be grasped independently
of analyzing their ways of being as experienced by us.

   Insofar as philosophical investigation proceeds according to fundamental ontology, it must formulate the question "of the meaning of
being."[26] To answer this question is, according to Heidegger, to undertake an inquiry essentially different from that "in which entities are
discovered," for example, in a manner scientific research. It means, in
fact, going beyond a study of entities divided into different "definite
areas of subject-matter," such as "history, nature, space, life, human
being, language." Scientific investigations take such areas, delineated
by regional ontologies, as their "respective themes" in order to "examine entities as entities of such and such a type."[27] In doing this, scientific investigations "already operate with an understanding of
being."[28] On the other hand, fundamental ontology, "from which
alone all other ontologies can originate," must analyze existentiality
"as the constitution of being of the being that exists"; it "must be
sought in the *existential analysis of Da-sein*."[29] From this standpoint, in

constructing his analysis of being Heidegger cannot consider it in an abstract and idealized way that aims at discovering the essence of things, because being "is no class or genus of entities."[30]

The focus on existentiality entails going beyond the tradition established by Aristotelian ontology (metaphysics), the aim of which is to investigate being qua being, or, in fact, beings qua existent, or insofar as they are.[31] To maintain this does not mean that one studies their existentiality; rather, their existence is presupposed as a universal and unproblematic predicate that needs no further analysis.[32] Aristotle admits that there are different ways of being pertinent to different categories of entities. However, this pluralistic perspective on ways of being is not foregrounded, since he believes that all have something in common—namely, what is grasped by the primary use of the term *to exist* as it refers to substances.[33] On the basis of this assumption, the question "What is being?" transforms into the question "What is substance?" or "What items exist?" Accordingly, to decide what it is for something to exist, we need to ask what sort of thing it is and discover its essence.[34] In this way, being becomes "substantialized": "the *is* of the thing is the *what* of the thing, not the fact that it exists, but that which the thing is and which makes it to be a substance."[35] Furthermore, the search for essences requires the analytic purification of entities by abstraction and idealization. This means that they must be deprived of all factual and empirical properties considered unnecessary from the metaphysical point of view.

The plurality of ways of being, articulated by Aristotle, was reduced to two fundamental ways by Saint Thomas's separation of existence from essence. There is God's being and the uniform being of finite entities, their existence given in the act of creation. The Aristotelian equiprimordiality of being and essence is preserved only in the case of God, otherwise essence becomes primordial. Moreover, the possibility of essences that do not exist spatio-temporally appears on the philosophical stage. In this light, the real issue of metaphysics becomes substantiality, with essence as its core, and not existentiality. Not surprisingly, in later thinkers, such as Duns Scotus, Suarez, and even in Descartes and Spinoza, existence is a modality of essence either in the sense that it belongs to "a completely individualized essence" or is "the complete actuality of essence."[36] The term *being* in its predicative usage—that is, as used to refer to being—came to be identified with "being" that refers to an abstract substance—that is,

Being. What emerged is a metaphysics of Being without being and a phenomenology of being without Being.[37] Metaphysics of Being has roots in Plato's conception of ideas and in Aristotle's view that everything is composed of form and matter. It is also present in the writings of those medievals who follow Aristotle in the Cartesian idea that the necessary property of entities is thinking or extension, in the phenomenology of early Husserl, and is even presupposed in analytic philosophy. Heidegger's early ontology belongs, undoubtedly, to the phenomenology of being. In surmounting the identification of being with Being and essence, he rejects the presupposition that essence precedes existence and denies metaphysical attempts to transcend immediate reality and ascend toward higher (absolute) Being: for instance, toward eternal necessity. He concentrates on factual human existence and studies its ontological structure. Clearly, to adopt a Heideggerian perspective on being is to reject the widely held objectivist view that the concept of being is "the most universal and the emptiest of concepts."[38] So understood, the concept of being can be equated with the propositional "is" and defined either as the value of a bound variable or as an entirely objectified presence-in-a-particular-space-time-region.[39] In both cases, it becomes a "colorless" predicate, a feature that can be ascribed to any entity whatsoever from the outside and thereby not a real attribute. For Heidegger, being is not a universal property understood as the actuality of essence, and the notion "is" lacks the neutrality and self-evidentiality that could make it independent of other predicates. Indeed, he cannot consider being in an abstract and idealized way, because "Being is always the being of a being."[40]

However, our perspective on being goes beyond Heidegger's view. Not restricting ourselves to his distinction between two ways of being, that of Dasein and that of all other entities, we want to work with a concept of being that has a complex internal structure and manifests itself through different ways of being that belong to various entities. Therefore, we want to replace Heidegger's dictum with this: "Being is always the ways of being of entities." What is central here is that clarifying the structure of being is an ontological issue, whereas clarifying the manifest ways of being is an ontic issue. Both are connected by an ontic-ontological circle of understanding: being and beings (entities) are understood only in terms of an analysis that moves back and forth between ontological and ontic levels. Ontological considerations refer to those features and structures of being that cannot be put aside in the course of abstraction or idealization applied to real entities; this is because without these features entities do not and cannot exist. Assuredly, to adopt Heideggerian being-ontology is to realize that being cannot be considered in a substantialist manner, as a substance, as a sum of given (empirical) features, structures, or relations, even if they are not treated statically but as dialectically related. Being is constituted in the course of being, existing, and our understanding of it is itself an event of being. From this standpoint, "to be" does not mean "to be this particular thing" or "to belong to a given category"; "to be" means "to act upon" and "to be acted upon," or "to constitute oneself/itself" and "to be constituted." Being cannot be separated from acting and from becoming this or that in the course of acting.[41] *Any entity is constituted by its ways of being, and the latter are established in the course of its ongoing activity.*

An important consequence of rejecting an abstract concept of being is the contextualization of being: entities exist in various ways in different contexts. As we will argue in more detail later, being is always situated and characterized by interrelatedness. A hammer is-a-tool in the context of everyday activity, when it is "in use, transparent in the swing of the carpenter intent on driving the nail"; in science it is-an-object, for instance, when it is in "the product-tester's laboratory, being weighed and scrutinized for flaws,"[42] unless, of course, it is used for hammering. Both these ways of being are related to people, who hammer, or for whom a hammer is present as an object and who study its physical or chemical properties as such. Only by neglecting these different relativizations of being can the natural (commonsensical or scientific) attitude lead to a realistic ontic view of what a hammer is, regardless of whether it is a tool for nailing, an object of scientific study, a murderous weapon, or a piece of art. In the ontological analysis we endorse, a concept of pure existence disappears: hammer is-in-hammering or it is-in-scientific-researching.

Given this contextualized concept of being, being-ontology—unlike metaphysics—is not threatened with failure when it approaches wholes and their parts. In metaphysics, in which the content of the concept of being cannot be changed from one context to another, an attempt to portray a thing as existing within one context while not existing in another, or an attempt to present something as being different within one whole from the way it is in another is a hopeless quest.

To be able to say meaningfully, as Latour certainly tries, that facts are produced by scientists and simultaneously not produced by them or by anyone else requires the concept of contextualized being. From this perspective, not only *what* the facts are within the context of scientific practice but also *that* they exist in this context—that is, that they come into presence as scientific constructs—is different from their existence outside science. To contextualize what entities are and that they are is to adopt a radically holistic and synergistic ontology, such as the Nietzschean, in which each thing is nothing but the sum of its effects.[43] A thing is constituted entirely "through its interrelations with, and differences from, everything else."[44] As Nietzsche puts the point: "If I remove all the relationships, all the 'properties,' all the 'activities' of a thing, the thing does not remain over; . . . thingness has only been invented by us owing to the requirements of logic."[45] In this light, objects are such as they are only through their interrelations with one another. Everything, human and nonhuman alike, interrelates dynamically as it enters into changing complexes.

The conception of contextualized being encounters a fundamental difficulty. There is no intrinsic warranty that the different ways of an entity's being compose one coherent way of being-this-entity and, furthermore, that all the ways of being of various entities form one universal being. As Frede rightly notes, different modes of being "do not form any unity that would constitute anything like *the* meaning of being."[46] Heidegger's solution to this problem is twofold: first, care unifies the different ways of being of Dasein (once a user or a carpenter, once a scientist or a philosopher); and, second, care is a common reference point for the various ways of being of entities that differ from us. More precisely, different forms of our understanding, "theoretical understanding, practical concern, solicitude, and the many ways of comportment toward one's own self,"[47] are unified by two forms of care: concern and solicitude.[48]

The existentialist position is not, however, an attitude we can or wish to adopt for analyzing science's way of being, if it is to be regarded as a social and historical enterprise. As others emphasize—particularly Rouse—scientific research is a form of practice, in Heideggerian terms, a circumspective activity, and it "must be understood as a concerned dwelling in the midst of a work-world ready-to-hand, rather than a decontextualized cognition of isolated things."[49] This provides an additional reason for introducing the social sphere as that which is not reducible ontologically to the sphere of Dasein's existentiality: in the relations between different spheres of social life — for example, everyday life and science — one can find dynamisms, such as the nexus of power-knowledge relations, that produce and secure the consistency (even if relative and partial) of different ways of being.

There is one important idea in Heidegger's existentialist standpoint that we want to preserve, namely, his belief that any fundamental ontology must begin with thematizing human being.[50] It must thematize our way of being without relativizing it to nonhuman beings or to Being, understood as Reason, Divine transcendence, biological life, or laws of nature. Otherwise, it will not be able to understand it as a primordial and unique (for us) way of being, and, consequently, it will be incapable of understanding itself as an event of human being, for it will not be able to clarify why it is elaborated from a human standpoint and why it cannot be elaborated from any other perspective, such as the God's-eye point of view. Unlike Heidegger, however, we believe that our way of being need not be thematized solely by contrasting it with nonhuman ways of being. This a secondary contrast and is revealed as a part of our self-identification as participants in the human world. And although self-identification is a crucial component of our self-understanding, it is not its constitutive element. Human self-identification, in terms of the opposition with the nonhuman, is not the only form of human self-understanding; it also has objectivist forms that assume the way we are is not different from the way other entities are, such as La Mettrie's view of *l'homme machine,* a biological account of human beings as organisms, or the idea that we are nonartificial intelligences. A similar ontic and, in fact, perceptual self-understanding of human being can be found in Merleau-Ponty's phenomenology and in philosophers of science who follow him. Heelan, for example, identifies Dasein with physical embodiment within the body itself. "I take the individual human subject to be identical at all times with a Body that he or she uses or experiences; that Body is inserted into its experiential setting, a World (for that subject), within which it is both a noetic subject, and an object through which physical causality flows freely without interference or pause. The human subject as Body, then, is an embodied subject connoting physicalities as well as intentionalities."[51] On this interpretation, being-in-the-world is identified with physical presence and the body's causal interactions with its immediate experiential context. This leads to the privileging of perception as the basic act of a sentient body situated within its ambient causal nexus. All cases of such ontic self-understanding fall short of an ontological understanding of our existence.

The ontology we want to elaborate is not existentialist in a purely individualist sense, and it does not elaborate our self-understanding in any naturalist (and, therefore, ontic) terms. Its main aim is our ontological self-understanding comprised of existentialist, social, and historical concepts understood ontologically. Furthermore, our ontology has to be fundamental (but not foundationalist!), hermeneutic, and historicist. Its *fundamental* character is limited and nonabsolute. We want to consider the grounds of the subject/object dichotomy; but we do not want to refer to nonhuman—that is, ideal, natural, or transcendental—Being or entities in order to ground science. We intend to find the basis of science and its social and historical nature in the human way of being, both individual and social, and to do this without presupposing any epistemological concept of scientific cognition or knowledge. We believe that such an ontology must be *hermeneutic:* it must emphasize meanings and see human activity as pervaded by meanings and their understanding, although, of course, not reducible to them. Finally, the ontology's *historicist* character means that we assume, contrary to Heidegger, that ontology cannot be closed in on itself. It is not a self-sufficient exercise; on the contrary, it requires concretization through purely ontic studies of historical phenomena. To search for such an ontology it is necessary to go beyond Heidegger, though not beyond his idea of elaborating fundamental ontology, in which being precedes beings.

INTERRELATEDNESS AND PARTICIPATION AS STRUCTURES OF BEING

As we have argued, Heidegger's ontology is individualistic in the sense that he solves the ontological issues of being by appealing to the existential structures of Dasein as it is in everydayness, since he believes it is Dasein's (one's own) everydayness. We believe that everydayness is not simply Dasein's everydayness; rather, it is our collective everydayness. Similarly, production is not simply individual concernful dealing with tools, nor is science reducible to objectifying thematization as performed by the scientist. Both are elements of our collective practice within the world. Their communal nature means that besides having an existential ontological structure they also have a social ontological structure.

What is missing in Heidegger's phenomenological analysis of human being is its relatedness understood in a proper sense—that is, not identified with self-referring. Heidegger separates nonhuman beings from human beings, emphasizing that the former are understood and that the latter self-understand themselves. Moreover, in Heidegger's conception, the understanding of other entities serves one general purpose—namely, Dasein's self-understanding, the revealing of the possibilities that constitute its way of being and are its relational "dispositions." Using a pen for writing, Dasein reveals its ability to write, which is a possibility that it already has, because it is-in-the-world. Dasein is not brought into the world by a pen or any other tool; Dasein does not begin to exist as a writing Dasein because of any particular act of using a pen.

Yet, contrary to Heidegger we want to claim that a human being and a tool are interrelated. Since a tool is constituted by being used for doing something, the use of a tool and Dasein's self-manifestation as a user are made possible by the complex of properties or, rather, meanings that already exist within a tool and which make it suitable for a given action. So considered, the relation between possessing a certain functionality (meaning) and being used, which, in fact, means a tool's existence, is a reciprocal, dialectical "conditioning." This reciprocal dependence between a tool's being constituted by an act of its usage, and a tool's possession of features which allow someone to use it, reveals an element of self-constitution in a tool's way of being. Assuredly, this element of self-constitution is not as obvious as in the case of Dasein, but it should not be ignored. A tool is not simply a passive object of constitution executed by Dasein, a mere potentiality for being-a-tool; it makes a contribution to the constitution of its being-a-tool.

The fact that a tool participates in its own constitution also reveals passivity in us: practical understanding, which "consists in an ability to use things appropriately,"[52] is possible due to an intelligibility residing as much in entities we encounter as in ourselves. Heidegger expresses the idea of Dasein's passivity when considering us as recipients of being, together with its entire ontological (existentialist) structure. Consequently, for him, Being is not our construction. He does not, however, introduce interrelatedness with other entities as an onto-logical feature of Dasein. Yet this is a necessary move if we agree that the content of different ways of being of a given entity is revealed when the entity is considered as it acts and interrelates with other entities.

Interrelatedness is the ontological structure of all being and this means that all entities, human and nonhuman alike, are within rela-tions with other entities: that they are and what they are is in no way separated from these interrelations. We cannot ask, for instance, what an entity is by presupposing in an idealized manner that it exists in iso-lation from other beings. Such idealization is ontologically un-justified. The interrelatedness of being is an ontological condition of the actual, ontic connections between different entities; in particular, it is a condition of the subject/object relation, of (mutual) relations between empirical things or actions, of causal dependencies, and of different forms of self-relating, such as modes of human conceptual self-understanding. These cases presuppose that what acts or under-stands and what is acted upon or understood are different entities or actions. They are not, however, simply different by themselves in vir-tue of their essential nature. Their actual, realized form of separability presupposes, but does not show, that each one is-in-relation to the other. It presupposes interrelatedness as its primordial ontological condition. Consequently, it is important to stress here that within the ontic-ontological circle causality is an ontic relationship derivative from interrelatedness.

Our stress on interrelatedness as the ontological structure of being has Nietzschean background. For Nietzsche, entities lack the feature of being unequivocally a "this" or a "that" in and by themselves: "The prop-erties of a thing are [its] effects on other 'things': if one removes other 'things,' then a thing has no properties, i.e., *there is no thing without other things*, i.e., there is no 'thing-in-itself.'"[53] Richardson aptly glosses this view: "Thus, being is not distributed individually, but ranges over net-works of interrelations, over quasi-unified wholes, so that it is never fully adequate to speak in either the singular or the plural voice."[54] Clearly, to embrace the ontological significance of interrelatedness is to replace the perspective of substance-ontology that concentrates on the what-ness of entities with a perspective that focuses on their forness. In speak-ing of the "whatness" and "howness" of an entity, Heidegger says that these features are constituted by its involvement. "The functionality that goes with a chair, blackboard, window is exactly that which makes the thing what it is."[55] So following Nietzsche and Heidegger, we reject two beliefs characteristic of traditional essentialist ontology: that the whatness of something is changeless and that thereby it is not relational. If entities are always already in mutual and dynamic interrelations, we must answer the question "what they are" by reference to their being-for, that is, to their forness.

Forness is a characteristic of both human and nonhuman individual identity and differs radically from whatness, which consists of the essential features of an entity conceived to be statically copresent. Forness has a dynamic structure. As Heidegger argues, the disclosing of a certain for-the-sake-of-which covers over, at the same time, other for-the-sake-of-whichs; that is, the revealing of one way of being closes off other ways of being. A razor used for shaving has a different forness than when it is used for trimming hair, or for cutting someone's throat, or when exhibited in a museum as a precious remembrance of a national hero. The same dynamism of forness occurs in the case of human beings. The forness of a person who is at one moment a teacher is different from that of the same person who is a parent, a lover, a shopper, a computer user, or a politician. Just as the identity of a particular razor may be constituted by forms of forness, so also may be the identity of a person. There are, however, important differences between human and nonhuman identification. What is specific to our identification is the fact that we constitute the who, rather than the for-what, that each of us is. What distinguishes *whoness* as a special case of forness is one's sense of self, one's personal self-awareness. Although it is the necessary ontological element of whoness, it is not the foundation of our whoness, since we are a synthesis of innumerable activities, relations with others, and interactions that go into making our self-identification.

An important element of the interrelatedness of being is its situatedness: that entities are means that they *are-situated* and localized. We are always somewhere and in a certain situation; whatever we do, we do it in a particular context; whatever exists, exists in certain circumstances; however it operates, it does so within a certain framework; whatever happens is local. Being, then, is always being-within, within a situation, site, place, or world. The situatedness of being does not simply mean its localization in a physical place and time. The being-within of an entity is always being-interrelated with other entities; for instance, relating-to them, being-toward them, having-been-assigned, being-for-the-sake-of them, and so on. Things are-in in the sense of being "in" the midst of doing something, in the thick of things, in language, etc. Spatio-temporal localization is nothing more than an ontic manifestation of situatedness, and the latter is an ontological feature of entities which is implied by their interrelatedness by the fact that all being is situated. Analogously, the situatedness of our being should not be reduced to being-situated-within-language. To extract the linguistic aspect of our being (or our linguistic practice from the entirety of our practice) is, ontologically considered, a dubious move. Furthermore, the situatedness of human being need not be seen, in Heidegger's manner, as Dasein's stretching along between birth and death. How situatedness is understood is an empirical, ontic matter. From the philosophical perspective it usually means more than simply having spatio-temporal localization and is connected to human finitude. The traditional metaphysical dichotomy between the finite and the infinite, sometimes equated with the separation of humankind from the absolute, is one of many historically constituted ways of conceiving situatedness. Heidegger's existentialist view is yet another, as is the hermeneutic idea of a horizon. We return to the idea of situatedness in chapter 5.

In applying the ideas of interrelatedness, forness, and situatedness to human beings we also follow one strand in Marx's thought, who—as Margolis puts it—believes that humans should be "identified with the entire structure of social existence taken holistically," itself "intrinsically subject to historical change."[56] Marx remains, however, within the ontic perspective. He conceives particular human beings as "points of social space," as "the ensemble of the social relations."[57] Moreover, he rejects individual conscious self-identification as a necessary condition of our whoness. Unlike Marx we do not identify a human being in terms of more primordial social relations; we conceive interrelatedness as the ontological structure of human being and social relations as its ontic, phenomenal manifestation: individually each of us is within social relations, within our interrelations with other human beings, and we each participate in various human communities, from informal groups, through particular societies, to the entirety of humankind; collectively, as communities, we are our relations.

Individual *participation* in communities and their *embracement* of their members are distinct human forms of interrelatedness. The use of the terms *participation* and *embracement* allows us to avoid those typical of objectivist (ontic) discourse concerning things and their relations. Within this discourse, the relation between an individual (or a subcommunity) and a community is a connection between two already existing objects and is characterized causally, or in terms of parts determined functionally by the wholes to which they belong, or as a relation between means and ends, or between the less important (individuals) and the more important (a community).[58] From our perspective, relations between individuals and a community are not instances of forms of dependence that obtain between self-existing substances. Individual participation in a community is simultaneously an activity of making this community. A community's embracement of its members as persons is also a process of shaping them into agents and actors. Ontologically speaking, social entities, namely, human beings and communities, are constituted within the ontological circle of participation and embracement. Put otherwise, participation-embracement is an ontological feature of our being, within which social being (both individual and collective) constitutes itself in the course of ongoing and mutual networks of influence.

Participation constitutes the social content of the ontological structure of an individual human being, its individual whoness: in virtue of this content it is always already a social being. Together with embracement they form an ontological condition of the social nature of all (human) being, both individual and collective. In this sense, these categories designate simultaneously existentialist and social ontological conditions. This is to say that the interdependence obtaining between participation and embracement is a relationship within which human being-together is constituted in both its forms: my-being-with-others and our-being-together. Conceived as participation, our social being, our being-together, is not just a matter of relating to others, as it is for Heidegger; it is decidedly a matter of our taking part in relationships and being embraced by human communities.

That a community embraces its members means, of course, that they are "encircled," influenced, and shaped by that community. However, as Elias emphasizes, what results is not "something merely passive, not a lifeless coin stamped like a thousand identical coins, but the active centre of the individual." In other words, an individual is not simply an actor, but a person as such. Within a community its members interact with each other and influence each other. "Even the weakest member of society has his share in stamping and limiting other members, however small."[59] Moreover, the community does not exist by itself. It is the participation of individuals that constitutes it in its social being.

In making participation-embracement a fundamental structure within which our being occurs, we give ontological priority to inter-relatedness and participation over the Heideggerian modes of self-understanding. This takes us, we believe, beyond Heidegger's perspective. Heidegger (interpreted in a slightly unfavorable way) thinks that Dasein's being is constituted as being-with through reflective self-understanding; whereas, in contrast, our conception of human being is socio-ontological, human being is constituted by participation in communities or, to be more precise, by one's actually participating in relationships and interactions with other people. What Foucault observes in reference to action, or Margolis in reference to dialogue, namely, that selves "are never fixed and finished entities within any exchange," but are always undergoing a process of reconstitution,[60] holds for all forms of participation. Only from such a perspective, I, thou, and others constitute a *we*; the world is *our* world; an action is *our* deed; my activity is a contribution to what we do, and so on.[61] We replace, therefore, the Heideggerian concept of Dasein as an entity for which being (in particular its own) is an issue, with a concept of us for whom our-being, our human-being-together-within-the-world is an issue. Human being together within the world has three inter-related forms: one's own personal being-together-with-others, or one's participation, our-being-together, and being-for-us.

Emphatically, to grasp the social nature of human being (individual and collective), it is not sufficient to say—as Schutz does following Husserl—that *we* includes you, me, and "everyone whose system of relevancies is substantially (sufficiently) in conformity with yours and mine."[62] Nor is it sufficient to appeal to coexistence in a common social world, or to Dasein's recognition that other selves exist, or to the conformity of individual systems of relevancies. None of these constitutes a *we*.

Ontically, being-together is realized as the net of social relations, and it has—as Carr argues following Hegel—its own internal ontic structure which does not reduce to the existentialist structures of the individual but consists in mutual recognition by individuals of their independence.[63] Mutual recognition is not, however, a purely cognitive acknowledgment of others' existence and independence; it is the struggle for recognition that underlies the relations of an authentic community and the relations of domination that permeate an "inauthentic" community. An authentic community in which participants mutually recognize their independence is genuinely plural: it preserves the individuality of its members yet always carries the potential for conflict. Domination, which violates the mutuality of recognition, is a distorted form of being-together.[64] Ontically considered, relationships and interactions, in particular the interrelations of recognition, constitute *we;* indeed, they are prior necessarily to both our participation and our reflective (self-)understanding. From the ontological standpoint, however, the condition of *we* is plurality, which is ignored by Heidegger but recognized and analyzed by Arendt.

Plurality, in Arendt's account, supplements Heidegger's notion of human mortality (temporality). For Arendt, in virtue of plurality, "to live" means "to be among men" and "to die" means "to cease to be among men."[65] Plurality is a condition of political activity "because we are all the same, that is, human, in such a way that nobody is ever the same as anyone else who ever lived, lives, or will live."[66] This is well and good. But plurality goes further than this and can be generalized as a condition of all human activity and not only of the political. If "to live" means "to be among men," the same may be said of any form of human activity; for example, "to produce" means "to act together" and "to cognize" means "to interact with others." Plurality and being-together form an ontological circle: being-together is conditioned by human plurality as it could not happen in the absence of more than one human being; but it also conditions human plurality since without mutual recognition of individual independence human beings could not be individualized and, what is more important, could not interact.

Recognition of the fundamental social character and structure of human being, together with the disclosure of plurality as its ontological condition, has an important consequence for our analysis: it requires modification of the Heideggerian idea that a fundamental ontology must thematize human way of being existentially, as mybeing. The proper perspective for our ontology is not the Heideggerian first-person perspective, but the hermeneutic perspective of the we; this is still a phenomenological first-person point of view but it is plural, not singular.[67] To adopt the plural first-person perspective is at once to overcome objectifying metaphysics and subject-centered philosophy established by Descartes and developed later, among others, by Husserl, Kierkegaard, and in a different way by Heidegger. These thinkers make an individual human being (either reduced to consciousness, transcendentally purified, or conceptualized ontologically as Dasein) the center for philosophizing: understanding the world (read: epistemologically or ontologically, as in the work of Heidegger) is in each case *mine* and presupposes that self-understanding is prior to other forms of understanding. From the first-person singular perspective, everything is "as it exists *for* the individual or figures in the individual's world";[68] and "the things under consideration can only be understood or investigated in their being-for an individual consciousness, that is, in their status, configuration, or position within a personal, individual order of consciousness."[69]

The first-person philosophical perspective rejects the objectifying standpoint because the position of an external observer is inadequate for a philosophical analysis of human self-understanding. However, the rejection of the possibility of understanding a subject as an object has been equated with the necessity of considering the subject from its own internal perspective. Consequently, both standpoints may be criticized for their individualism, an artifact of ignoring the "social relatedness of the individual, his natural dependence on a life with other people" and neglecting historical forms of the emergence of the self and the phenomenon of individualization.[70] Two other positions view the subject in a nonobjectivist way: the perspective of another human being and the perspective of a participant in a community who expresses (consciously or not) its common viewpoint. The first appears as a theoretical perspective in Habermas; the second is adopted by Rorty in his ethnocentrism and by Kolakowski in his Eurocentrism.[71] The idea of replacing the singular first-person perspective with the plural first-person perspective is more convincing than Habermas's ideal view of social interactions in which a participant reflects from the perspective of the other from that of a second person.

The crucial difference between the singular and the plural first-person perspectives is visible in their ways of treating collectives. From the singular first-person perspective, groups are entities external to the philosophizing subject. The subject can perceive these groups as collectives in which it participates, but it does not need to perceive them in this way, since it does not understand itself in social terms as a member of a community, a viewpoint decidedly different from understanding itself in terms of being within the social world. This is why

Heidegger's conception of the ontological content of Dasein's being-with-others is in danger. Dasein does not necessarily understand itself as participating in human collectives and as engaging in social interactions. From the plural first-person perspective, on the other hand, the I is always already a participant. One's perspective is the perspective of a group, namely, that in which one participates. Indeed, the perspective of a participant is a primordial and inescapable attitude toward self, others, and the world. In other words, we ascribe priority to participation over knowing and understanding, even though we conceive these in Heideggerian terms. So understood, being becomes being-for-us, I becomes an I-participating-in-a-community, and my-being is truly a social being-with.

This move, which gives ontological priority to self-understanding-in-participation over Heideggerian individualized self-understanding, allows us to see that the post-Kantian idea of the *intentionality* of consciousness should be supplemented by the idea of the *sociability* of consciousness. Moreover, sociability and intentionality form an ontological circle of mutual conditioning. The intentionality of human consciousness means that it is always consciousness-of-something.[72] Consequently, the out-there-ness of the world does not require justification. From the Husserlian perspective, the intentionality of consciousness is purely subjective: intentionalities are "the subjective conditions of possibility of the presence (or absence) of objective structures within human experience."[73] On the other hand, for Heidegger, intentionality is conditioned by the ontological structure of human being—namely, by the fact that we are always within-the-world. However, one should realize that since human being is also being-together, individual consciousness is not simply my own, entirely idiosyncratic consciousness of something, but my consciousness as a member of a community, my social consciousness of something that finds expression according to existing patterns of thought. To speak of the sociability of consciousness reflects the shift from focusing on issues of individual intentionality to focusing on being-together-in-the-world. This view overcomes the standard approach to intentionality that theorizes it within the subject/object opposition and exclusively in reference to the individual.

Intentionality of consciousness exists within the interaction of mutual recognition that joins different persons.[74] Both my consciousness and the consciousness of other people become individualized and "placed out-there" through mutual recognition. The sociability of consciousness is the ontological condition of our self-identification as participants: we are able to conceive ourselves as members of communities because our consciousness is social, both in form and content. The sociability of consciousness establishes a link between us and a community even prior to our recognition of being connected with this community, and before we begin to thematize the intentionality and sociability of consciousness. In everyday life we see the world of things we use and the world of social interrelations in which we participate as existing out there; they are as they are and do not need to be constituted by the work of our consciousness. This is not to claim that we are privy to a nonconceptualized given; rather it is to see that we are within a communal world before we actively recognize it for what it is.

The sociability and intentionality of consciousness condition each other within an ontological circle. Without the intentionality of my consciousness, if it were not consciousness-of-something, the consciousness of other people could not be recognized by me as external and different from my own consciousness. Consequently, my own consciousness could not become individualized and recognized by me, because my self-recognition requires the contrast of my consciousness with that of others. On the other hand, without sociability, the intentionality of consciousness could not be realized; it is only because our consciousness is "socialized" that it is able to be consciousness-of-something. If we and the world exist in the interrelation of our being-within-the-world and the world's being-for-us, it is not simply the world that "gives itself to me as something that exists out there."[75] In Husserl's view, only a self-existing and self-determining world could "give itself to us," could impose itself and imprint its internal structure on our consciousness. This is the traditional view of the relation between mind and the world. If, however, the world does not exist by-itself and does not self-determine itself, but exists-in-interrelation-with-us, we must initially learn from others that it exists and in what comprises its internal structure. To say that the intentionality of consciousness can be realized only to the extent we learn from others is not to deny that we contribute to establishing knowledge that refers to the existence and internal structure of the world.

If the way of being of each of us is ontologically being-together and our consciousness is always sociable, we cannot understand ourselves otherwise than in social and—as we will argue later—historical terms. This means that participation in a community cannot be described, as an act of participating, *objectively;* that is, it cannot be described by someone who does not participate in a given community. It can be grasped only through the acts of self-awareness of the participants. Hermeneutics "clearly pursues the course of understanding a society from within."[76] It means, in the second place, that our self-understanding is constituted in communities, even if what is immediately present to us indicates that self-understanding is prior to social participation. Long before we understand ourselves and the communities to which we belong, we already belong to them. So our consciousness and self-consciousness are necessarily socially effected in the hermeneutic sense.

> Long before we understand ourselves through the process of
> self-examination, we understand ourselves in a self-evident way
> in the family, society, and state in which we live. The focus of
> subjectivity is a distorting mirror. The self-awareness of the in-
> dividual is only a flickering in the closed circuits of historical
> life.[77]

Thus, our self-description is limited both by the language of our community, culture, and epoch, and by the web of relations within which we participate. Language and social relations are the equiprimordial structures of our individual being. Language conditions who we are by limiting how we can understand ourselves; but at the same time it opens up possibilities for who we might become "by offering more than one possible 'true' linguistic account" of ourselves.[78] Also the network of social relations that encircles us and is experienced by us conditions who we are and opens up possibilities for projecting our life. This does not mean, however, that the relation between us and the communities in which we dwell reduces to the fact that our understanding is conditioned by communities. We and our communities are always within an ontic-hermeneutic circle: as we cannot understand ourselves other than in a socially formed way, so communities cannot build their self-understanding other than through the individual activity of their members.

Human beings and human communities are ontologically interrelated, and so neither can be reduced to the other. Communities are not simply sums of individuals; their way of being cannot be reduced to individuals' being if participation in a community is to be constitutive for an individual way of being. Furthermore, the being of a community does not reduce to the fact that I and some others say "we." A community does not exist merely as something perceived by individuals or because people identify themselves with it. Moreover, even if we agree with Carr that a narrative account of a group's existence and activity, according to which "the group achieves a kind of reflexive self-awareness as a 'subject,'" conditions its existence and activity, it alone does not constitute the existence of a group.[79]

Given the presupposition that a society is not a sum of individuals, we reject individualism and consider communities (social groups, societies, and finally, the entirety of humankind) as real and as having their own way of being. They are wholes ontically described by sociologists in terms of social structures, or networks of social functions (roles) of individuals, or in terms of social interactions that have stable regularities of their own. Societies are, however — as Elias points out — very particular and special wholes. First, a society is neither a supra-individual substance nor a field of forces considered in analogy with natural forces.[80] Second, it is neither a harmonious unity without "contradictions, tensions, or explosions," nor a "formation with clear contours, a perceptible form and a discernible, more or less visible structure."[81] Openness, changeability, and internal tensions characterize its being as a whole. A society is a whole in the constant process of (re)constitution and (self)reproduction, and this process should be understood historically as leading to new forms of social organization and ontologically as the process of establishing its being and perpetuating it. What needs consideration now is the way of being of communities.

PRACTICE AS THE WAY OF BEING

That the way of being of communities cannot be readiness-to-hand is obvious: they are not instruments that we manipulate (though there are politicians who live such an illusion!). Communities' way of being is also not that of objects (though they are treated as objects by the objectifying social sciences). Nor can the being of communities and their public life-worlds be understood literally as identical to Dasein's way of being. Communities, cultures, languages, science, and so on, are not Daseins.[82] Clearly, Heidegger uses the concept of Dasein in reference to a human being: "Da-sein is a being which I myself am, its being is in each case mine."[83] Although this is an ontic characterization of Dasein, its ontological analysis does not change the reference of the concept of Dasein. It reveals only that Dasein is not "I myself" as a self-constituting and closed monad; but rather that it is a being open and directed toward what it is not, toward tools, objects, others, and finally, toward the future and its own ultimate death.[84]

The fact that individual being is ultimately being-toward-the-end, toward death, is the reason why we think — contrary to Haugeland — that the concept of Dasein should not be expanded beyond human beings, so that one cannot say "a person is a *case* of *Dasein*."[85] Moreover — contrary to Dreyfus — we think that the way of being of cultures, science, etc. cannot be the same as Dasein's way of being.[86] Supra-individual, sociocultural systems might be considered to exist like Dasein, if it were shown that their being is, by their very ontological nature and not merely by historical contingency, a stretching out in time in the Heideggerian sense, a process directed toward death. This is not the case, however, even though it happens that societies, cultures, social institutions, and so on, cease to exist, this is a fact conditioned by historical circumstances and not by their ontological structure. Although the way of being of communities cannot be the same as that of Dasein, they are, of course, similar. A basic ontological characteristic of Dasein — namely, that being is an issue for it — is also a feature of human communities. Human self-understanding and, in general, human understanding of being has two forms, or two sides: an individual and a communal. They are not, by any means, identical. Activities, through which individuals understand (reveal and project) their own being, are different from the practices of communities. Cognition, understood in individual terms, as an activity composed of sensual experience and the exercise of one's intellectual capacities, may serve as a good example of how understanding is attributable to an individual but not to a community. On the other hand, politics is an example of understanding that, by its very nature, cannot be attributed to an individual. It is a communal activity and, to the extent it contributes to the establishment and maintenance of the existence of a society, it may be considered as an activity for which the being of a given society is an issue.

So as to conceptualize social, communal activity we begin with the ancient concepts of *praxis* and *poiesis*, concepts not foreign to Heidegger's analysis. The three fundamental modalities of being that concern Heidegger — namely, Dasein, handiness, and objective presence — can be seen as developments from the Aristotelian concepts of *praxis, poiesis,* and *theoria.* "Heidegger explains the Aristotelian determinations of *praxis, poiesis,* and *theoria* as if they were only modalities of being, thereby rigorously excluding any understanding of their ontic significance."[87] It seems, however, that the most fundamental concept for Heidegger is "poiesis." Indeed, his conception of human everyday activity bases itself primarily on this concept. His reappropriation of the concept of praxis is, if anything, partial: he omits the plural and political dimensions of praxis.[88]

In Aristotle's reading, both poiesis and praxis are individual activities, but they are fundamentally different.[89] Poiesis, or productive activity, is defined by "(a) its beginning, the plan elaborated by the producer, (b) its goal, the completion of the product, (c) the means available for its implementation, (d) the capacities required of the producer, and beyond itself, (e) by a specific use of the product."[90] In poiesis a subject, objects (resources, tools, products), and action are ontically separated and ordered into a stable time sequence. In later appropriations of this concept the Aristotelian structure of production is extended to all rational, goal-oriented forms of human activity.

Praxis, on the other hand, is not defined by any such univocal characteristics. Its end is internal to the activity itself. Moreover, "at the most elementary level (the very life of somebody in relation to and among others), every *praxis* is indeed inscribed within a preexisting network of relationships and of verbal exchanges which create multiple and constant factors of ambiguity."[91] Praxis is "the only activity that goes on between men without the intermediary of things," so it reflects the *condition per quam* of all political life, namely plurality. In the Aristotelian view, praxis is, therefore, in contrast to "production based on knowledge, the *poiesis* that provides the economic basis for the life of the polis."[92]

What is typical of praxis, but not of poiesis, apart from its ambiguity, is boundlessness, unpredictability, and individualization. As poiesis is locked up within the limits of a pre-given schema, so praxis is boundless in the sense that it has "an inherent tendency to force open all limitations and cut across all boundaries."[93] Praxis is unpredictable because its "preexisting network is constantly renewed as newcomers arrive."[94] This unpredictability is closely related to the revelatory character of praxis, "in which one discloses one's self without ever either knowing himself or being able to calculate beforehand whom he reveals."[95]

Finally, it is individualized because it is not executed by an anonymous agent, by anybody, but by a community of individuals that differ among one another and are bound by social relations.[96] In praxis we reveal who we are; we appear to one another in the sphere of public affairs as subjects: in (political) action and speech "men disclose themselves as subjects, as distinct and unique persons, even when they wholly concentrate upon reaching an altogether worldly, material object."[97]

Furthermore, because of the differences between them, praxis and poiesis relate to time in fundamentally different ways. In the case of poiesis, time is an external dimension, along which elements of an action (the end, plan, the means, and capacities of a doer) are ordered; whereas praxis does not have an end, but is an end, as in acts of seeing and knowing, which, for Aristotle, are examples of praxis. By its nature praxis is complete at every moment, and there is no distinction between beginning to activate a potential for action and the completion of this activation, so that no duration need be involved in the action. This does not mean that praxis does not take time, but rather that it is completely what it is at each moment of its duration; at each moment it unifies its past and its future.[98] For example, seeing cannot be analyzed into stages toward its actualization, since it is complete at every moment. That is, there can be no time at which that potential is not fully actualized. Thus, duration or temporality need not be involved in the structure of such actualities as seeing.[99]

For Aristotle, poiesis and praxis are forms of individual activity, constitutive for human ways of living, although it is praxis alone that is unique to humans. Also for Arendt, who follows him, human activity is individual and divided into (1) labor, connected with "the biological process of the human body"; (2) work, providing "an 'artificial' world of things"; and (3) political activity, "preserving political bodies" and composed of action and speech.[100] Although Arendt considers human activity as individual, she notes its social aspect. Action, in particular, goes on between humans, and its condition is human plurality. Action and speech are either concerned with an objective in-between, with "the matters of the world of things in which men move, which physically lies between them and out of which arise their specific, objective, worldly interests," or with the "subjective in-between," the web of human relationships that is intangible though "no less real than the world of things we visibly have in common."[101]

Two points concerning this Aristotelian perspective may be criticized. First, it presupposes the distinction between the natural and the cultural components of human beings and reflects this in separating poiesis, as the reproduction of the life cycle (with various degrees of artificiality, of course), from praxis, as the productive creation of political life. Since nature is regular and deterministic, poiesis is determined, routinizable and predictable, whereas praxis is not. Also, as Arendt emphasizes, labor is "subject to the necessity of life," and work depends "upon given material,"[102] whereas public life is a matter of human creation and is consequently omnipresent, frail, and unpredictable. Human reproduction of the life cycle, however, is culturally differentiated in almost all its elements, so in this sense it is not purely natural. On the other hand, political (cultural) activity should be seen—from the Aristotelian perspective—as influenced by the natural component of the human constitution, if human beings are to remain consistent in their natural and cultural constitution. Moreover, as Arendt herself recognizes, in the modern world there is a tendency to substitute making for acting, which results in praxis becoming instrumentalized and routinized.[103] Finally, rationalization, as a process, seems to be a universal tendency of human interactions.[104] These qualifications justify—in our opinion—the rejection of the opposition between poiesis and praxis at the ontological level: if there are differences, no matter how deep and significant, in human activity, they are historical, ontic products, not ontological structures; their emergence is conditioned by processes internal to human ways of being and not by a pre-given human nature or by nonhuman realities. Rejecting the opposition, we will use the term *praxis* (or *practice*) in reference to the entirety of human activity.

In the second place, to conceptualize the in-between as the public world of things, facts, and artifacts tangibly lying between humans, and as a web of social relationships, carries the danger of objectifying and stabilizing it as an ontological condition of "action and speech," or of human activity in its entirety. Things and social institutions are not, however, absolutely prior, independent, and conditioning structures. They do configurate us; but they are themselves the outcome of human activity. Thinkers who talk about the relationship between humans and artifacts, and, in particular, about the phenomenon of alienation, stress that "things that owe their existence exclusively to men nevertheless constantly condition their human makers,"[105] and that the structure of human life "is both medium and outcome of the reproduction of practices."[106] However, even such a dialectical view privileges beings over being and presupposes the conception of praxis as the activity of human agents possessing a predetermined essence, or, at least, a persistent yet historically modifiable nature. In other words, agents are definite entities, even if historically changeable, and practice is their activity.

In opposition to this approach, we consider *practice,* the entire human cooperative activity, the activity that proceeds in the course of social interactions and refers to things, as the way of being of human communities, societies, and ultimately of humanity. Since a human way of being is always being-in-the-world, practice is also the way of being of the world and of every single thing that participates in the world.

To regard practice as a way of being of human communities does not mean that they are subjects of practice, formed prior to it and independently of it. What they are cannot be separated from their activity, which, in turn, is their being. As individual being cannot be ontologically defined without appeal to activity, so the being of communities cannot be ontologically conceptualized without an appeal to collective activity because, according to the anti-essentialist perspective, we are what we do. It is not sufficient, however, to appeal to political activity; if communities are what they do, if the sense of their being is constituted in the course of their action, their activity does not reduce to purely political forms.

Thus, for us, practice is social and—following Aristotle's notion of praxis—individualized, unpredictable, and boundless because it is ontologically circular—that is, self-referential. It is individualized due to human plurality, as it establishes personal idiosyncrasies that relate us to each other and bind us together into collectives. The individualization of practice does not mean, however, that it is simply a sum of individual activities. As communities do not reduce ontologically to idividuals, so practice does not reduce to actions of individuals, though each of us does, of course, engage in activities that compose practice. Certainly, discursive, cognitive, and practical acts cannot be abstracted from their individual carriers. But also, to act is to have one's actions supported by others and integrated together with their actions in ways that are recognizably appropriate. The idea of the individualized character of practice allows us to illuminate the structure of our participation in communities: we participate in a community not only in the sense that we are within social relations, but also in the sense that we are within interactions during which practice is realized. Thus, as practice is the way of being of communities, so participation in practice is the way of being of individuals.

To consider participation in practice as our individual being goes beyond the statement of several authors, such as Brandom, Dreyfus, and Rouse, that "human beings become subjects only through their participation in practices"; and even beyond Rouse's correction of this view, that to be a subject is already to be "situated historically and culturally."[107] The statement that humans become subjects (even though they are held to be situated culturally and historically) presupposes substance-ontology: it presupposes that participation in practice adds new features to what we essentially are — namely, human beings. It is a view that takes our sociocultural nature to be a "second nature," laid upon our "thinghood" and a continuance of our biological nature. From our perspective, in contrast, participation in practice constitutes us: we are, or we become, humans, beings, organisms, subjects, agents, doers, actors, etc. by participating in practice.

Understood as a way of being, practice is indelibly *holistic:* due to its sociality and historicity, it is not simply a sum of individual deeds or particular subpractices. And its features — namely, individualization, unpredictability, and boundlessness — are not features of individual activity. Only by conceptualizing practice in a holistic way can we realize the idea of an ontic-ontological hermeneutic circle: an ontic analysis of a particular subpractice or individual deed presupposes an ontological view of the entirety of practice; and, on the other hand, an ontological view of practice presupposes an ontic understanding of our own actions.

Practice is *unpredictable* because it is located within an ontic-ontological circle in three senses. First, it perpetually constitutes the network of factual social relations, sociocultural institutions, systems, and so on, as an ontic structural web within which it proceeds. This network is an ontic condition of practice and its actual product. The mutual dependence between practice and the network gives relations and sociocultural systems the status of the *correlatives of practice*.[108] Second, practice constantly objectifies itself and converts into *realities,* into facticity, that, in turn, influences it. Within this circle the world is always a life-world, a world of a given community's practice. Finally, individual activity is interrelated with social structures. This circularity is referred to by Giddens when he talks about the "duality of structure" or the essential recursiveness of social life. The duality of structures "relates the smallest item of day-to-day behaviour to attributes of far more inclusive social systems: when I utter a grammatical English sentence in a casual conversation, I contribute to the reproduction of the English language as a whole."[109] Since any such act is performed within a given language, the relation between an individual act and a supraindividual structure is reciprocal. Polanyi similarly characterizes every individual scientific discovery as dialectically connected with science as a whole: "every scientific discovery is conservative in the sense that it maintains and expands science as a whole, and to this extent confirms the scientific view of the world and strengthens its hold on our minds; but no major discovery can fail also to modify the outlook of science, and some have changed it profoundly."[110] In general, therefore, to perform individual as well as collective activities keeps sociocultural structures, institutions, and systems alive; they would not exist by themselves, independently of our activity. This activity is, in turn, informed by already existing sociocultural structures.

Practice is *boundless* in the sense that it embraces all human and nonhuman ways of being. Society, Nature, and supranatural Transcendence are not realities that are prior to it, but correlatives of practice. They are not, of course, created *ex nihilo* in the course of practice but are products of practice in the sense that they exist relative to its different domains. In particular, Society, Nature, or Transcendence are correlatives of particular (metaphysical) points of view that conceptualize them as stable, independent, and prior realities. Such perspectives belong to practice. To emphasize that society, nature, or the absolute become visible from a viewpoint that is internal to practice is, of course, redundant: there is simply no viewpoint that is external in reference to the entirety of human practice. This does not deny, of course, that a viewpoint may be transcendent and objective in reference to this or that particular subpractice.

Finally, practice is *self-referential in the sense that it is our self-making*. As Fackenheim rightly notes, a nonessentialist (historicist) view of human history requires the rejection of the distinction between our being and our acting. If there is no eternal, pre-given and nonhistorical, human nature that is capable of acting, our being is in fact self-making or self-constitution. What we are, our nature, is itself the product of our practice, and hence—as Fackenheim stresses—it is not "a proper nature at all."[111] We constitute ourselves and our world when we outrun in practice the past constituted by our ancestors, and in turn we constitute future conditions for our descendants.

The self-referential and self-transcending nature of practice, conditioned by the situatedness of being, is—as we will argue in chapter 5—constitutive for historicity, temporality, and consequently for time. Time is within practice; it is the temporality that is "inside" and "encompasses" individual life. Consequently, "the ontologist, inquiring into human being *qua* being and *qua* human should turn historian."[112]

Clearly, to view ourselves as participating in self-making is at odds with the traditional view, according to which we are substantial selves or mental substances, or countable individuals localized in space and time. There are three commitments that constitute the substantialist view of human beings: we possess an unchanging essence, unrelated to anything else and independent of accidental changes; we are self-conscious and disembodied subjects that persist self-identically in time; and we, as individual substances, are agents that act causally as autonomous seats of change in virtue of directed ends.

The identification of human being with practice and the understanding of practice as self-constitution not only preclude essentialism in reference to human nature. They also debar a modified, weaker version of the idea of human predetermination—namely, the Aristotelian opposition between possibility and actuality, according to which actuality is the realization of preexisting possibilities. Practice does not actualize preexisting possibilities; it is not a process of embodiment or materialization. It is the way societies are in virtue of their activities. Moreover, insofar as practice is boundless, it embraces possibilities; indeed, they are constituted during practice. Together with possibilities, practice constitutes impossibilities and thereby constraints. Their way of being is also practical: impossibilities exist as the limits of our plans, actions, and thoughts.

The belief that the self-referentiality, and not the rationality of human action, is its ontological feature, the "existentiale" of practice, leads to the conclusion that poiesis, goal-oriented, rational action, is nothing more than a constituted, derivative form of practice. It is a form elaborated in the course of the repetition of certain deeds that become routinized, arranged into protocols, and subsumed under clear goals and rigorous control. Its routinization is recognized and expressed in different instrumental conceptions of the intentionality and rationality of action. These conceptions are, in turn, the source of the belief that rational action is repeatable and predictable. The idea of poiesis, of a "conduct based on a preconceived project,"[113] such as production, is established from an ontic perspective that presupposes the subject/object dichotomy. It is an objectified and standardly individualist description of practice: having separated an action, a doer, the aim of action, its means, results, rules, etc., it reifies and stabilizes both humans and nature. It conceives action as bodily movement caused by mental states, such as desires, beliefs, goals, or an intention to perform an action. A separation of the mental from the bodily and worldly things is presupposed, and priority is given to deliberate intentional acts of consciousness. Any consideration of thought in action, of thought-action is excluded or, at least, marginalized.[114]

Although, as Nietzsche argues, the separation of the doer and the deed is engraved in the subject/predicate structure of many languages, it is, of course, merely a metaphysical postulate.[115] Nietzsche's attack centers on cause and agency, understood as independent of effects and activity. Believing that everything is interrelated, he rejects causes and agency as standardly understood in philosophies that emphasize individual subjectivity and claim that the self acts *sui generis* in making its own individual world. Following his criticism, we reject any need for the concept of goal-oriented action as an ontological category that defines practice. Thus, unlike many writers, we do not consider rationality as an ontologically necessary structure of human action.

Heidegger's ontological picture of human acting contrasts significantly with rational choice views of human action. For Heidegger, the individual is a potentiality for its possibilities either to be or not to be itself, and not an agent as that term is ordinarily understood. The key to this view is his claim that our situatedness roots itself in our resoluteness and is thereby disclosed as such. It is through resoluteness that an irreducible link is created between the futurity of our unrealized possibilities and the ontological structure inherent in the thrownness of our past. In coming to an awareness of not having achieved what it might have been, the authentic self becomes resolute. Although it is always already directed toward the future, its resoluteness derives its drive from responding to unrealized possibilities inherent in its past and connected serially with its present. Always already we are in the fray, dynamically thrown into the midst of things, our lives directed to outcomes whether or not we commit to them objectively. Within this hurly-burly, the resolute self seizes a moment to take charge of affairs; but this is not the overt act of an agent taking stock of its immediate environment. So the "term *resoluteness* can hardly be confused with an empty '*habitus*' and an indefinite 'velleity.' Resoluteness does not first represent itself and acknowledge a situation to itself, but has already placed itself in it. Resolute, Dasein is already *acting.* "[116] Thus, to see modes of human behavior as deliberate responses to objectively present physical situations, or action as simply the manifestation of an individual's intentional landscape, does not capture the resolute human situation. As we have argued, that situation find its home in the fact that willy-nilly individuals are dynamically interrelated within sociohistorical circles and structured in webs of practice and action that they constitute and by which they are constituted.

That we do not consider rationality as an ontologically necessary structure of human action aligns us with the later Wittgenstein's rejection of the regulist view of the status of rules governing human performances. This view states that the application of a rule is guided by another, higher order rule, or by a justifying reason and holds, moreover, that those rules "form an autonomous stratum of normative statuses."[117] Contrary to the regulists, Wittgenstein locates the correctness of application of rules within practice itself, in social habits and training, not in rationally explicit forms of justification.[118] "A rule, principle, or command has normative significance for performances only in the context of practices determining how it is correctly applied."[119] Criticism of the regulist view can be taken further. It presupposes that rules have an either/or structure: that it is a matter of following this rule or, in fact, another. Accordingly, if actions are understood as entirely rule-governed, there is no space for capturing the endless openness of action, for accommodating the circulation of the infinitesimal differences that constantly accrue to actions within practice.

The concept of practice is the fundamental ontological category necessary for "activating" the ontic-ontological circle within which the dialectical juxtapositions of what is possible and what is actual, of acting and being acted upon, of understanding and being understood, operate and constitute the ontic oppositions of subject and object, of the doer and the thing done, of cause and effect, and so on. Our conception of practice supports the conclusion that the concept of participation has a broader sense than the purely social: it also refers to the content of our being-in-the-world. Our being is not only our participation in communities; it is always our participation in the life-world of a community, the world that is a correlative of our practice. Participation in the world becomes recognized as personal, as, for instance, the involvement of our bodies in nature and our selves in a supra-individual (cultural or sacred) reality, and as communal, namely, as the involvement of our community in its life-world. Our individual participation in the world of a given community is realized always through participating in practice, and it is an ongoing unity of acting, experiencing (feeling), understanding meanings, and evaluating situations. It takes various forms, from acts of practical coping with things, producing and manipulating them, to mythical rituals and rites, mystical acts, ecstasies and trances, religious mysteries and ceremonies, to scientific cognition. This means that our being-in-the-world is not just a matter of using particular tools or understanding the totality of meaningful equipment or simply living in the worldly context. These ways of reading our being-in-the-world are subject-centered: they see the individual as the center of the meaningful everyday world and, moreover, as having the world. On the other hand, each of us, considered as a participant in practice, is decentered; we refer to the world indirectly, through the mediation of practice. We have a world, but simultaneously we are in the world. As a result, for us pieces of equipment are as mediated by practice as are the objects of science. In other words, none of our individual ways of being, using tools, perceiving, thinking, engaging in religious or mystical acts, etc., provides a direct access to the World, to entities as if they were in-themselves.

The choice of the concept of practice as an ultimate horizon for our ontology of science seems unavoidable if we are to articulate a post-transcendentalist ontology which takes seriously the task of transcendentalism, though in a modified form. This modified task is no longer one of carrying out a foundationalist criticism of reason from its own perspective, in the manner of Kant; rather, it follows the Young Hegelians in their recognition of the situatedness of reason within practice.[120] In agreement with the transcendentalist rejection of the search for absolute Being, we locate the conditions of the possibility of our being, acting, and, in particular, cognizing within practice. Our concept of practice is close to the Hegelian and Marxist ways of understanding praxis, although there are significant differences between us and them. Like Hegel and Marx (and like Heidegger after them), we believe that people are what they do. Like Hegel, we aim at sublation *(Aufhebung)* of the idealist/materialist dichotomy, but we reject the idea that Spirit is hidden beneath and behind human practice and history. Also, like Hegel, we want to unify theory and practice, but we do not want to conceive philosophy as the comprehension of the logos embodied in praxis.[121] For us, rather, it is an attempt at human hermeneutic self-understanding, which now disguises itself as representationalist theory of the empirical world and now as a speculative metaphysics of the Spirit. Like Marx, we want to unify the natural and the social within practice; but we do not understand practice as an interchange between the human species and nature taken to exist prior to practice, even if only as "a totality of possible satisfactions" of human needs.[122] We also refrain from following Marx's understanding of the essence of social practice in terms of the goal-directed reproduction of material life, division of labor, class struggle, and revolutionary practice that overcomes the alienation of authentic human nature. For us, practice is the way of being of humankind and nature, so that both are constantly (re)constituted in its course. Next, we agree with Marx that, as Margolis formulates it, "man is alienated insofar as he fails to grasp that his own nature and the world's (the world in which his labor—his praxis—is effective) are the products of his praxis, through history";[123] but we stop short of advocating Marx's revolutionary recipe for overcoming this alienation. Furthermore, like Marx we consider language as a human product; but for us—unlike for him—language, or more precisely, communicative activity, is productive (constructive) just as much as the action of human hands and minds. Finally, unlike Marx (or Heidegger, for that matter) we do not presuppose, within ontological consideration, any internal hierarchical structure of practice. Moreover, for us history is the sphere of the unity of theory and practice, whereas, for Marx, history gives rise to the dualism of theory and practice that it must overcome.[124] Similarly, it is not a task of ontology to divide practice into spheres, even as Arendt did, by distinguishing labor, work, and action. Unlike her, we think that it is practice that constitutes us as an *animal laborans, homo faber,* and *homo politicus* and not our relation to nature and to life processes that are prior to our practice and constitute us as nourishing life, fabricating artifacts, etc.[125] Any division of practice accepted in ontological analysis would be an ahistorical separation requiring concepts more primordial ontologically than the concept of practice. From our perspective, there are no such concepts.

Crucial for our concept of practice is its involvement in the ontic-ontological circle, which means that (1) from the ontological stand-point, practice is the structure of our communal ways of being, and (2) from the ontic standpoint, it is a totality of local (historical and so-cial) subpractices realized by different communities. Moreover, since practice is the structure of the being of communities, it cannot be di-vided ontically into—to use Schatzki's phrase—different *X-ings*. A practice of X-ing is, for him, a set of doings and sayings "linked prima-rily, usually exclusively, by the understanding of X-ing." This under-standing is composed of knowing how to carry out acts of X-ing, how to identify X-ing, and how to respond to X-ing.[126] Schatzki's concept of dispersed practices, identified with shared activities, is construed from a first person perspective, even though it allows him to talk about prac-tices performed by groups. On the other hand, for us subpractices are not X-ings but complexes (composed not only of activities) that main-tain the life of societies and constitute their relative self-containment.

In this light, our concept of practice is radically different from the pragmatist concept of practice as a rational, goal-oriented, and indi-vidual struggle to control and dominate the changing circumstances of one's life. It is also different from the notion of praxis as an in-grained habit or a bit of tacit knowledge;[127] although, practice does contain—in the sense in which we employ it—habits, tacit knowledge, and goal-oriented actions. It also differs from Rouse's view, although we are in accord with many of his ideas regarding scientific research. He distinguishes his view from those of Dreyfus, Winch, Brandom, Turner, the Marxists, and Bourdieu. According to Rouse, practices are "temporally extended events or processes,"—that is, "not just pat-terns of action, but meaningful configurations of the world within which actions can take place intelligibly."[128] He construes them as hav-ing been brought about in part by agents. Composed of situated pat-terns of actions they must be "teleological, holistic, instrumentally mediated, and socially regulated." These are patterns of "ongoing en-gagement with the world," of flexible responses to situations, regard-less of whether agents are taken individually or collectively.[129] We agree with Rouse when he claims that "Violations of a rule that previ-ously encompassed all and only correct instances of [a] practice may nevertheless mark the continuation of a significant pattern of correct performance."[130] This makes room for the real possibility that activities develop and change over time. As will become apparent below, we can accommodate his desiderata within our picture of science, although for us the prior issue is to provide an ontological account of science. We are not offering, as does Rouse, a discussion concerned only with the ontic status of particular research activities that involves an ontic account of their epistemic credentials.

THE CULTURAL AND THE NATURAL WITHIN PRACTICE

We are elaborating an ontology that is neither essentialist nor based on the separation of different sorts of entities, such as subjects and objects, physical structures, social relations and ideas, the ideal and the real, the immanent and the transcendent, the mind and body, things of nature, and things of culture.[131] Nor does it presuppose any natural classification of entities into genera and species. Such dichotomies and classifications pertain to the world of practice and all are correlatives of practice: they are the products and simultaneously the conditions of human activities, in particular of language and cognition. Consequently, they belong to the sphere of ontic, not ontological knowledge.

Unlike the natural attitude of everyday life, we do not believe that the world is pre-given "as the universal field of all actual and possible praxis, as horizon."[132] The relation between our practice and the world is mutual: as the world is the sum total of the correlatives of our practice, so practice occurs *in actu* within the world.[133] Like Nietzsche and Foucault, we reject as fictional the metaphysical idea of the world as a determining ground or *hypokeímenon*. We also accept their idea of the primacy of interpretation over signs, although we part company with their claim that there are no facts, only interpretations, that every *interpretandum* is always already an *interpretans*.[134] We depart from their view since we believe that facts are correlatives of practice, which includes, of course, interpretation but does not reduce to it.

Individual being, either personal, natural, or social, occurs always within-practice. So the very being of entities is relative: they are products of practice, its means, circumstances, agents, and conditions. We reject the idea of the thing that is in-itself (and by-itself); or, more precisely, we reject the entire opposition between the thing in-itself and the thing for-us. Since existing, acting, and interrelating are inseparable, nothing can exist in-itself and nothing can exist by-itself. In other words, the existence of entities is "derivative" in respect of their taking part in practice and entering into interrelations and interactions. From this perspective, gravitation, cats, quarks, or stars exist exactly in the same sense as toasters, scientific theories, friendship, or pictures of Dutch painters. All exist within the world of practice, or, in fact, within one or more particular regions of the world of practice in its entirety. So quarks, gravitation, or stars exist in the world of physics as correlatives of physical research practice; though gravitation exists also in other spheres of the world—for example, in the world of everyday life, the world of space travel, or the world of science-fiction literature—and stars exist, additionally, in the world of poetry. That entities exist simultaneously in different spheres of the world of practice, although they are different in various spheres, means that they are disclosed in their particular ways of being within particular configurations of practice. Thus, within particular practical situations we disclose ourselves as producers, or artists, or subjects of cognition, and the nonhuman discloses itself as a physical body, a tool, an object of theoretical description, or a sacred entity. Since this disclosure happens within interrelations and interactions, it is always a unity of active self-identification and a passive state of receiving identification.

When entities exist simultaneously in different spheres of the world, or move from one situation within practice to another—for instance, from the everyday life world to the world of science, or from the latter to nature mobilized for production—they change their way of being. This change does not, of course, simply happen; nor is it a result of what an entity does by-itself; it is always a process that is worked out practically. Our practical activities, in particular our modes of interpretation (theoretical, hermeneutical), are necessary for establishing the relations of similitude or identity between heaviness and gravitation, between the heart of poets and the heart of anatomy, between the Sun that soars in the firmament and the Sun of Copernican astronomy. In the case of the objects (correlatives) of science, social constructivists describe the ontic manifestation of this process by analyzing the social networks, processes, and activities through which this materializes; and the phenomenological philosophers of science discuss the issue in terms of perceptual profiles and invariances, horizons and hermeneutic interpretation, cultural praxis-laden meanings and "naturalized parts of the furniture of the lifeworld."[135] We discuss this issue in the next chapter.

Unconvinced adherents of substance-ontology will here argue that the being-ontology we outline does not really differ from their position because behind all these changes in ways of being there are entities that undergo transformations. They are partially right: but they see only the ontic half of the ontic-ontological circle. Every particular ontic process of transforming a way of being, such as that which leads from being a tool to being an object of physics, or from being a scientific artifact to being an element of nature, happens in the world already populated by entities. To admit this is not, however, to acknowledge entities as the ultimate ontological units that exist independently both from social cognitive activity and from the entirety of practice. To see entities as the ultimate ontological units requires stripping away all traces of construction they continue to carry, even when taking on a life of their own. From our perspective, such a move is an ontologically unjustified, impermissible idealization that should not be attempted even in reference to the natural sciences and the objects of nature. Scientists do not work in a world populated with things-in-themselves; they work in a world that is "already infused with meaning, intelligibility, and familiarity."[136] So from the ontological standpoint, the separation between the nonhuman (the natural) and the human cannot be accepted as primordial. Indeed, it must be shown as constituted in the course of practice. This, in turn, requires that several concepts be introduced that allow us to go beyond Heidegger's account of the relation between a human being and the world of equipment. Two of the most important of these, natality and meaningfulness, relate to the ontological features of our being.

As we have argued, on the basis of Heidegger's ontology, an ontic image of human beings who become—for example, warriors, courtiers, or users of computers as a result of historical changes in the world of practice—cannot be analyzed adequately for two fundamental reasons. First, Dasein is not constituted by the world as the totality of useful things; and second, the world does not possess historicity independent of the temporality of Dasein. In our opinion, if the Heideggerian concept of *mortality* is supplemented by Arendt's concept of *natality*, and a hermeneutic reading of the relation between us and our world is adopted, the idea of us beginning something and becoming someone can be embraced and clarified.

Arendt's concept of natality restores the balance disturbed by Heidegger's emphasis on human mortality. She clarifies the notion of natality in a phenomenological manner as both an ontological condition and an ontic faculty of beginning something new. Natality interrupts the death orientation of human life, which carries inevitably "everything human to ruin and destruction"; it is inherent in human action "like an ever-present reminder that men, though they must die, are not born to die but in order to begin."[137] The primary ontic, empirical manifestation of natality is human birth, with its potentiality for a beginning. The factual birth of a human being is "the installation of his capacity to initiate."[138] Other manifestations of natality emerge—according to Arendt—in the sphere of politics, where natality is an "ability to create a public space between oneself and others so that freedom can appear," and in the sphere of thinking in its relation to ethical experience.[139] Although action (political activity) and thinking have the closest connections with natality, natality and mortality are the ontological conditions of a large range of human activities. Arendt admits: "Labor and work, as well as action, are also rooted in natality in so far as they have the task to provide and preserve the world for, to foresee and reckon with, the constant influx of newcomers who are born into the world as strangers."[140]

We want to use this broad concept of natality as an ontological (existential) condition of the phenomena of beginning, becoming, creating, and novelty with respect to practice and its world. The concept prompts us to think about ourselves as *beginning* something new and as *becoming* whom we have not been, through our activities that constitute novelties. It also allows us to think about other, nonhuman entities, events, or situations as coming-into-being, in virtue of our beginning something new and so becoming this or that. Seen in this light, natality "gives the world its worldliness" in the sense that it is a condition of the world as a human construction, as the embodiment of our activity of beginning, becoming, and creating. In this sense natality is an ontological (existential) condition of both the social world and the self-transcendence of practice.

On the other hand, what the world gives to natality "is worldly reality," as it is "a human context for the manifestation of its factual, political and theoretical experience."[141] In this sense the world is an ontic structure within which human beginning and becoming occur. We are born into the world, we always begin something new within the world, and we also became whom we have not been in the world.

The concept of natality exposes the mutuality of the relation between us and the world of practice. To account for this mutuality we must go beyond an ontological analysis of the phenomenon of our being and mobilize a hermeneutic understanding searching for *meanings* that may be *interpreted* and not for essences that may be revealed in philosophical speculation, for senses that may be discovered in phenomenological analysis, as in the manner of Husserl, or for tangible features that may be experienced.

As Taylor rightly notes, the hermeneutic concept of meaning as applied to things, actions, situations, or whatever, is different than the linguistic concept of meaning: it has a three-dimensional not a four-dimensional structure. Hermeneutic meaning is always the meaning of something for a given subject or for a community and exists in "a field, that is, in relation to the meanings of other things." Linguistic meaning is also a meaning of something (a signifier) for a subject and exists in a field of meanings, but, additionally, it links a signifier and a referent.[412] So linguistic meanings may be considered as a subclass of hermeneutic meaning. There is, however, one difference between Taylor's conception of meaning and the hermeneutic way of conceiving it. Taylor says that as meaning is always of something "we can distinguish between a given element — situation, action, or whatever — and its meaning," though not in the sense that they are physically separated; in other words, "there can be no meaning without a substrate." We can, however, separate two descriptions of a given element; one, in which "it is characterized in terms of its meaning for the subject,"[143] and another, which is free from meaning. This is exactly the point where hermeneutic ontology differs from Taylor's view. From its perspective there is no substrate, a bearer of meaning that is there to be studied prior to meanings. Whatever is, is through the complex of meanings it has for us.

Read in this way, hermeneutics is an ontological interpretation of human experience, practice, and its correlatives saturated with meanings; that is, in terms of understanding (interpreting) and being understood (interpreted). Its ontological employment means that it seeks the basis of understanding and meaningful structures as they pervade both the objective and the subjective. This search may proceed in different ways. Heidegger, having connected hermeneutic analysis with existentialist ontology, locates the basis of understanding and meanings in Dasein and its finitude; Gadamer situates it in language and the necessity of its use. According to us, it is practice that accounts for the existence of meanings and for their understanding. This does not mean, however, that practice is a meaningless activity that produces meanings; if it were, it could not create them. Practice and meaningfulness are involved in an ontological circle of mutual interrelatedness and reciprocal conditioning. Indeed, meaningfulness is an ontological feature of being. From the hermeneutic perspective the situatedness of entities as correlatives of practice means that they are situated within reference frames of meanings, in axiological frameworks, such as the ethical, aesthetical, mythical, religious, or practical. Through these axiological frameworks their being is contextualized and woven into meaningful networks. "Nothing is by itself ordered or disordered, unique or multiple, homogeneous or heterogeneous, fluid or inert, human or inhuman, useful or useless. Never by itself, but always by others."[144] By regarding meaningfulness as an ontological feature of being we can deepen our idea of forness. Unless meanings come into the picture, forness seems purely objective, as if derived from the utility and functionality of entities. In fact, however, forness consists of meanings and valuations that go beyond the merely utilitarian.

The meaningfulness of being and of the interrelations among different ways of being does not mean, of course, that meanings form an autonomous sphere of reality prior to practice, anymore than it implies that they are mental entities. They too are correlatives of practice: they are established and when in use are always within the realm of practice. Even understanding another culture is practical: it consists in sharing "its know-how and discriminations rather than arriving at agreement concerning which assumptions and beliefs are true." Seen in this way, understanding is not a matter of "translation, or cracking a code," but of "what Heidegger calls 'finding a footing' and Wittgenstein refers to as 'finding one's way about.'"[145] Clearly, the hermeneutic approach to practice allows us to realize that although practice is not—from an ontological perspective—a goal-oriented action, it is not simply the exercise of technical skills that can be described in purely objectivist—for example, physical—terms. Practice, taken as a whole, and together with its elements, is *meaningful* and significant for us. In particular, it has an *axiological* dimension—that is, it is valued by us. The axiological dimension of practice consists of different sorts of values (moral, aesthetical, practical, epistemic, religious, etc.), as well as of norms, rules, and standards. As MacIntyre stresses following Aristotle, there are goods, and virtues as conditions of achieving those goods, as well as standards of excellence internal to different spheres of practice, and the "human conceptions of the ends and goods involved."[146] And, of course, all forms of human action are symbolic, not just those that are linguistic. This means that symbolicity is not reduced to linguistic structures alone, but characterizes the human condition of dwelling in a universe of cultural meanings. Given this holistic picture of human symbolicity, all forms of human intentionality are derived from practice within which human beings pursue free and active lives. This is not to deny that individuals act intentionally in the sense of acting for a reason in order to achieve an end that they desire or want. It is to recognize, however, that human intentional action derives its meaning and significance from the cultural contexts that embrace those actions.

In virtue of their meaningful and axiological nature, our actions are not directed simply toward things, or objects that exist by-themselves, but are directed toward meaningful pragmata—for instance, the tools we use, things we manipulate, products we make, artworks we admire and take pleasure in or consider valuable, things we cognize in (scientific) cognition or refer to in mythological thinking. What is constitutive, therefore, for the being of meaningful pragmata is interrelatedness within the network of meanings. Also our (individual and collective) ways of being are realized within meaningful networks, and when we interact with others we do so within meaningful frames of reference, regardless of whether those interactions involve reference to the other as a person or only as an actor performing a social role.

The ontological meaningfulness of practice implies that it is always *linguistic* and *cognitive*. There is no human activity devoid of a linguistic and a cognitive aspect, and there is no language or cognition, no matter how abstract, not embodied in practice. Like many writers, we believe that "the constitution of language as 'meaningful' is inseparable from the constitution of forms of social life as continuing practices."[147] The same should be said, however, of forms of social life, namely, that their constitution is inseparable from language. In other words, as practice cannot be reduced to purely linguistic activity, so likewise it is not purely material and nonlinguistic.

The same mutual dependency links practice and cognition. We assume, like Habermas, that "the symbolic reproduction of the lifeworld and its material reproduction are internally interdependent";[148]

and, like Foucault, that thought is interrelated with action and power with knowledge. Ontically speaking, therefore, we integrate linguistic, cognitive, and practical aspects of our being. But from the ontological standpoint, we consider this integration in terms of humankind's self-making and our ongoing participation in this process. However, considering cognition as an inevitable aspect of practice, we do not want to conceive reason as simply a situated reality or an embodied activity. Cognition is a dimension of practice consisting in making being intelligible; it is, in its various historical forms (e.g., representational or metaphysical), a phenomenon that entirely pervades practice. This does not mean, however, that every entity correlated with human practice, in one way or another, has its clear and distinct representation or is identical with its conceptualization. As "we are more than we know,"[149] so entities are within the world of practice to an extent that surpasses our immediate acts of knowing; our being does not reduce to knowing, practice cannot be reduced to cognition, and the being of entities other than ourselves cannot be reduced to their passive presence for our (sensual) perception or thinking.

CHAPTER 4

Fundamental Ontology: Science

IT IS NOT UNCOMMON NOWADAYS for philosophers of science to think "our theoretical representation of the phenomenon may not be unique and the aspects of nature to which we attend will be culture-relative. But so what?"[1] This captures the standpoint of those forced to abandon foundationism and to acknowledge openly the social and historical nature of scientific justification and rationality. Over the past few decades the combined efforts of philosophers, sociologists, and historians of science have conjured up a specter that haunts the philosophy of science: its object of study, scientific knowledge and cognition, has lost logical clarity and methodological sharpness; the supremacy of scientific knowledge has become problematic; and cognition is now threatened with reduction either to computational models of how the mind works or to mere social practice. Confronting this situation, philosophy of science can either ignore relativist, historicist, and sociological approaches to scientific research and sustain a neo-positivist picture of idealized science, or it can wholeheartedly accept a purely empirical and sociological view.

To opt for the first choice is to insist that science is an enterprise that aims at truth, or empirical adequacy, or at some other definable epistemic goal, despite the complexities that need to be overcome. From this perspective, interactions among scientists and the practical manipulation of things can be seen either as a context and background for scientific cognition or as a phenomenal sphere through which scientific cognition actualizes itself. Popper and Kitcher, in his minimalist social epistemology, adopt the first approach;[2] Heidegger and Dreyfus, following him, opt for the second.[3] In both cases, however, the social and practical aspects of science are not thought to constitute science as a cognitive enterprise and become irrelevant to what makes science a subject matter for epistemology. Indeed, the social and the practical become nothing more than merely contingent and incidental circumstances that attend the production of knowledge.[4]

To opt for the second choice is to conceive science as a social activity that is not directed toward any particular configuration of cognitive aims. From this standpoint, science involves the manipulation of things, the construction of laboratory micro-worlds, social relations, and interactions with equipment and can be analyzed in ways similar to other types of practice without explicit reference to epistemological concepts. As a result of being "socialized" or "politicized," science becomes disenchanted; it is deprived of its special status, its specificity is clearly endangered, and its epistemological characteristics are either marginalized or threatened with extinction. So making this choice entails that epistemological content must be added to these descriptive features of scientific activity. Otherwise, no justice can be done to the truly epistemological aspects of science, none of which can be eliminated from any perspective that hopes to understand science as a form of cognitive activity.

In our view, philosophy of science is not thrown unavoidably into this *tertium non datur* situation. The possibility of establishing a third position emerges once we realize that questions such as "What is science?" "What is scientific cognition?" or "What are the relations between science and the world?" have been liberated from standard epistemological presuppositions. They show themselves anew for those who believe that traditional epistemological terms "like *observation* and *representation* reveal little about the various epistemic activities that can be associated with those names."[5] We agree with Lynch's judgment, though not exactly for his reasons. If scientific activities are no longer epistemologically obvious, we should rethink the concept of cognition.

One step toward such rethinking has been executed by the hermeneutic phenomenology of science that abandons a purely epistemological way of understanding concepts such as cognition, knowledge, perception, experimentation, etc. It emphasizes their interpretive character and studies "the common hermeneutical structure of reading, perceiving, and observing with the aid of scientific instruments."[6] Even if it adopts—as Heelan proposes—a version of realism (horizon realism), it does not simply presuppose a representational function of a scientific theory or a model; rather, it wants to know "how the model functions empirically; how can it come to play a semantically descriptive role in empirical situations?"[7] What is necessary for securing the descriptive role are "practical empirical procedures such as processes of measurement."[8]

However, even a phenomenologically and hermeneutically transformed concept of cognition is ill suited to grasp the collective nature of scientific cognition, since the knower is still viewed as an individual. Further rethinking of the concept of cognition in reference to science means, in particular, that we ask and answer questions such as the following: does scientific cognition, considered as a form of the sociocognitive, simply amount to a form of social practice, to a collective activity that does not refer in any epistemological sense to anything external to itself? Or, is scientific cognition a communal way of referring to the world so that it can be appropriated in a cognitive way? In our view these questions can only be answered on the assumption that science is, strictly speaking, *collective* and *cognitive*. Needless to say, the interrelation of these two aspects must be carefully established in order to avoid any misleading personification of cognitive communities as subjects of cognition and to avoid the deconstructionist reduction of cognition to forms of social practice seen solely in terms of noncognitive goals and interests.

PRACTICAL EMBODIMENT OF COGNITION

As we mentioned in the previous chapter, ontological considerations need not presuppose a hierarchical structure internal to practice: neither the primacy of practice over theory nor an internal nonhistorical hierarchy of subpractices. Practice is not a "space" within which various concrete spheres, domains, and subpractices assume a stable and well-defined place. It functions, for the most part, as a multidimensional whole in which each subpractice exists in interrelation with other subpractices. Cognition is an aspect of practice that undergoes the historical process of gaining autonomy and of becoming a relatively self-governing subpractice. It gives birth to scientific cognition and becomes embodied in science as a sociocultural system that produces knowledge. Most importantly, the emergence of science marks the transition from individual to social cognition. This is not recognized, however, by traditional epistemology and philosophy of science because they adopt an individualist perspective.

Individualism is undoubtedly still a dominant position in modern epistemology, philosophy of science, and history of science, although recently it has become a target of severe criticism. The belief that cognition is an individual enterprise is the core of Cartesian and Kantian epistemologies. To be sure, Kant revolutionized the traditional concept of cognition, but he eliminated neither its individualist nor its representationalist orientation. For Kant, it is our transcendental epistemological equipment, forms, categories, etc, that condition us as cognizing subjects and make possible both cognition and cognizable objects. If we turn to Heidegger, he too remains clearly within the ambit of epistemic individualism. Epistemic individualism is also embraced by those intellectual historians who believe that ideas are essentially mental constructs and that their history is ultimately a succession of predecessors and successors who proliferate these ideas in varying contexts. Epistemological individualism must be overcome, however, in order to understand the collective nature of science as a cognitive practice.

It is now beyond doubt that the social character of science, embracing, as it does, both the communal context of scientific practice and its interrelation with broader social contexts, cannot be ignored by any adequate philosophical account of science. It is impossible to describe how scientists are situated without considering them as thrown into local scientific communities that comprise social relations, institutions, and socially constituted habits, values, preferences, prejudgments, and biases. Moreover, a sound understanding of the social nature of science is critical for grasping both its cognitive and historical character. Even those who see the history of science as a process that selects and replaces hypotheses, theories, and explanations, itself governed by (transcendental) rules of rationality, must admit that these items are constructed, and then, *objectified*. Couched in Popperian terms, their objectification means that they must become elements of the third world; or, as seen from Lakatos's perspective, that they become public. Of course, they achieve public status only through processes, institutions, and relations that constitute the sphere of what is public. There would be no science as an empirical phenomenon without social, supra-individual networks of relations, processes, etc. Let us direct our attention, therefore, to issues squarely connected with the social nature of science. Epistemological and historical matters will be considered later.

Emphasizing the practical embodiment of (scientific) cognition means that, unlike Heidegger, we do not want to separate science from (everyday) practice. If scientific knowledge is not the product of detached, disinterested, and objective reflection; if scientific cognition is conditioned by social structures; and if scientific facts are "the products of culturally and historically specific human interests and achievements,"[9] then science cannot be abstracted away from the entirety of practice, so that it stands in contrast with it and is subjected to different mechanisms.[10]

There are, of course, Marxist conceptions in which science is treated as part of social practice—for instance, Althusser's structuralist conception, Nowak's idealizational theory of science, or the historical epistemology of Kmita. They are based, however, on ontological foundations that differ from the ontology we articulate here. Nowak and Kmita give, first, a methodological description of scientific acts and their interscientific functions, and, second, they describe functionally the social role of scientific practice. They characterize the relation between science and practice in accordance with the Marxist "layer cake model": science is functionally determined by more fundamental spheres of praxis, in particular by the economy. According to Kmita, science "codifies and deductively systematizes predictive elements of the direct, subjective social context of the basic practice, and of subjective contexts of the remaining types of social practice functionally subordinated to the basic practice."[11] Moreover, scientific knowledge is genetically and functionally determined by demands of social practice.[12] According to Nowak, science adapts to demands of the socioeconomical basis and contributes to the development of this basis by projecting new means of production, new social and economical relations, new political and legislative systems, and so forth.[13] In so far as they advocate an ontology of the social world, these positions leave it unclear just how material things determine nonmaterial ones or, indeed, how social practices embrace cognitive characteristics.

To begin an elaboration of our sociohistorical ontology of science we must first note that to understand scientific cognition as belonging to practice is to abandon a theory-dominant view of science; this orientation is found in most philosophies of science, in the theoretical hermeneutics of Hesse and Rorty as well as in the work of Heidegger.[14] Having assigned a preeminent role to theories, they consider experiments and observations as "significant only within a theoretical context."[15] What is opposite to a theory-centered view of science is an experimentation-centered one adopted in the hermeneutic phenomenology of science, represented by Heelan, Crease and Ihde, and in Rouse's pragmatic view of science.

Heelan emphasizes that the thrust of the hermeneutic phenomenology of science is centered on experimental phenomena and their constitution as perceptual objects. For him, scientific phenomena "are 'dressed' for the world by standardized scientific instruments used as readable technologies."[16] Hermeneutical analysis, concentrated on empirical acts such as perceiving, observing, experimenting, and measuring, conceives them as cases of understanding and looks for their common hermeneutical structure.[17] Perceiving, for instance, is "understanding a perceptual object, recognizing it, and naming it"; to perceive something is "to be able correctly to interpret *directly*—that is, to 'read'—the conditions of its presence; these conditions are textlike structures in the World ('texts')."[18] From this perspective, experimental phenomena are praxis-laden rather than theory-laden because theory-ladeness means nothing more than the fact that theoretical terms are used to name observations and any theory "has become embodied and hides itself in a public praxis."[19]

For Rouse, "laboratories can no longer be regarded as merely incidental to the achievement and assessment of referential success," since reference to the world "is achieved only through the mediation of the constructed world of the laboratory."[20] This means that "scientific research is a circumspective activity, taking place against a practical background of skills, practices, and equipment (including theoretical models) rather than a systematic background of theory."[21] It seems, however, that although Rouse replaces a theory-centered view of science with one that is practice-centered, he remains within the traditional epistemological approach. He simply states that scientific research is "a kind of practical activity, one that reconstructed the world as well as redescribed it."[22] Clearly, he presupposes that scientific research is a cognitive practice, an activity that has epistemic (representational) functions: "some theoretical models realistically describe some concrete situations (and not merely the 'observable' ones)," and experimental micro-worlds are "local reconstructions of the world."[23]

We opt for a practice-centered view of science that goes beyond simply equating scientific subpractice with perception, experimentation, and measurement. Both theory-centered and experiment-centered views are endangered by an attempt to decontextualize "practical empirical procedures" and by a tendency to take for granted the cognitive character of scientific activity. From their standpoint, the cognitive character of science does not need to be viewed as constituted in history, during the development of science; science simply (and somewhat mysteriously) carries that character. To avoid this outcome, we begin not by presupposing epistemological concepts that allow science to be understood as cognition, or as an information process, or as a composition of theorizing and experimenting; we begin with a nonepistemological view of science, understood in terms of practice. In the course of our ontological analysis we will elaborate its epistemological meaning—namely, the cognitive character of scientific activity. This is, in our opinion, a proper way to clarify what Crease calls an "antinomic" character of scientific knowledge that on the one hand is "a social product, the outcome of a concrete historical, cultural, and social context" and, on the other hand, "has a certain objectivity or independence" from this context.[24] Our analysis of the social, and later of the cognitive character of science (while temporarily suspending the analysis of the historicity of science), will indicate that the social and the cognitive can be constructed only in an ontological and not in an epistemological or an ontic way, which, by its very nature, requires factual particularities. Assuredly, this does not imply a search for universally valid concepts, but rather a search for concepts that refer to our ways of being and which are filled with various contents at different times. Our aim in reconstituting the concept of scientific cognition is to find those ontological structures that are conditions of (scientific) cognition as a sociohistorical enterprise.

SCIENTIFIC RESEARCH AS A FORM OF PRACTICE

We want now to concentrate on considering science not as cognition but as a *sociocultural system* that is connected with a certain subpractice. We do not deny or ignore its cognitive character but merely defer its analysis for later. That scientific cognition does not exhaust the totality of scientific activities is readily apparent. Scientists undertake teaching, administrative, and organizational duties that are directed, in general, toward maintaining the institutions and functions of science; they also solicit funds, present their results to the public, and oversee their practical applications; they look for the support of the media or politicians or try to protect their activity from external scrutiny. Moreover, if we take into account the results of sociological studies devoted to contemporary science, we readily see how much time scientists spend on these types of activities. It can indeed be wondered whether cognitive activities are really the dominant type of activity that engages scientists.

The idea that scientific research is not individual but collective and that scientists are shaped in social circumstances is commonplace in contemporary reflection on science. However, standard ways of elaborating this idea usually privilege one side or another of the interrelation between an individual cognizer and a cognitive community: either individual activity or collective epistemic structures; either the social or the cognitive character of scientific practice. Recognition of the sociality of science, on the part of philosophers of science, has sanctioned a view of the scientist as a socially formed individual and has produced views of global epistemic structures, such as Popper's third world or Kuhn's paradigms. None of these views has been successful, however, at clarifying what it is to say that cognition is a collective activity.

The idea that the subject of (scientific) cognition is socially conditioned replaced the earlier epistemological concept of the subject as "the marble statue which thinks for itself, untaught by others," characterized by Elias as *homo clausus*.[25] This phrase refers to "a single thinking mind inside a sealed container, from which each one looks out and struggles to fish for knowledge of the 'objects' outside in the 'external world.'"[26] This metaphor is no longer acceptable. The subject of cognition is now seen as a mind in a container that is open to social influences usually considered in objectivist (rationalist or empirical, sometimes almost quasi-physical) terms. These influences are conceived in various ways. For example, as (1) the impact executed on the mind by objective, public knowledge and the logic of its development (for example, Popper); (2) the determining impact on scientific cognition made by global scientific structures (for example, Kuhn); (3) hermeneutically understood tradition; or (4) the influence of know-how that functions in a given setting (for example, constructivist sociology of science). Such views presuppose a fundamental assumption, expressed in Elias's conception of *homine aperti*, that knowledge does not originate autonomously in the individual mind.[27] Moreover, neither the natural (physical and biological) nor the transcendental features of human beings (should they exist at all) constitute them as subjects of cognition. Individuals become subjects as a result of a constitutive process, which is thoroughly social. The traditional epistemological concept of an unconditioned and disencumbered subject, "a knower guided only by content-and value-neutral methodological rules"[28] is now largely rejected; and subjects of cognition are seen as conditioned by "various aspects of their social location—from their dependence on government agencies and industry for funding, to their location in an intellectual lineage, to their position in the race, gender, and class grid of their society."[29]

Although analytic philosophy of science no longer views the subject as a mind in a sealed container, to *be* a subject of cognition is still to think and observe in accordance with explicit rules or tacit know-how. In other words, the opening of the container, within which the cognizing mind was sealed by traditional epistemology, has changed the *situation* of the subject of cognition but not its way of being. This is still defined individualistically: either in psychological terms, when the reference is to mental acts and contents, or in epistemological terms, when cognition is defined either as a relation of representing between a subject and an external object or as an involvement in a hermeneutic circle. This individualized view of cognizing alerts us to three dangers faced by any philosophy of science that would approach cognition in nonindividualistic terms. The first danger is that of personifying cognitive communities and treating them as individuals writ large who perform cognitive activity. The second is the possibility that aspects of the relationship between a cognitive community and its members which cannot be objectified—namely, the personal and axiological—can go unnoticed. The third is an oversocialized conception of the subject: the relation between cognitive communities (and their knowledge, rules, ideals, etc.) and particular subjects of cognition is conceived as a unidirectional influence; this can be the case even if supra-individual structures are recognized as human products. This stance disregards the circular structure inherent in a subject's way of being and the hermeneutic character of the relation between supra-individual structures and particular subjects of cognition. Moreover, as we emphasized in discussing constructivism, the appeal to communal cognitive structures as a matrix shaping scientists in a uniform and harmonious way makes it impossible to understand the communal character of scientific subpractice.

There are, therefore, two philosophical-sociological dilemmas that we seek to avoid. The first is the danger of personifying cognitive communities or of deconstructing cognition into forms of social activity governed by noncognitive goals or interests.[30] The second is the possibility of oversocializing the subject of cognition or of understanding cognition in individualistic terms. To avoid both dilemmas, we propose to regard social relations and interactions as constitutive for cognition.

The social character of science, in particular that of scientific research, cannot be understood simply as the existence of social influences impacting upon cognizing (experiencing) subjects. This understanding is trivial and does not go beyond an individualistic view of cognition. The nontrivial senses of the social character of scientific research are best displayed by a critical analysis of three basic ways of understanding cognition. It can be viewed (1) as a relation between a subject of cognition and an object, (2) as an activity of discovering objective knowledge, and (3) as the individual, intellectual, and sensuous activity of creating knowledge.

To understand cognition in the first way implies that the subject is an entity equipped, either in virtue of itself or as a result of social shaping, with cognitive resources, such as powers of cognition, cognitive structures, forms, or inherent ideas. It implies also that an object of cognition is an entity independent of both cognition and social influences, devoid of cognitive content, and either purely natural (that is, nonsocial), or social but independent of the subject. However, there is no object of cognition deprived of all traces of the cognitive or, more broadly, of all traces of the social. Scientific research proceeds in a life-world that is thoroughly social and permeated by cognition since all forms of practice have cognitive aspects. What is socially and cognitively preconstructed is not only earlier scientific knowledge, the theoretical background for research; everything found in scientific settings, for instance, in laboratories, is a product of practice and of cognition involved in it:

> It is clear that measurement instruments are the products of human effort, as are articles, books, and the graphs and print-outs produced. But the source materials with which scientists work are also preconstructed. Plant and assay rats are specially grown and selectively bred. Most of the substances and chemicals used

are purified and are obtained from the industry which serves the science or from other laboratories. The water which runs from a special faucet is sterilized. 'Raw' materials which enter the laboratory are carefully selected and 'prepared' before they are subjected to 'scientific' tests.[31]

Knorr-Cetina concludes her description of the artificiality of laboratory settings by stating that "nowhere in the laboratory do we find the 'nature' or 'reality' which is so crucial to the descriptivist interpretation of inquiry."[32] Indeed, it does not matter whether an entity, event, process, etc. that becomes an object of scientific research enters science from the world of everyday life or from the world of philosophy, or whether it occurs in science as an instantiation of theoretical postulates or as an incidental discovery. What happens in every case is either a modification of its previous conceptualization or a case of establishing its conceptualization. Even in the case of accidental discoveries, such as X-rays, the event of discovery is simultaneously and inseparably the event of conceptualization. Clearly, we do not discover something, we discover something as-something. Hence, there is no setting within science where the subjective, in the absolute sense, encounters the objective in an equally absolute sense. Scientific research is done by already socialized subjects, proceeds in an already socialized world permeated by meanings and knowledge, and refers to already socialized and knowable objects. Boundaries between the human and nonhuman, the subjective and objective, the known and unknown are prior to scientific activity only in a relative sense: scientific research is done within existing boundaries and contributes to the establishment of new ones.

Understanding cognition as an activity of discovering objective knowledge, a view accepted for example by Plato or Frege, presupposes that it is a human activity directed toward the achievement of subjective knowledge through the process of discovering objective knowledge. The latter is viewed as nonsubjective and so not produced by humans. This is an illusion, however. Knowledge we discover, say, through learning, has been constructed by other people. It is always—Fleck and Popper are right—social knowledge.

For Fleck, in particular, cognition is not a process that proceeds in any one individual consciousness but is "the result of a social activity, since the existing stock of knowledge exceeds the range available to any one individual."[33] This indicates that three factors are involved in cognition, "the individual, the collective, and objective reality (that which is to be known)." They can be reduced, however, to two elements: social collectives "composed of individuals" and ideas, since "objective reality can be resolved into historical sequence of ideas belonging to the collective."[34] Fleck's approach is based on a collective concept of the subject of cognition and contains an epistemological element—namely, the notion that ideas compose the stock of knowledge. In other words, Fleck replaces the traditional view of cognition as a relation between the subject and object with a relation obtaining between cognitive communities and socially created ideas (little wonder that he calls those communities "thought collectives"). Moreover, for him, it is this relation, and not interactions between individuals belonging to cognitive (scientific) collectives, that is constitutive for cognition.

For Popper, cognizing is a relation between cognitive subjects and socially constructed knowledge; it connects subjective knowledge (searched, acquired, and possessed by an individual) and intersubjective knowledge, itself once equally subjective (searched, acquired, and possessed by other individuals). To adopt such a picture, however, stops short of the proper conclusion of the analysis: behind these two types of knowledge there is a relation between people, there is *dialogue.*

Using the hermeneutic concept of dialogue to describe scientific cognition amounts to more than simply recognizing the existence of conversational exchanges considered by sociologists. For Latour and Woolgar, activity in scientific laboratories is essentially a matter of different conversational exchanges that refer to (1) known or recently established facts (recently published papers); (2) practical activities, in particular, to "the reliability of a specific method"; (3) theoretical matters (often related to other issues, for example publishing papers); and (4) other lines of research, often containing an evaluation of a paper.[35] In virtue of such conversational exchanges, beliefs are changed, statements enhanced or discredited, and reputations and alliances between researchers are modified. Notice, however, that Latour and Woolgar's conception of conversational exchanges already presupposes the cognitive nature of scientific activity.

The hermeneutic concept of dialogue, on the other hand, is prior to any epistemological view of cognition. It is an ontological conception of dialogue able to uncover the ontological structure of cognition.

As dialogue is—for advocates of contemporary hermeneutics—an ontological structure of cognition, so language, as it exists in communication, is the ontic condition of cognition. Neither language nor cognition exists outside and prior to dialogue; all forms of conceptual understanding are constituted during communication; discourse and communication are "the natural context in which formation and developments of meanings and rules of thinking take place."[36] For Gadamer, "language has its true being only in dialogue, in coming to an understanding"; and coming to an understanding "is a life process in which a community of life is lived out."[37] But even philosophers who are not exponents of hermeneutics emphasize the fundamental role of dialogue. For Wittgenstein, "the meaning and justification of forms of speech must be found within the world of human discourse, not in independent reality beyond language."[38] Habermas also contends that communication is "the only truly constraining source of truth, if not of salvation, in the midst of a desacralized world," and that communicative understanding "comes to substitute for the transcendent authority formerly assumed by myth, the sacred, or religion."[39] For Rorty, too, and in his opinion for pragmatism in general, "there are no constraints on inquiry save conversational ones—no wholesale constraints on inquiry from the nature of the objects, or of the mind, or of language, but only those retail constraints provided by the remarks of our fellow-inquirers."[40] Finally, Heidegger not only locates discourse at the center of his analysis of Dasein but also emphasizes its ontological, existential status. Dasein "is essentially determined by its ability to speak";[41] and since discourse is an existential state, within which Dasein is disclosed, it is constitutive for Dasein's being: indeed, the human way of being displays itself in discourse. Clearly, discourse is not Dasein's monologue, but a conversation that reveals its being-with-others: "Being-with is 'explicitly' *shared* in discourse, that is, it already *is*, only unshared as something not grasped and appropriated."[42]

The mutuality of relations between humans is revealed in discourse as the interaction of listening and talking, and as such it becomes recognized and thus appropriated. Furthermore, discourse is the articulation of Dasein's capacity to understand and interpret its being-in-the-world. "Discoursing is the 'significant' articulation of the intelligibility of being-in-the-world."[43] This means that talking is always about something, so that an assertion, a command, and even a wish are also about something. Clearly, dialogue is always ontologically and hermeneutically *open* in the sense that every person who participates in dialogue must be open to what others are saying and manifest this openness through listening and being directed toward what is being talked about. As Grondin rightly notes, the adherents of hermeneutics "are unanimous in emphasizing the virtues of dialogue in the search for objectivity,"[44] although hermeneutics in general takes an antifoundational outlook. In particular, the hermeneutic view is based on the belief that dialogue

> rebels against the idea of ultimate foundations because any such foundation can be called into question in the course of dialogue. Those who already have a definitive foundation do not really need communication. They have already found the answer. To put it in thesis-form, we have dialogue because we don't have any ultimate foundations.[45]

Dialogue is necessary for those who need communication to confront their views with the views of others and, possibly, to elaborate an intersubjective, common position.

Moreover, it allows us to approach, in a nonobjectifying way, the relation between individual scientists and the reservoir of communal epistemic structures that comprise existing knowledge, cognitive rules, and ideals. These epistemic structures that each of us discovers, as it were, in social space, and which appear as "something 'given' that stands over against us,"[46] rarely informs scientists in a relatively uniform and harmonious way. As we have previously mentioned, from a socio-ontic perspective, to account for the transmission of knowledge, skills, values, etc., a common background need not be presupposed that is prior to, and nonchanging in, the course of conversation, or, more broadly, social interactions. "Social action is interconnected not because of what is shared, but because of what is transmitted from one locus of action to another, transformed, and reintegrated, or because of a continued *process of conversion* which consists of the circulation *and* transformation of social objects."[47]

The convergence of knowledge and activity is achieved through the continual incorporation of someone else's results into one's research.[48] What happens during discursive interactions is the fusion of opinions, skills, procedures, and interests, so that a simple consensus is not the immediate result; "fusion always means transformation of what is fused."[49] Authors, who believe that shared knowledge, rules, and procedures are constitutive of the social character of scientific practice remain decidedly inside the subject-centered way of considering science. For them the shared content functions as no more than an external context for individual actions, a context that allows them to explain the uniformity and cohesion of either individual or group activities in science.[50]

Seen from a hermeneutic perspective, the reservoir of global epistemic structures is a *tradition* that is involved in the process of understanding. And the relationship between scientists and a tradition in science is an instantiation of the hermeneutic circle;[51] in fact, it is an ontic socio-hermeneutic circle connecting individual understanding and social knowledge behind which there are other subjects. As a given, scientific tradition is an ontic condition of understanding, performed by scientists who work within this tradition, so their understanding is an ontic condition and the source for the existence of this tradition. Scientists constitute the tradition, revitalize and confirm it as a tradition, in the course of absorbing, affirming, invoking, and interpreting it.

That human dialogue is central within science is—we contend—acceptable to anyone who admits the social character of science. The view has, however, one fundamental drawback: to say that scientific cognition, and not just those parts of scientific practice that result in cognition, is, in its ontological structure, a dialogue among scientists (a linguistic game) seems counterintuitive if not ludicrous. The view not only ignores entirely the experiential and manipulative aspect of scientific research, but it is endangered by hermeneutic panlinguisticism. Viewed in this light, we cannot simply see reason "as thoroughly embedded in language."[52] If cognition is dialogue, it is human interaction: there is someone who is addressed and someone who addresses. Moreover, there is mutual influence and changes conditioned by the exchange. Dialogue may have a direct or indirect form; it may involve dead or living participants; and it may be oral or written. The view that ontologically (scientific) cognition has a structure of dialogue and, hence, of social interaction leads us to the analysis of a third understanding of cognition.

This understanding of cognition, the view that it is an individual, subjective activity, can be criticized by making clear three main ideas: (1) scientific research is a highly collective activity based on division of labor and cooperation, exchange of ideas, skills, and on mutual trust; (2) the local settings of collective activities are connected by social networks and processes; and (3) those networks are openly directed toward other spheres of practice and connected with them through various technological links. Scientific research is collective in two senses: it is teamwork, that is, it is additive, "as when a number of people join together to lift something heavy"; and it is collective work in the strict sense: "the coming into existence of a special form, comparable to a soccer match, a conversation, or the playing of an orchestra."[53] This second sense of the collective character of scientific research is particularly important. Although research is undoubtedly composed of the circumspective acts of scientists, it is not simply a haphazard mélange. The division of labor, myriad interrelations that merge specialized operations performed by individual scientists into complex systems, interactions that many people engage in jointly, etc., form the ontic structure of scientific research that is a network of activity, a subpractice. Analogously, although experience "is always actually present only in the individual observation,"[54] it is more than simply an activity of an individual scientist; it is social scientific experience (experimentation), a network of activity that forms an integral part of a scientific subpractice.

In general, then, *scientific research* is one of the various ways of human being, but it is a way of being of communities and only derivatively of individuals. It forms the basis of *science* understood as a sociocultural system, which (1) has achieved relative autonomy from other sociocultural systems and the subdomains of practice connected with them, and (2) has reached a position of domination over the entirety of historical practice and as such informs a way of being of Western societies.

Social relations among scientists generate an ontic structure necessary for scientific subpractice: communities, social organizations, and the network of social institutions. Scientific communities play a crucial role in scientific research so regarded. They are neither abstract structures that determine their elements nor aggregates of similar individuals who share beliefs and norms, but social entities whose way of being is constituted by social relations, in the course of which research is realized. The circle of participation and embracement that works in the case of scientific communities, as it does in other communities, means that every individual activity and every interaction among members of a given community is performed within it and, at the same time, contributes to the making of the community. From this standpoint, it is scientific subpractice, embodied in historical communities, with their traditions, ways of proceeding, rules, norms, and values, that constitutes the condition of possibility for being a subject of any scientific activity—in particular, a subject of scientific research. Scientists, as scientists, are constituted by processes of social formation (education, training within a particular scientific community, etc.), and by mechanisms of coercion, which compel conformity and are part of power-knowledge relations. These processes and mechanisms constitute *the social role of the scientist.* This role should not, however, be understood as an abstract and universal scientist but in terms of a variety of situated and specific roles. People become scientists and engage in specific assignments in a given place and time within a particular sociocultural context, in a given scientific community, and in a certain laboratory. Hence, although in our practice-centered approach, individuals and their (psychological) activity are decentered, they do not disappear: they are in no way abstract individuals who perform a social role according to uniform rules, but are concrete persons engaged in specific tasks.

The communal nature of scientific research means that it is always situated: it proceeds in particular settings according to protocols that direct scientific collectives. Consequently, individual scientists always enter certain existing settings. Localized research may be, and is, transcended due to networks of communication and cooperation. Within these networks the proliferation of new empirical knowledge, experimental techniques, new instruments, units of measure, and practical skills takes place through standardization, replication, transformation, and adaptation of the means and products of research to other settings.[55] Finally, the networks within which scientific research proceeds are open in the sense that the boundary between them and nonscientific networks—for instance, other sociocultural systems—is purely conventional, as is the boundary between scientific and nonscientific applications of technologies elaborated by science.

From our ontological perspective, the *activity of the scientist* is not an internal, psychological act or a relation between a subject and objective knowledge, or even between the subject and the object, but, as in the case of any individual activity, it is participation in scientific research. In other words, to adopt a non-epistemological, practice-centered perspective on science is to reject both individualism, which gives priority to the individual as a subject of cognition, and the epistemological perspective, according to which the scientist realizes cognitive activity that is understood as producing knowledge that represents the world. Being a subject of scientific activity is to participate in a particular subpractice that belongs to the entire practice of certain historical societies. This participation is a form of being-together-in-the-world: through participating in research an individual takes part in social relations that form the structure of being-together and interacts with other scientists and with the life-world of science. Participation in scientific research is not, however, a unidirectional dependence but an ongoing interrelation. As scientific research constitutes scientists, so the involved activity of scientists conditions scientific subpractice. The nature of this conditioning may be revealed by considering the *creativity* of scientific subpractice.

The fact that scientific subpractice establishes novelties means that social relations and interactions within which it proceeds are not duplicative or directed toward the preservation of its given global state or its particular elements—for example, an existing body of knowledge. The creativity of scientific collective activity clearly implies that it cannot be entirely routinized and patterned. As Rouse notes, it must be radically open: "whether a subsequent action counts as a continuation, transformation, deviation, or opposition to a practice is never fixed by its past instances."[56] The openness of certain scientific activities means that relations and interactions among scientists are not depersonalized, as are those between a customer and a shop attendant or between a lawyer and a client. They are highly individualized and personalized exchanges (mainly discursive), in which all the varying idiosyncrasies come into play among participants, differences in the knowledges they accept, in their skills of reasoning, persuasion, and the ability to perform experiments, in their perceptivity and imagination, in the value-systems they espouse, and in their subjective preferences. Scientific communities embrace scientists as persons: the need for scientists' distinctiveness and idiosyncrasy stems from the creativity of scientific research. Participation in research is constituted, therefore, within an ontic circle of *individuality* and *conformity*.[57] Each element of this circle, the cultivating of one's individuality and being shaped by a given scientific community, is necessary for individual scientific activity, for research, and for the existence of research communities. Moreover, unlike the reference to shared content, the reference to an ontic circle of individuality and conformity can explain controversies, the emergence of discordant ideas, or scientific revolutions without referring to something external to scientific interactions—for instance, to the rebellious personalities of scientists or to empirical irregularities that accumulate despite the unifying power of a shared paradigm. An interaction-centered view is able to account both for uniformity and nonuniformity by referring to the character of the processes of transmission and circulation within which fusion and transformation constantly take place. It is true that the shared content of a certain scientific discipline, with its numerous skills and resources, is necessary to transform novices into adept scientists who can then practice this discipline. What is common, however, is only a partial condition of cognitive interactions, since all controversies, negotiations, and teaching require another, opposite and no less important condition: the difference between participants and, to that extent, the lack of consensus.

From the perspective of the circle of individuality and conformity, the activities of scientists cannot be treated in an objectifying way that ignores the emotional and ethical aspects of their work. Contrary to the imperatives of the objectivist approach, scientific knowledge is not acquired by detached, indifferent, and dispassionate (rational) cognizing subjects. Scientists are involved and committed in research, and that cannot be grasped without taking into account their *emotionality*, an element ignored for the most part by rationalist conceptions of the cognitive subject, but strongly emphasized by practicing scientists: "The concept of absolutely emotionless thinking is meaningless. There is no emotionless state as such nor pure rationality as such."[58] Clearly, scientific research cannot proceed without individual intellectual passion, excitement, deep emotional involvement, and even obsession, or without the commitment of scientists.[59]

Emotional involvement in research, however, is not simply a psychological, subjective factor that informs individual activity. As Polanyi argues, it conveys the *axiological character* of scientists' participation in scientific activity. Intellectual passions and excitement saturate objects and issues approached by scientists guided by values: they become interesting, attractive or repulsive, worth studying or being disregarded, important or insignificant.[60] The axiological (in particular, the ethical) dimension of scientific research constitutes its human, cultural character, since the presence of values distinguishes the human from the nonhuman.

Polanyi derives existentialist conclusions from his analysis of the axiological dimension of scientific research. Scientific inquiry, which proceeds through axiological choice and concludes with discovery, creates in the discoverer "a new existence, which challenges others to transform themselves in its image."[61] Polanyi's conception remains, however, predominantly psychologistic, despite the fact that he refrains from portraying the subject as a self sealed within its solitary container, but views it as influenced by social context; nor does he consider the subject as purely rational, but as one that is both rational and emotional. Intellectual passions are forces driving human personality, but Polanyi does not pursue their sociocultural source or their ontological foundation. However, they clearly refer to the phenomenon of natality as an ontological condition, and they are revealed as social products because they belong to the circle of individuality and uniformity. They belong to the individuality of scientists and are elements constitutive of their idiosyncrasies that underlie creativity, as it is conditioned ontologically by natality. Simultaneously, they are social products in the sense that a certain type of social context is necessary for shaping human psychological forces into scientific, intellectual passions that are at the same time sociocultural and historical phenomena.

It is not only individual involvement in research that is the locus of the axiological and, in particular, the ethical character of research. Relations among scientists and their interactions also have an axiological aspect: they are based on *trust.* As Shapin contends, the role of trust "in the constitution and maintenance of systems of valued knowledge" has been largely ignored by modern epistemology, much of which "has systematically argued that legitimate knowledge is defined precisely by its rejection of trust."[62] Disregarding the cognitive role of trust is connected with attempts to treat scientists as agents who simply fulfill the rules of scientific method and to marginalize the sociocultural relations between them; "the role of trust is rendered invisible by positing a solitary knower as the sufficient maker and possessor of scientific knowledge."[63] Consequently, it is the individual subject of cognition that is the locus of authority, and individual cognitive activity gives warrant to the epistemic validity of knowledge. Shapin, on the other hand, insists that interactions between scientists should be considered as the locus of authority. Following Schutz, Wittgenstein, Putnam, Polanyi, Barnes, and others, he notes that the collective character of science, the fact that systems of knowledge are handed down by others and that scientific research must be done cooperatively, means that the members of any scientific community cannot fruitfully pursue their goals without mutual trust.[64] Relations between members are morally textured, and social norms identify "who is to be trusted, and at what price trust is to be withheld."[65] Relations between laymen and scientists are also based on trust. Science is not simply a collection of facts verifiable by anybody; it is highly professionalized, and laymen accept scientific results if they can rely on the authority of scientists,[66] or, even more abstractly, on the authority of depersonalized science. In fact, it is characteristic of science today that trust "is no longer bestowed on familiar individuals; it is accorded to institutions and abstract capacities thought to reside in certain institutions."[67] Let us refer to a seventeenth-century example.

Shapin provides an account of experimental activity in Boyle's London laboratory. There Boyle lived and worked experimentally with various paid assistants until his death in 1686. Because of the character of his research and the accessibility of skilled technical assistants, he engaged "others' hands, eyes, and minds." Shapin believes that Boyle's "laboratories were among the most densely populated workplaces of seventeenth-century experimental science."[68] The careers of Robert Hooke, Denis Papin, and Thom Huyck illustrate this claim. Hooke was Boyle's paid assistant at Oxford from 1661 to 1662, a curator of experiments at the Royal Society, and a natural philosopher and experimental genius of the highest rank. As Boyle's paid assistant, he enjoyed a high degree of collegiality and probably constructed and designed, together with another remunerated assistant, Ralph Greatorex, the air pump known as *machina Boyleana,* with which he executed experiments on the air.[69] Papin came to Boyle from Huygen's laboratory in Paris. While with Boyle (1675–79) he designed a remarkable air-pump, operated it in experimental trails, and recorded his strategies and operations with Boyle's concurrence. Huyck, though not possessing Hooke's or Papin's skills, knowledge, and status, was a valued and knowledgeable assistant.[70]

For Shapin, the role trust plays in collectively making experimental knowledge can be rendered visible by showing how skill "functioned *as attributions*" and how such attributions are "related to distributions of social value" and embodied in interchanges between a knowledgeable individual and technicians.[71] Of course, this means considering what technicians did and clarifying how their work was regarded within the experimental culture, and how it related to "the identity, the legitimacy, and the credibility of knowledge."[72] In his role as author-employer, Boyle recorded Papin's technical activities and personal views in a published tract, thus providing a clear picture of how "technicians and knowledgeable agents defined each other and their respected capacities in the course of mundane interaction and within institutionalized systems of cultural practice." So skill, while not necessarily embracing manual labor, rested at one pole of an evaluative scale with knowledgeability at the other.[73] Clearly, in Boyle's laboratory trust resided in the hands and eyes of others, in interpretative and manipulative skills, and in the cognitive abilities of minds. Certain forms of work were done interchangeably and stood in an inverse relationship to cognitive authority. Other forms required great skill and dexterity: lens-grinding, metal-boring, glass-blowing, and various types of chemical operations. Seen in this light, knowledge making does not stand outside situated practical activity, but is made and sustained by that activity. But what counted as knowledge, and as skill, or perhaps not even as skill? Shapin delivers an unequivocal answer. Technicians had routine experience of how operations tended to go, and an unverbalized, perhaps unverbalizable, appreciation of how to reproduce them and to assess results. But it was Boyle, with his knowledgeability and status, who effectively subsumed the technicians under the spokesmanship of his authoritative voice.

The discussion presented in this section is aimed at exhibiting the social structure of science, a structure ignored by early work in the philosophy of science and overly simplified in much of the recent literature. Since the social character of science is usually buried by philosophers of science under the cognitive, we have tried to excavate the former by showing that the traditionally understood relation of cognition is underlain by social relations. None of the traditionally recognized versions of this relation, a relation between the subject and the object, a relation between the subject and objective knowledge, or the individual activity of constructing knowledge, exists by-itself. Behind each there are social interrelations and human interactions. Yet, it is important to stress that it is not the network of social bonds and institutions that constitutes science as a sociocultural system, that identifies it and allows it to be separated from other systems; science is constituted by scientific research that forms a particular sphere, a subdomain of practice.

Certainly, we do not want to culminate our analysis with an excavation of the social that underlies the cognitive. As we have emphasized, any adequate account of science cannot ignore the fact that it is a cognitive enterprise. Moreover, we do not intend to reduce science to the social anymore than we identify concepts with patterns of social behavior. What we maintain is that thought and knowledge inhabit action and circulate among subjects. Both scientific knowledge and patterns of scientific cognition come to be within the nexus of social relations and are sustained by the networks of power within which they are carried. Seen in this light, the view of scientific cognition as a subpractice, and of science as a sociocultural system, must be supplemented by a careful analysis of this particular type of practice as cognition—that is, as a cognitive form of practice. If cognition is an interaction, it is not the presence of the other subject that requires explanation but rather the presence of an object of cognition. Moreover, both views, the view of scientific cognition as practice and the view of a certain subpractice as scientific cognition, must be integrated.

A FORM OF PRACTICE AS SCIENTIFIC COGNITION

The hermeneutic-ontological reconstitution of the epistemological dimension of science begins by putting aside those intellectual presuppositions (either psychologistic or realist) deeply rooted in epistemology. Yet, it must bear in mind that scientific cognition should not be reduced to purely linguistic phenomena since it bears nonlinguistic characteristics, even if they are manifested through language. It is not just tacit knowledge and practical skills—acknowledged finally by analytic philosophers of science—that have a nonlinguistic way of being. Personal, emotional, and axiological (ethical) aspects of scientific subpractice, the circle of participation and embracement, as well as the observation and manipulation of things, also have a way of being that is not purely linguistic.

Even opponents of the dialogical view of cognition would probably admit that in certain disciplines, particularly those with a humanistic standpoint, dialogue between scientists, especially those representing different intellectual traditions, takes place and contributes to the establishment of knowledge. A discipline that appears (at least to many of its critics) to be nothing more than dialogue is philosophy.[74] To say that philosophy from the historical perspective is nothing more than dialogue among philosophers is, of course, an exaggeration; but, at the moment, this oversimplification need not be argued against. After all, an idealized picture of this sort shows something important about philosophy. It unmistakably shows that cognition does have a *dialogical structure* (someone who believes there is no cognition in philosophy should not read this book in the first place!). Moreover, it shows that cognition does not disappear even if abstracted from relations to whatever is external to it, as well as from its involvement in noncognitive activity, whatever form that may take. Let us now digress and consider the ontological structure of philosophical cognition as a dialogue, a form of activity not related to anything outside itself.

What is constitutive for philosophical dialogue, what makes it a dialogue, is its *self-referentiality.* Philosophy is directed toward itself, and the cognitive activity of philosophers refers directly to work of other philosophers.[75] As we said in the introduction, we see philosophy, following Kolakowski, as an endless dialogue between the priest and the jester, or, put more adequately,[76] between two distinct philosophical attitudes: the *conservative* and the *innovative,* or between two trends in philosophical thought: the absolutist, dogmatic and foundationalist, on the one hand, and the critical, skeptical and relativist, on the other. The priest, searching for the absolute, is the guardian of the past and its traditions, although the priesthood is not simply a cult of the past; its aim is "a survival of the past intact in the present, an outgrowth of itself."[77] The jester is the critic of tradition. Assuming, correctly, that "our thoughts about reality are also part of reality, no less important than other parts," Kolakowski concludes that neither the priesthood nor the jesterhood is simply an intellectual attitude toward the world. Both are certain forms of the world's existence: the priesthood is "a factual continuation of a reality which no longer exists"; the jesterhood is an intellectual realization of possibilities before they become facts.[78] The world would have been different from what it is if the guardians of various traditions had not maintained the continued existence of the past; moreover, the world will be different from what it is to the extent the jesters succeed in elaborating intellectual alternatives to existing systems.

Philosophical self-criticism and self-preservation compose *philosophical self-transcending,* the ability of this particular subpractice to surpass its own existing stage or state of development. Self-transcending does not mean or entail the ability to rule out situatedness and to overcome the local contingencies that shape philosophical activity and thereby allow it to reach an absolutely objective perspective. Nor is self-transcending the essence of philosophy or of philosophical cognition in any traditional sense of the term *essence*. Self-transcending, the internal and ongoing dynamism of philosophical cognition, is its way of being; it is what happens in the course of philosophizing. It requires the existence of both those who participate in the dialogue and the occurrence of the dialogue itself. Because of ongoing interaction between claims to absolute truth and disrespectful criticism, philosophy is not simply a search for answers to perennial questions, which may be seen as a never-ending quest that abrogates the possibility of conclusions.[79] Philosophers do not simply seek for answers to questions posed by other philosophers and, in particular, to those posed by their predecessors. They problematize existing issues and critically elaborate new problems; they transcend inherited levels of philosophical discourse and often reach new, insightful, transcendental, historicist, etc., perspectives. The nature of philosophical self-transcendence means that philosophy's cognitive character is not attributed to it by any authority external to it, human or otherwise; philosophy attributes this meaning to itself when it develops (as cognition) and recognizes its cognitive characteristics within the epistemological frameworks it elaborates for defining cognition. It constitutes itself practically as cognition both in-itself and for-itself. Self-constitution does not mean that anything goes in philosophical cognition. There is a *horizon* beyond which philosophy cannot move without threatening to convert itself into scientific, ideological, theological, or commonsensical discourse. This horizon cannot be regarded, however, as something that escapes change: and it is, of course, largely settled by philosophy, but not entirely. It is also determined from the outside; and codetermination must be assumed from outside since the dialectical existence of the outside is required by the very concept of horizon. The idea of a philosophical horizon prompts us to restore an undistorted picture of philosophy as something that is involved in practice. It is, in fact, practice in its entirety that constitutes philosophy as a particular subpractice that cannot be reduced to the dialogue of philosophers. It now becomes clear that philosophy's own self-determination is partial, not global.

Science, unlike philosophy, does not have a horizon settled by itself, despite its philosophical roots, because science is not self-referential: indeed, its modes of development are not through self-criticism and self-preservation. The lack of self-reflectivity means that although science permanently constitutes itself as a subpractice, it does not constitute itself as cognition for-itself. Even the existence of ideals, indicating those activities that are scientific from the point of view of scientists, does not mean that science, like philosophy, constitutes itself— cognitively—as cognition. Science does not contain activity such as would constitute it-as-cognition-for-itself. In its case, external authorities, such as philosophy, theology, politics, ideology, are necessary for constituting it as cognition (-for-them). For this reason, we believe that reflection on the cognitive character of science cannot be advanced by any approach elaborated solely within the objectifying (natural) sciences, or by a perspective that finds its methodology and basis uniquely in science. If science is not self-reflective cognition, then from within science itself its cognitive character cannot be grasped and apprised. This means that the cognitive nature of science can be analyzed only from a philosophical perspective, either from the viewpoint of traditional epistemology or from an ontological stance from which it is clear that

> [t]he world of physics cannot seek to be the whole of what exists. For even a world equation that contained everything, so that the observer of the system would also be included in the equations, would still assume the existence of a physicist who, as the calculator, would not be an object calculated. A physics that calculated itself and was its own calculation would be self-contradictory. The same thing is true of biology.[80]

Manifestly, the natural sciences are products of human activity, which they themselves do not study and, therefore, do not and cannot describe or explain in their own terms. They marshal only practical and methodological self-reflectiveness: scientific discourse includes considerations of *how* the research is to be done or how it should be executed. Science may be self-critical—as postempiricist philosophers emphasize—and it may be able to correct its own mistakes, "be these mistakes in erroneous data or false theories, or even in the scientific method itself."[81] Such self-correctiveness does not mean, however, that the process of theorizing is simply a matter of criticizing earlier scientific theories. Scientists, when acting as scientists and not as philosophers, do not problematize existing questions; they simply abandon them together with the rejected or useless theories to which they were

related. New theories typically do not emerge from a purely conceptual analysis of existing issues. They are generated when scientific
knowledge is confronted with empirical reality, new theories, or with
new mathematical models; or they are inspired by practical needs
addressed to science; or they stem from discoveries of unknown
spheres of reality. If this is so, the ontological structure of scientific cognition is clearly different from that of philosophy.

This is not, of course, to deny that there is dialogue in science,
both in the natural and the social disciplines. Acknowledging this, together with the idea of the idiosyncrasies of scientists, emphasized in
the previous section, allows us to understand the way of being of *scientific knowledge*. It is neither accumulated thoughts existing in individual minds nor an objective, ontologically self-sustaining system; it is
sustained in the sphere in-between subjects, and it circulates among
real persons within power relations.

> Thoughts pass from one individual to another, each time a little
> transformed, for each individual can attach to them somewhat
> different associations. Strictly speaking, the receiver never
> understands the thought exactly in the way the transmitter in-
> tended it to be understood. After a series of such encounters,
> practically nothing is left of the original content.[82]

From this observation Fleck supports the view that knowledge belongs
to scientific communities and not to any single individual. His conclusion is justified; communities are the primary bearers of research; nevertheless, individuals, among whom thoughts circulate, cannot be
eliminated. The idiosyncratic character of the understanding that
real, differentiated individuals possess and express is an important
condition of the existence of nonsubjective scientific knowledge, of
how it changes, and of the unpredictability of scientific research. If
individual differences in understanding ceased and there were no
transformations of thoughts passing from person to person, there
would be no knowledge that could be external and objective in relation to the receiver. Clearly, individuals contribute to the establishment of socially embodied knowledge, which, once created, must be
maintained and learned.

At this point it is important that our position not be mistaken. Scientific research is not pure dialogue. Although scientists in all disciplines
talk to each other and interpret existing traditions, they do more than

that. What constitutes scientific research is, in fact, *dialogue*, together with *experience*, and the *technological involvement of science*. For us experience is not simply individual receptivity or individual acts of perception but the collective empirical (in particular, experimental) activity of scientists. Scientific experience is understood holistically as comprising language, understanding, manipulation, methodological protocols, and social relations. By referring to the technological involvement of science we mean its interconnections, through technologies, with other subpractices, in which scientific knowledge is applied and the techniques, procedures, and equipment elaborated in science are implemented. We understand technologies very broadly as networks of social activities and equipment, based on well-established and regimented precepts or algorithms that guide processes of production in order to achieve uniform results. This concept of technology is applicable not only to the material production of goods but also to social and political activities.[83]

None of the elements that constitute science in its sociohistorical being can be separated from one another: scientific research does not exist once as dialogue, once as pure experience, and once as practical application. They are dynamically interrelated, and scientific research exists as an interaction of dialogue, experience, and technological involvement. It is not only scientific experience that is theory-laden, impregnated with the contents of scientific dialogue and shaped by the technological involvements of science; most importantly, dialogue is open to what is experienced and to what is technologically manipulable; and their joint results are projected beyond science through its interactions with nonscientific subpractices such that the practical success of science can be measured and presented in scientific terms.

In virtue of the way these three elements interact, scientific research is a distinctive type of cognition: it is the collective elaboration of a particular form of the understanding of being through the synthesis of human experience. From the ontic perspective of the individual human being, cognition—that is, building knowledge—is the activity of experiencing and synthesizing individual experience of what there is. From the ontological perspective, this activity is conditioned by the ontological features and structures of human individual being— primordially, by the fact that for us being is an issue. Cognition is, however, more than that: by virtue of the intentionality and sociability of individual consciousness, the synthesizing of the experience of being is a simultaneous contribution to the cognition of the world of practice and to the elaboration of communal self-understanding. Seen in this way, the synthesized experience of being, the cognition of the world of practice, and the elaboration of communal self-understanding come together. As we argued in the previous chapter, it is the sociability of consciousness that conditions its intentionality, and this conditioning relation shows that social relations, interactions, and, in particular, social knowledge, underlie our individual understanding of the world. As a result, an individual contribution to the cognition of the world is always performed in a particular sociocultural and historical way. In turn, because the intentionality of our consciousness underlies its sociability, individual contributions to our common self-understanding presuppose distance between oneself and a community. Consequently, communal modes of self-understanding may not differ in kind from modes of cognizing the world; they too can treat the community objectively, as existing out-there. The objectifying way of building human self-understanding is, we will argue in the next section, typical of the social sciences.

Cognition of the world of practice and the elaboration of communal self-understanding are also conditioned by the ontological features and structures of our social being: plurality, interrelatedness, etc. They are those elements of practice that constitute the world as *objective*, existing out-there, for any individual subject and for particular spheres of practice, such as scientific research. So it is collective practical activity, containing regulated acts of manipulation, rather than thinking or sensual experiencing, that turns scientific research into objectifying thematization.

Ontically speaking, scientific research—in particular, laboratory activity—is a spontaneous, creative process of fabricating order, imposing it on "an initially chaotic collection of observations," and demonstrating that the process of fabrication has been done correctly and is objectively valid.[84] Fabrication is a continuum that is composed of several processes. There is a process of coming into existence for scientists during which external reality is deconstructed and "melts back into a statement."[85] Then, what begins to exist is crystallized, stabilized into scientific facts in the midst of experimental activity, and descriptions narrating how those facts and objects were discovered are elaborated.[86] In turn, this new factuality influences the production of further knowledge: facts become "embodied in mechanisms which (are said to) enable further work (experiments, inferences, measurements, data collection). Facticity is thus enshrined in terms of instrumental value."[87] Finally, scientific facts are transferred outside scientific experimental settings, so that scientific factuality achieves an objectified, thinghood type of being.

> Up to a certain point on this continuum, the inclusion of reference to the conditions of construction is necessary for purposes of persuasion. Beyond this point, the conditions of construction are either irrelevant or their inclusion can be seen as an attempt to undermine the established fact-like status of the statement.[88]

At the point of stabilization, what was artifactual at the beginning of the process of fabrication becomes a fact since all traces of production are now extremely difficult to detect. Before this point scientists deal with statements; at this point "there appears to be both objects and statements about these objects";[89] and beyond this point objects and facts are treated as prior to statements and function as their constraints. They are no longer relativized to the process of their construction; they take on a life of their own and become reified. In a sense, "facts refuse to become sociologized. They seem able to return to their state of being 'out there' and thus to pass beyond the grasp of sociological analysis."[90] Scientists "rewrite" the history of their discovery, so that "the involvement of the agent of representation appears merely peripheral and transitory."[91] Final research reports are realist in their treatment of facts, since references to their construction are abandoned. What happens in the laboratory constitutes a constant swing from fact-like status to artifact-like status. "While one set of agonistic forces pushes a statement towards fact-like status, another set pushes it toward artifact-like status."[92]

This constructivist picture can be juxtaposed to a phenomenological one. The latter sees empirical procedures as a mediation between science (scientific knowledge) and the world. "Each practical empirical procedure is a humanly planned process in which Nature is made to 'write' in conventional symbols a 'text' from which scientific information is 'read' by the experienced scientist using the resources of scientific language within which this is then expressed. The 'pages' on which Nature 'writes' its 'text' is a scientific instrument used as a readable technology."[93] Due to scientific experimental procedures that involve instruments or readable technologies—that is instruments that are able "to communicate information *directly*."[94] — scientific phenomena present themselves to us. Standardized instruments and readable technologies "can define the perceptual profiles and essence of a scientific entity."[95]

Constructivists stress the role of networks within which facts, scientific statements, data, and instruments move and influence each other and change their ways of being. As Latour conceives it, everybody and everything is active in the actor-actant network. Within this network the ontic conditions of cognizability are practical and social in the sense that to become a cognized object requires active participation in the process of transmission from nonscientific spheres of practice to science, then participation in the process of the fabrication of scientific facts, and finally in the process of objectification outside science. On the other hand, phenomenological philosophers of science stress epistemic conditions of cognizability. "To be known, a phenomenon must fall into the horizon of intentionality, and fall into it in a certain way. This is what the instrument makes possible."[96] What is crucial within this horizon of intentionality are "the intentionality-structures that underlie cognitive and deliberative acts of persons";[97] cognitive acts, in particular, perceptual experience and interpretation; and finally, *noetic-noematic* structures. What happens in the case of science, within the horizon of intentionality, is the (self-)presentation of (perceptual) objects through their different profiles; the appearance of new noematic features arising from the use of scientific instruments; and the specification of new horizons of reality due to instrumental embodiment in readable technologies.[98]

It seems that according to both views, entities, processes, and phenomena studied by science are active elements of the world that are not simply given to scientific observers but give or self-present themselves to observers. What is different, however, is the fact that from the phenomenological perspective it is the way-of-presentation-to-perception that changes when an object becomes visible through different profiles; whereas, from the constructivist perspective, it is the way of being of things as they travel within social networks that changes.

SCIENTIFIC RESEARCH AND OTHER COGNITIVE SUBPRACTICES

Scientists, as likewise philosophers, religious prophets, and artists, try to achieve and express human experience of being, to elaborate human self-understanding, and to make the world of practice intelligible. They differ from prophets, artists, or philosophers only insofar as scientific research is organized and proceeds in a specific way.[99] Clearly, science is a cultural system in which human cognition and self-understanding are realized, as they are in philosophy (metaphysics), mythmaking, and religion. Our aim in comparing science with other cognitive subpractices is not to show that science contains camouflaged myths and degenerated rituals, or that it ultimately supplies us with demystified knowledge about reality as it is in itself. The aim is to show that each subpractice is a form of cognition and to reveal the specificity of those forms science takes. In short, as Krohn says, comparison "is not meant to cast aspersions, but rather seems essential to lay a comparative base, a prerequisite to taking up the question of possible specific, unique features of scientific practice and institutions."[100]

Mythmaking, religion, and scientific research are, as are all subdomains of practice, ways of our-being-in-the-world. Superficially they have nothing in common; advocates of science believe that within myth and religion, knowledge, if it exists at all, is distorted, since within these systems cognition is an integral part of religious activity; whereas in science, cognition is seen as independent and governed by objective rules of Reason. Undoubtedly, the cultural functions of mythmaking, religion, and scientific research are far from identical. Nevertheless, from a hermeneutic perspective, they share two features in common: they contain forms of cognition (understanding) and they are totalizing cognitive subpractices. They are cognitive subpractices because they situate us in the world in a reflective and meaningful manner. All three offer an understanding of our being-together-in-the-world, although they do so in different ways. They facilitate our knowledge of the world and ourselves, as individuals and as humankind, as well as the knowledge that we are in the world. They synthesize human experience in terms of its symbolic or conceptual interpretation, constitute its meanings, and explain it in teleological, causal, or functional terms. None is composed, however, purely of cognitive activities, such as thinking or passive observation. All are forms of active participation in practice and indeed manifest this participation through rites, ceremonies, liturgical acts, theoretical disputes, experiments, and so forth. Mythmaking, religious faith, and scientific research are different ways of synthesizing human experience as a whole, and there are no absolute reasons, judged from this broad perspective, for privileging one over the other. Hence, scientific desacralization of human activity and the world is not an objectifying, decontextualizing act that reveals their true nature (as the adherents of scientism believe), but simply an act of recontextualization.

All three are totalizing forms of cognition in the sense that they entirely permeate the way of being of societies in which they dominate; second, that they tend to eliminate other totalizing forms of cognition and strive to marginalize nontotalizing cognitive activity—for example, (modern) philosophy; and third, that they connect the subjective and the transcendental in characteristic ways and generalize their distinctive approaches to all ways of being, as in the case of science, which treats everything as objectively present. It is not only myth that underlies "a compact existential engagement between man and world" and touches upon all aspects of human life;[101] the same may be said of religion and science. That mythmaking, the ritual observance of religious faith, and scientific research can permeate the entirety of practice means that the way of being-in-the-world of societies in which they predominate may be characterized as mythical, religious, or scientific. The fact that for modern, nonreligious man "a physiological act— eating, sex, and so on—is in sum only an organic phenomenon, however much it may still be encumbered by tabus," and not a sacred act, "a sacrament, that is, a communion with the sacred,"[102] means that such an act is experienced as it is informed by science and not by myth or religious faith. Each totalizing form of cognition claims exclusiveness: it offers a complete system of the world, so that no element of the world requires an appeal to symbols, concepts, or explanatory strategies belonging to another totalizing discourse. In other words, each constitutes itself as an ultimate authority, as the source of apodictic truth whose aim is to offer a sufficient context for understanding any and all forms of human experience. Other discourses can only spoil, corrupt, or profane this understanding; and for this reason alone, if none other, they must be completely eliminated. Clearly, totalizing cognitive subpractices cannot be reconciled with each other; they may nevertheless be reconciled with nontotalizing forms, such as philosophy,[103] but their cohabitation never amounts to equality. Totalizing cognition tends to push nontotalizing discourses to the margin and devalues their claim to cognitive importance. When neopositivists proclaimed metaphysics nonsense, they acted as spokesmen of science; when sociologists or historians of science insist that the only perspective for considering science is "from inside," as it allows science to be characterized as historical actors saw it, they also contribute to marginalizing or even to eliminating philosophical discourse about science.

Crucial for understanding the cognitive character of mythmaking, religious faith, and scientific research is the way in which they connect the objective and the subjective. Their manner of connecting them determines the content of the human self-understandings they elaborate, and it decides how the world is to be made intelligible. Although they appeal to different fundamental oppositions—the sacred/profane, in the cases of myth and religion, and the objective/subjective in the case of science—it is taken for granted that the sacred or the objective exists. Moreover, they believe the sacred or the objective has a definite structure, which manifests itself through *hierophanies* or *ontophanies* or through empirical manifestations, or through phenomena that are created in the divine act or are discovered in scientific experiment. As in a hierophany, "something sacred shows itself to us," so in an ontophany a being shows itself.[104] As a hierophany, "a personal manifestation of deity—the sacred, the taboo, or the holy,"[105] is an interpretation of what is experienced by people as power, so scientific objects and phenomena are the interpretation of what is experienced as determining and confronting us. On the assumption that the transcendental exists—namely, the sacred and the objective—these social, cognitive activities make the world intelligible: they build their systems of the world and establish their relation between oppositions as the dependence of the profane on the sacred or of the subjective on the objective.

There is, however, a fundamental difference between the sacred and the objective: the sacred order is often composed of personal gods, whereas the objective is by definition depersonalized. Consequently, human being is different in a world constituted by myth or religion than it is in the world constituted by science; accordingly, these ways of being establish different forms of human self-understanding. In myths the world has not yet been thematized as that which opposes humans.[106] Gods condition human existence by establishing "paradigmatic models for all human activities"; by imitating the gods humans participate in the sacred and reenact the time of origin, and rituals, ceremonies, and festivals periodically reactualize the divine act of the world's creation.[107] In short, myth and religion underwrite a periodi-

cally renewed (mythical) participation in the sacred and our experience of the sanctity of human existence as the result of divine
creation.[108] What is crucial for mythical participation is its *individual*
and *direct* character: every individual participates personally in the sacred, and every single ritual is modeled by the gods. There is, in other
words, the individual and unmediated dependence of human beings
on the sacred as the direct source of meaning and order. Moreover,
human openness toward being is experienced as listening to what the
gods and the world say and understanding the content, the sustaining
meaning of their message. Science replaces participation with *determination:* humans become objectified, determined entities that exist, in
an objectified way in the objective world, a world free from meanings
and values. So construed, the being of a determined entity cannot be
experienced as a participation in the meaningful order that speaks to
us. Science attempts to understand humans by placing them "in a pre-
meaningful, prehuman world (the history of the species, the unconscious body)."[109] It indeed fulfills "a desire to step outside oneself into
an order in which one treats oneself as a thing." Such an approach "is
not only suprapersonal—that is, not only one which destroys me in
my origins—but also suprahuman; for it is not only I who surrender
to the objectifying gaze in which I identify with that transcending
movement—I surrender the whole humanity to that gaze."[110] The personalized and direct character of the relation between an individual
and the source of meaning and order, typical of mythical and religious
forms of human self-understanding, disappears. Relations between
persons and the objective source of order are mediated: persons are
nothing more than instantiations of an abstract human being, taken to
be related to the objective by equally abstract processes of causal
determination.[111]

The comparison of science with myth and religion allows us to understand that if science is to be "certain about entities by methodically
organizing its knowledge of the world," it must liberate itself from everyday experience of the world so as to fulfil its claim to "serve the
growing domination of being."[112] For Heidegger and Gadamer, science is

> primarily a response to *Fremdheit,* the condition of being no long-
> er at home in the world. To be at home means to belong, to live
> in surroundings that are familiar, self-evident, and unobtrusive;

its contrary, *Fremdheit,* consists in the schism between past and present, I and others, self and world. Method derives from this sense of living among objects to which one no longer belongs.[113]

Science indicates human homelessness; it is alienated from our everyday activity, and it deals with the alienated, objectified world. What distinguishes science from other forms of human understanding is that it "responds to alienation with alienation." The concept of scientific method drives a wedge between the object and the subject of scientific research; so instead of "homecoming from the condition of *Fremdheit,* method strives for dominion over the world. It aims not to understand the world but to change it, to recreate it in the image of consciousness."[114] The sociocultural situation this epitomizes can be characterized in this way: valuable and powerful new truths can be achieved by universalized methods of studying an already dehumanized and disenchanted nature. Notice here the link between power and the disenchanting of nature. Historically, Bacon, Galileo, and Descartes each strove to impose his methodology on science and to make it the unique means for securing this new picture of the natural order.

What has pivotal importance for the ontological view of scientific research as cognition, a feature central to Heidegger and other thinkers, is *objectification.* Scientific research constitutes an objectified (depersonalized and desacralized) form of conceiving both our being in the world and the way the world is. "Science disregards the intentional movement which directs us toward things; it takes its objects as simply given and does not know that they reveal themselves originally in encounter and coexistence."[115] By means of objectification, science reveals (projects) possibilities for ways of human being that are absent in societies dominated by other forms of cognition. In the light of scientific objectification, the world becomes—as Heidegger says—a picture, more precisely, it is "conceived and grasped as picture."[116] Scientific objectification changes the relation between the subjective and the objective from one of mutual interaction into a unidirectional conditioning, determination, or subordination. The objectification of the subjective results from the primordial assumption that the subjective must be defined in such a manner that the influence of the objective may be understood, described, measured, and explained.

Objectification proceeds by means of the projection of conceptual frameworks, horizons, which allow scientific objects to appear.[117]

Scientific facts "are isolated by a special activity of selective seeing rather than being simply found," and they "are not merely removed from their context by selective seeing; they are theory-laden, i.e., re-contextualized in a new projection."[118] Projection constitutes scientific research since it opens up a sphere in which scientific procedure may move, and it discloses those modes of the being of entities that it can thematize.[119] In the case of the natural sciences, nature is projected in a mathematical way. In terms of mathematical projection alone, "can anything like a 'fact' be found and set up for an experiment regulated and delimited in terms of this projection."[120] The fact that scientific research is based on projection does not mean that it creates its subject matter — that is, (scientific) facts. Projection is not construction, it is rather an act of *revealing* or disclosing.[121] The view of scientific research as projection that discloses allows Heidegger to claim that the traditional conception of truth, according to which it is an *adaequatio intellectus et rei*, is based on a tacit assumption that being has already uncovered itself, that its disclosure is purely cognitive, and that the human subject is only the source of mistakes and falsity.[122] For Heidegger, on the other hand, although scientific research *"discloses something that is a priori,"* that is, entities, it projects their states of being — for instance, what it is to be a Newtonian body — since it always proceeds on the basis of the subject/object dichotomy.[123] In other words, scientific research does not give us access to entities as they are in themselves; it presents them as objects and in this sense objectifies them. Objectification is not, however, a construction ex nihilo; it is a discovery. It discloses a certain way of being: to be an object is included in entities' ways of being. Yet, although projection is disclosure, a total, fully objective revelation is impossible. The disclosing of a certain way of being closes other ways of being. Since scientific cognition is not the only way of understanding nature, when it discloses being in its distinctive ways it simultaneously veils other aspects of being and closes other ways of understanding being. We return to relations of truth and disclosedness in chapter 6.

Scientific projection is possible since in scientific cognition, as in any form of understanding, there are specific forestructures: *forehaving,* shared by scientists, technical assumptions and skills as they are explicitly accepted or embodied in research practices; *foresight,* scientific projection that allows scientists to plan and organize their research; and *foreconceptions,* initial hypotheses that determine ways of collecting empirical data, which will either confirm or falsify them.[124] Each forestructure is an a priori condition that makes all scientific disciplines, natural and human, possible.[125] Together they set up a horizon within which scientific inquiry proceeds. "When the basic concepts of the understanding of being by which we are guided have been worked out, the methods, the structure of conceptuality, the relevant possibility of truth and certainty, the kind of grounding and proof, the mode of being binding and the kind of communication— all these will be determined."[126]

In contemporary science, at least in the natural disciplines, objectification is manifested through epistemological features that distinguish scientific research from the other forms of cognition mentioned above. Scientific knowledge is constructed in a discursive, argumentative way and is based on experience submitted to depersonalizing rigor. Moreover, what is typical of scientific cognition in most contemporary disciplines are: (1) an *intention to construct universally valid knowledge,* (2) the *search for explanatory determination,* (3) the *ideal of strictness and consistency achieved through mathematization,* and (4) the possibility of the *practical application of knowledge.* The existence of such features of scientific research have been noted and examined by historians of science as well as by philosophers in their discussions of the role of deduction and induction in science, the theory/observational dichotomy, the existence of scientific revolutions and incommensurability, the universality of scientific laws, the measurement problem in quantum physics, and so forth. These features should not be regarded, however, as absolute or essential attributes in the traditional Aristotelian sense of the term. First, none is typical of every scientific theory or instance of scientific research, contemporary or historical. Second, they are distinguished from essential attributes by the fact that they are processes within the historical development of science. We need to embark, accordingly, on a further discussion of objectification as a historical phenomenon. However, before fully discussing objectification, which we will do in chapter 7, we want to consider causality as an example that throws light on objectification.

CAUSATION AS AN ELEMENT OF SCIENTIFIC OBJECTIFICATION

How we conceive causal relations plays a key role in the constitution of the *object* of study within the subject/object distinction. For example, substance-ontologies bind individuals separated in space and time by positing causal relationships. Causation becomes "the cement of the universe"; and as cause and effect are distinct conditions, they demand an "ontic glue," an agency able to produce effects or a regularity that links their instantiations. A venerable notion of causation holds that effects derive from their causes, that there is some relation of resemblance between what is "effected" and its cause. The idea that causes produce effects got superseded by the view that posits relations of necessity, regularity, and universality between causes and effects.[127] An extreme case of the ontic separation of cause and effect is ancient atomism. It viewed nature as a composition of immutable and non-productive atoms whose relations are accidentally altered by chance combinations in space and time. Atoms are never touched by real change; their relations alone arrange them into spatial configurations. On this view, causality is simply alteration in terms of spatial contact, and in the Humean schema, an instance of this notion, causes are prior to effects and provide a template for ordering time and change into a unidirectional framework. Entropy in thermodynamics provides a different causal schema for linking the world together. If a lower entropy state is always earlier than a later, it provides an irreversible process that defines the objective direction of time.[128] Given this picture, the causal order of entropy states coincides with the temporal order and defines it. These asymmetrical relationships between causes and effects accordingly set the direction that natural processes can take.

In causal pictures, nature is conceived as a self-regulating system indifferent to human purpose and plan. Consequently, both individuals and society are typically divorced from nature. The emergence of this divide can be described as one that disenchants, dehumanizes, and depersonalizes nature. Human and nonhuman, mind and matter, and values and facts are opposed: on one side, there is the social and human, on the other, the nonsocial and nonhuman. When these irreconcilable "realities" became culturally rooted during the seventeenth century, human alienation from an indifferent physical world became a datum of educated society. Science's great technological successes during the early modern period and its impact on eighteenth-century epistemology and metaphysics entrenched objectified causal categories within other forms of human understanding.[129] As scientific causalism grew, its models became paradigms for emulation elsewhere. This encouraged attempts to subsume humans under objective causal regularities similar to those of the regional ontologies of science. Science and scientism joined hands. To understand the historical emergence of objectifying strategies from the causal standpoint, we will briefly analyze its development in Descartes, Leibniz, Newton, and Hume, then show how these classical pictures are problematized in late-nineteenth and early-twentieth-century physics.

For Descartes the world is constructed from endless configurations and reconfigurations of passive extension. Bodies interrelate according to the principle of the conservation of motion, whose total quantity is (re)distributed among bodies with no net gain or loss. If "bodies" are simply passive extensions, they lack agency and self-activity and mutually affect one another only through spatial contact directed by their conserved motion. True causation is either an expression of God's infinite creative power or arises from the self-active intentions of created beings who act according to preconceived ends. This restricts productive causation to God and to the intentional acts of conscious agents. Descartes posits two created substances: mind is a nonextended thinking thing and matter is an extended nonthinking thing. This stark ontology incorporates two causal assumptions: that the sole source of causality is intentional acts and the Neoplatonic notion that causal efficacy is from higher to lower entities. Given these assumptions, de facto regularities arise governed according to the instituted laws of nature. Ultimately, productive causal change lies in divine action as it receates the world from moment to moment. Descartes conceptualizes changes in extended things according to his laws of impact. These apply to bodies considered in abstraction from their surroundings. So considered, bodies simply exchange spatial relations and move with determinate speeds in definite directions, but no account is available of how they themselves produce change among themselves. Descartes's position is typical of substance-ontologies: dispersed individuals are linked through superadded causal relations such as spatial contact or by the instantaneous exchange of modes of action, and the modes of change invoked are far from clear.

In Newton's metaphysics we have, probably for the first time, the notion of transcendent objectification grounded in divine nature. In the course of composing his *Principia Mathematica* (which he began in 1684), he articulated a critique of Descartes's philosophy known as *De Gravitatione*.[130] There is little doubt that *De Gravitatione* provides the metaphysical picture that animates the *Principia* itself.[131] Taken together, the *Principia* and *De Gravitatione* show how space and time relate to divine presence: they coexist eternally with God's necessity as causal consequences of divine presence.[132] Accordingly, space and time are infinite, uncreated, and objective actualities prior to the created world and to human existence. The *Principia* articulates a strong concept of objectivity: space and time constitute a boundless structure within which all natural processes take place. Indeed, they provide an independent and ordered set of positions existing prior to things that occupy them and with respect to which the latter play out their interrelations. Newton's view is consistent with substance-ontologies, according to which things have separate positions in space and time and are connected by extrinsic relations of causality. So viewed, real change takes place with respect to absolute space and time and is ontically independent of the relative measures necessary for human cognition. Within absolute space, gravitational action at a distance transpires simultaneously among all bodies. This underwrites an innovative notion of a "physical object" as what stands in causal relations to other physical objects. From this it is a small step to the notion that causal relations are constitutive of what physical objects are. The idea that objects relate causally becomes, according to the *Principia*, a primitive fact that pertains to their defining nature. And it relates to another conception of physical things posited by Newton that replaces the Aristotelian notion. For Aristotle, to be a physical object entails the inability to move in the absence of a mover. But in Newton's world something is a physical object just in case, if it is moving, it continues to move in the absence of a mover: once moving, always moving.

Leibniz rejects Descartes's causal picture, as well as Newton's idea that transeunt causation links spatially separated individuals. In his monadology, monads act synergistically as they unfold their life histories according to a preestablished plan programmed by the divine actualization of the best of possible worlds. Monads are self-sufficient bundles of activity and are endowed with inner drives enabling them to enact the processual drama of their histories. In this picture, transeunt causation does not express what is really real, and Leibniz substitutes the relation of sine qua non causation. For example, if the copresent details of the unfolding histories of monads x and y were not juxtaposed, they could not be individually expressed from their own points of view. Accordingly, the sine qua non causal relation of these two monads includes the inner act of x representing the act of y

representing it as *y* represents *x* representing it, and so on. In this way, Leibniz reconceptualizes causal relations in terms of an epistemology of generalized representationalism. What is ultimately real are self-active monads, the source and subject of their changes.[133] Leibniz also criticizes Newton's conception of action at a distance. In his view, it is merely another version of transeunt causation, the positing of causal relations among things spatially juxtaposed and isolated from one another. Moreover, for Leibniz the phenomenal world's temporal direction does not turn on a relation of before and after defined by the directional action of an agent. There is simply no agency, causal or otherwise, at the level of phenomena. Time is merely the order of succession and space of coexistence.

Leibniz's metaphysics reveals a basic theme in Western philosophy. Do existing individuals possess an inherent "ownness," an integrity that is independent of all relations and relata? His answer is emphatically "yes." In fact, each monadic individual is an autonomous *individuum ens positivum,* the self-sufficient ground of all that it is, has been, and will be: "a monad, like a soul, is a world by itself, having no intercourse of a dependent nature except with God."[134] The opposing tradition—for example, that of Heraclitus, Hume, and Nietzsche—conceives individuals as a nexus of relations and relata; they are what they are in virtue of their relations with other processes, entities, and properties. In this tradition, individuals lack autonomous integrity or "ownness." This view contrasts vividly with Leibniz's fundamental ontology, for which it is impossible that relations have this role. Monads possess individually and separately an inherent ownness and are such that they lack nonformal relations to one another such as objective causality or spatial relations that provide an objective nexus.[135] Their harmonious interrelation is a direct consequence of their being a compossible world of unique individuals that together express the principle of the best.

Hume's philosophy is starkly opposed to Leibniz's. In the Humean world, "things" neither inhere in substrata nor presuppose substances as bearers of their properties. They are simply collections of sense-impressions colligated together by associative acts of perception. The term *substance* merely refers to impressions bundled together by extrinsic relata and stabilized within perception. In Hume's ontology nothing possesses essential "ownness." Things are what they are in virtue of effects produced by the shifting relations they bear to one another. In Hume's view, their characteristics exist apart from "that unintelligible chimera of a substance."[136] Moreover, causality is predicated on such relata as spatial contiguity, temporal succession, and the constant conjunction of sense impressions. These are the glue that combine what appears as separate, distinct, and distinguishable. So individual efficacy is a myth, together with the notion that individual "substances" stand apart from the relational nexus. Seen in this light, causality cannot be posited as an objective "cement of the universe." Indeed, Hume denies that a logical connection, which ties cause and effect together, exists in "nature"; given this stance, it is not contradictory to suppose a cause to occur and the effect to fail. We are the source of substantial unity and causality: and it is we who impose unity on disparate facts. Ironically, from this chain of reasoning he is lead to preserve the connection between necessity and causation by transferring it to the mind's determination.[137] As a result, objective causation, as established by Descartes, Newton, or Leibniz, is clearly de-objectified.

Although in the late nineteenth and early twentieth centuries scientific pictures of the world remain causalist, classical views begin to get modified or abandoned. By the end of the nineteenth century physical science developed various conceptions of ethers and fields that view causation as implicit in physical processes and as no longer a "glue" imposed upon spatially separated things. Moreover, the term *cause* functions often as an abstract variable or "placeholder" to be filled in by particular mechanistic stories or in terms of interconnected processual mechanisms.[138] In 1884 Poynting identified the physical field with omnipresence energy. Unlike Maxwell, Fitzgerald, and others, he denied the necessity of a substratum to support its existence.[139] In 1889 Heaviside speculated that the ether was not a type of matter and suggested that matter itself might be explained solely in terms of an ether-ontology. In the early 1890s Larmor contended that matter is indeed constructible out of the field. For him, electrons are mobile singularities that move through the ether "much in the way that a knot slips along a rope."[140] In Larmor's conception, the ether-field is not a species of ordinary matter; rather, it is prior ontologically to matter, and, unlike Lorentz, Larmor does not separate matter and the field itself. Larmor was not alone: many physicists conjectured that various forms of physical action might be reducible to the motions of massless electrical centers in an omnipresent ether-field.[141] Their ultimate aim was to reduce mechanics and optics to electromagnetic theory. What emerged from an objectified standpoint is an extreme form of reductionism: the dematerialization of matter with change subsumed under one encompassing set of field equations. From 1892 to 1910 further developments occurred. Lorentz established an electromagnetic view of physics that dealt with problems of interaction between matter and the field and denied that classical concepts such as mass and momentum applied either to the field or to matter that passed through it.[142] Clearly, these worldviews problematized the classical picture of separate entities causally interacting in a "container" of space and time.

It remained for Einstein in his general theory of relativity to deprive space and time of their independence from energy.[143] Space-time, conceived as a manifold of events endowed with a metric field, now also contains the energy of the gravitational field. In this picture, the distinction between space-time, the container, and energy, the contained, breaks down. There is no longer a gravitational field in space-time that generates gravitational attraction across space-time. Gravitational effects arise from the curvature of space-time and are merely " warps" in the energy of the field. If general theory radically transforms causal concepts, quantum physics pushes causal intuitions to the limit. Since it is a fundamental theory, it applies to micro interactions among electrons, to Schrodinger's cat, and to us. This picture trades on an unclarified causal assumption: that interactions among electrons are the same as those among us. However, we are aware of correlations we set between ourselves and states of the theory, a state of affairs irrelevant to correlations among electrons themselves. Yet, any account of the theory's causal features must consider our awareness. For us, at a given time, Schrodinger's cat is alive or dead; but, according to the theory's standard interpretation, the cat's state is a state neither of life nor of death. For some, this indicates that the measurement problem is fatal to quantum physics and that the theory must await its successor. Certainly quantum theory challenges the micro/macro distinction and (no matter how the measurement problem is interpreted) the intuition that common causes operate at all levels of interaction allowed by the theory.[144] Moreover, quantum field theory (relativistic) bankrupts standard notions of physical objects and classical notions of the temporality of causal interactions among separate entities. Electrons and photons lack paths and violate the notion that particles move self-identically from state to state. And since the theory fails to satisfy Boltzman statistics, the number of electrons and photons present in the field under certain conditions is not determinable. Clearly these consequences, taken conjunctively, undermine criteria of identity demanded by substance-ontologies[145] and seriously problematize the subject/object opposition.

Certainly, ontic models of causation articulate many explanatory features important to human cognition. Moreover, they embody successful attempts at understanding the processes of the physical world. Nevertheless, none is conceptually unproblematic even in the physical contexts to which they apply. When applied, mutatis mutandis, to human phenomena they are more deeply problematic. The fact that we make ontic claims in terms of objectified models of reality cannot be abstracted away from the ontic-ontological circle. That circle involves dynamic interaction between the ontological features of our situated existence and conditions necessary for making ontic claims concerning our objective experience of the world. If we apply ontic models of causation directly to human action, we surely obscure the ontological conditions that are presupposed by the possibility of their articulation. Seen in this light, causation is not a primordial feature of our situated existence.

CHAPTER 5

Historicity and Becoming

OUR APPROACH, INSPIRED BY HEIDEGGER, Gadamer, Nietzsche, and Foucault, goes beyond both metaphysical and ontic accounts of history. Its aim is primarily ontological and hermeneutic: to understand history as our history and to uncover its historicity, its ontological features, structures, and conditions. We have no intention of establishing foundations or ahistorical invariances. Indeed, the ontological analysis we began in chapter 3 in no way presupposes that historical contingency can be tamed. As Foucault puts the point, if we listen to history, we find that "there is 'something altogether different' behind things: not a timeless and essential secret, but the secret that they have no essence or that their essence was fabricated in a piecemeal fashion from alien forms."[1] To set the scene for our approach consider Foucault's characterization of genealogy: "The search for descent is not the erecting of foundations: on the contrary, it disturbs what was previously considered immobile; it fragments what was thought unified; it shows the heterogeneity of what was imagined consistent with itself."[2]

At the core of our ontological considerations of history there is an ontic-ontological circle. Whatever is drawn into human experience becomes a historical condition of our existence. History owes its existence to human action, but that action does not straightforwardly make it.[3] From the standpoint of participants in practice, the claim to construct history is, of course, easily understood: if the past exists this is because of human effort, and if historical events are meaningful, the work of interpreting has been done. Given the hermeneutic character of our analysis of history, its ontological results are not ahistorical—that is, not rendered necessary suprahistorically. Any ontological analysis is situated within history, and its validity is therefore limited. What one age takes as categories that unify the self, or invariant features of experience, another will dismiss. Undoubtedly, one mode of cultural ordering is as transient as any other.

HISTORICAL UNDERSTANDING VERSUS EXPLAINING HISTORY

The perspective we want to elaborate is *historicist* and *hermeneutic* and opposes objectivist, naturalist, and essentialist approaches to history.

According to the *naturalist* approach, human history is an instantiation of general historical processes realized alike in natural and social life. It differs from other kinds of temporal sequences "only in that it happens in or to man."[4] The naturalist "disjoints having a history and possessing historical meaning."[5] On this view, historical processes are merely a succession of occurrences that "can stand to each other only in the same kind of external relations of time and space, contiguity, resemblance, and causation, as the facts of nature."[6] Within such processes, specific facts may accumulate or successions of emerging and disappearing states of affairs may happen. Only if temporal events are subsumed under (a combination of) universal laws can they be described, explained, predicted, and—it is hoped—manipulated.[7] Seen in this light, the idiosyncrasies of historical events are insignificant, and what matters is the search for invariances, for repeatable patterns, stable dependencies, and permanent determinants.[8] Moreover, the sole matrix within which historical patterns of change occur is simply physical time and space. Historical time is either indistinguishable from physical time or completely determined by the latter.[9] An *essentialist* approach to history accepts many of these assumptions, although it does not necessarily emphasize the unity of human and natural historical processes. What is crucial to any essentialist approach is the search for the essence of historical events. A famous example of essentialist history is Augustine's *City of God,* in which time and history march forward, in a linear two-step, to their inevitable fulfillment in the city of God. History is manifestly going somewhere. Two assumptions that both approaches share are ontologically and epistemologically naive: (1) history is an objective passage of events in the flow of time, and (2) historical episodes are independent atoms of history.

Nietzsche, Foucault, Heidegger, Gadamer, Ricoeur, and Margolis, among others, severely criticize these views. They steadfastly "aim at destroying the image of history as a resting reservoir of past facts."[10] Their anti-naturalist and anti-essentialist view presupposes that: "(a) history is concerned with unrepeatable, singular past events not subsumable under universal laws, (b) history is specifically concerned with the contingent rather than necessary doings of specifiable human agents, and (c) there is a gap between the event, and any invariant or constructed model."[11] In this light, the configurations and reconfigurations of history are dissolved continually under the weight of contingency: things appear in history when they can and disappear when they must.

Moreover, contrary to objectivist approaches, we believe that the historical does not speak for itself. It must be reenacted by an interpretive act. To make our point more intuitive, consider Mink's objectifying view of history. "There are hopes, plans, battles, and ideas, but only in retrospective stories are hopes unfulfilled, plans miscarried, battles decisive, and ideas seminal."[12] To say there are hopes, plans, battles, or ideas, without qualification, identifies them as separate from their localized appearing-in-history, and this deprives them of their historicity. It means, for instance, identifying the battle of Waterloo as a timeless event, as a complete instantiation of the battle that occurred in a certain place and at a certain time. Under such identification, the battle is an event nonidentical with Napoleon's decisive defeat that exists—according to Mink—only within the story about Napoleon's life; or, of course, in another historical story—for example, the history of France. To account for the battle of Waterloo as a historical becoming, on the other hand, is to see it as a process involving many events beginning with the initiation of the battle, through the battle in progress, to the battle lost; but it is also to treat it as a historical complex of meanings involved in this particular battle and in no other. These meanings appear within the interrelations of a given event with other and related becomings. Thus, the battle of Waterloo is the defeat of Bonaparte and the maintenance of the sociocultural balance in Europe; it is a direct cause of Bonaparte's second abdication; it is a decisive turning point in the history of Europe; it is an entitlement to fame for Wellington and Blücher; it is a topic of numerous stories told much later in world literature; or it is a tragedy for those who took part in it. It is all these meanings taken together that identify the battle of Waterloo.

Our historicist-hermeneutic view has affinities with German *historicism* to the extent it emphasizes the uniqueness of historical events and claims that everything emerges as a historical product. From the historicist perspective, history is productive in the sense discussed in process philosophy: "Process is productive, not only merely of new examples in accordance with patterns, but of new patterns as well as of new examples. There is not only change in accordance with laws. There is also a change of laws themselves."[13] Only if this is assumed "can man's very being be considered historical. Only then can the ontological inquiry into what man is reduce itself to a historical inquiry into what he has become."[14] Notice, however, that neither historicism nor the historicist-hermeneutic approach to history is simply a version of "the ancient doctrine of an Heraclitean flux, which encompasses man along with all else."[15] Both positions admit that patterns exist within history and both search for structuring conditions. There is, however, a crucial difference between them. Historicism presupposes a qualitative distinction between the natural and the historical—that is, between determined events and free action, between natural facts and cultural facts, themselves the vehicles of value.[16] The historicist-hermeneutic approach goes beyond these oppositions; its aim is to undertake an ontological analysis of the concept of historicity itself. Moreover, contrary to historicism, the historicist-hermeneutic approach does not exclude historians from history, since they are not treated as external spectators or as transcendental subjects of historical cognition.[17] Given that we cannot transcend our historical horizon, situated as we are within culture and society, our historicity informs our understanding of any historical situation. It is therefore an illusion to think that we freely interpret history. Historizing the subject of historical research overcomes historicism: the fixed point, from which historicism historizes the past, is transformed into a dynamic hermeneutic process involving the interaction and fusion of horizons. Arguments that support the historicist-hermeneutic view center around several ideas crucial for delineating its way of understanding history and historicity, since they refer to its narrative and hermeneutic features.

When considered ontically, historical events are *unique*, or, at least, we are interested in history, as history, in so far as it is composed of unique episodes. So, as Weinsheimer stresses, any historical understanding must see "every moment of history, including its own, as ineluctably factical and particular, immersed in having been, and never finally determined as an instance of a general concept under which it could be conclusively subsumed, but rather always awaiting interpretation and always exceeding it." Historical understanding "does not surpass the variety of the past because it does not surpass the past at all." Historical events must be seen as unique since any attempt at depriving them of their singularity is an attempt—paradoxically— at suppressing their historical character by "subsuming temporal particularity into atemporal generality."[18]

Hermeneutical knowledge that refers to history, and reflects its temporality, is not a generalized explanatory theory leading to prediction, but a *hermeneutical narrative,* since "the historicity of human experience can be brought to language only as narrativity."[19] There is a reciprocal dependence between narrativity and temporality. Many writers, such as Ricoeur, MacIntyre, Olafson, and Carr, argue that time becomes a human reality to the extent it is articulated through narrative. Ricoeur says, for instance, "I take temporality to be that structure of existence that reaches language in narrativity and narrativity to be the language structure that has temporality as its ultimate referent."[20] Thus, human existence and experience are irreducibly temporal, and narrativity brings them to consciousness; they are narratively enacted and unified. There is no distance, alienation, or discontinuity between the world of our experience and that of our narratives.[21] Historical narratives are both *about* the structure of human experience and *of* that structure. Human understanding, as it is present in narrative, is not a disembodied mental act; it is a sedimentation internal to practice. The products of understanding arise from temporally situated actions always engaged and enacted in ever-changing contexts. A narrative approach to history denies that we can bracket off the diachronic shape of our historical lives and retreat into timeless realms of abstract intelligibility and order. In historical narration the differences between the past, present, and future are not blurred, since past, present, and future events are not subsumed under timeless laws. Narrative structure provides the historian with an ontology of human action and worldly occurrence.

Certainly, narrative can be viewed as a structure imposed on events from without; but equally it can be viewed as what intrinsically structures historical events. As Ricoeur emphasizes, an event is historical insofar as it is incorporated within the plot of a story and in this way contributes to the development of a plot. The plot is "the intelligible whole that governs the succession of events in a story."[22] Narration is a diachronical representation whose end is implicit in its beginning, and certain events are excluded as irrelevant to the story's unfolding. Clearly, narrative structures are teleologically oriented and lead to closure. "Looking back from the conclusion towards the episodes which led up to it, we must be able to say that this end required those events and that chain of action."[23] So as narratively structured, events are not simply "beads" on a string ordered diachronically. They are joined together, given meaning, enabling us to see the inner connection between the ending and beginning of a story.[24] In other words, narration "constructs meaningful totalities out of scattered events."[25]

Hermeneutically conceived, narration rejects the view, often adopted by historians, that a distinction exists between a narrative representation and the events to which it refers. The situation is in fact otherwise; events are not pre-given but constituted historically in the very act of constructing a narrative. Understood in this way, *narratives are not instantiations of realistic description;* they possess cognitive significance not reducible to the cognitive significance of laws and the initial conditions of nomological explanation. They are not imperfect explanations awaiting improvement; the pattern of narrated events is not cognitively incomplete and in need of explanatory supplementation by underlying causes or regularities. To conceive all forms of the cognitive in causal or lawlike terms is an essentializing move. To assimilate a narrative understanding of x leading to y under a generalization of the type "whenever x then y" privileges the synchronic and nomological and masks the diachronic features and meanings of events that narratives express. The anti-essentialism involved in the narrative approach to history rejects "permanent natures, distinct from the processes in which they are involved."[26] Invariances or legitimizing necessities need not be presupposed, "strict epistemic or ontic invariances that cannot, on pain of incoherence or contradiction, be denied." Invariances that appear in history, such as death or taxes, "can be instantly relativized to the historical horizon within which they do appear."[27] To use Foucault's language, they are, at best, instances of the "historical a priori."[28]

A paradigm example of narrative is myth. Every myth is a story that embraces both *universals* and *abstract particulars* (the World, the Holy People, Darkness, the Center of the World, the Flood, the Land without Sin, etc.). The history of European thought may be seen as an ongoing elimination of abstract (non-empirical) particulars. They are present in ancient philosophy (Cosmos, Logos, the Soul of the World, etc.), and in medieval Christian philosophy. Modern philosophy and science attempts either to overcome or to banish them. Positivism and analytical philosophy of science, in particular, have formulated the strong requirement that genuine knowledge must be liberated from the grip of abstract particulars. Their ideal is to view scientific laws as strictly universal and, as such, without reference to unique objects. Modern science, however, still refers to unique processes (the process of the evolution of the universe, biological evolution, or the natural history of humankind).

What differentiates narrative understanding of history from its (causal) nomological explanations is the role *meanings* play in historical understanding and in history itself. Hermeneutic understanding, operating within historical narration, does not treat historical events as objective occurrences but as becomings inscribed with meanings. These meanings constitute the relative character of historical events: they relate to people who participate in them and to concurrent events; they relate to earlier generations and earlier facts as well as to generations yet to come and later occurrences; they relate, moreover, to us who engage in their hermeneutic interpretation. So an event may carry very different meanings, since later interpretations may add new meanings and change the layers of meanings conveyed through time. Unlike the objectivist account of the temporal order of historical occurrences, according to which it is always unidirectional, in hermeneutic understanding meanings move simultaneously in both directions: from the past toward the present and from the present toward the past. Ontologically conceived, meanings may be seen as a medium within which historical events exist and through which they acquire sociocultural and historical being. In saying this, we have no intention of reifying meanings; as we have argued in chapter 3, they are correlatives of practice and, in particular, of its various interpretative activities. So although continuance and rupture occur within the sphere of meanings, they are not driven by the internal dynamism of meanings themselves.

This character of meanings allows us to understand that relations between historical events and historical knowledge, as well as between an act of interpretation and an interpreted occurrence, are reciprocal. Objectifying approaches to history "must necessarily overlook the knower's share in making history."[29] Indeed, they are unable to recognize that the interpreter's participation in making history in the very act of interpretation is the reason why history is neither a merely re-petitive process, like the trajectories of certain natural phenomena, nor a process of reiterative cumulation. For instance, the temporal se-quence of events that take place in a piece of metal, as it undergoes cu-mulative change due to fatigue, does not embrace the scientist who is studying it. Also, his explanation of the sequence neither belongs to the sequence itself nor is responsible in any way for generating it. On the other hand, studies of history belong to history, and hermeneutic historical understanding clearly recognizes this fact: it sees itself as shaped by tradition and by interpretations that will emerge in the fu-ture, and as contributing to making history; "precisely because under-standing is itself an event in history, it extends, furthers, and carries on history. Understanding makes history. Even if it is understanding of the whole, it adds itself and integrates itself with the whole, which therefore always remains to be understood."[30]

What happens in history is the process of "preservation and at the same time surpassing what is preserved."[31] History is a shared way of be-ing of both subject and object, the historian and the historical event, the interpretation and interpreted fact. What is to be known histori-cally cannot be wholly set over against the knower. At bottom, our historical reenactments of the past participate in the historical mean-ingfulness of events whose unfolding still makes traces upon the life of society. From the hermeneutic perspective, we are-in-history in the way we are-in-the-world. The temporalities of history are the horizon and background against which the meaningfulness of our experience un-folds itself through the concordance of a beginning, middle, and end.[32] Thus, within the human experience of being historical the sub-ject/object opposition is undermined.[33] The hermeneutic attitude to history allows us to recognize that history is "something which the sub-ject knows only by being that something itself. . . . In such historical knowing the subject brings about the object of knowledge in and through itself. It understands its history by being that history, and it is that history in understanding it."[34] The historicist-hermeneutic posi-tion can be seen, in other words, as an immanent approach to history.

That studying history is necessary for our own self-understanding is obvious from the perspective of hermeneutics, since one of its basic as-sumptions is the idea that "who we are is a function of the historical cir-cumstances and community we find ourselves in, the historical language we speak, the historically evolving habits and practices we

appropriate, the temporally conditioned choices we make."[35] So in
studying history we comprehend the past as what brings the present
and our historical selves into being, precisely because we are guided by
an interest in the present.

Since the very act of telling history interpretatively belongs to
history, history cannot be objectified, subsumed under general laws,
and fully known. It cannot be subsumed under the umbrella of perfect and self-contained understanding. The primordial condition of
objectification—the autonomy of subject and object—is impossible to
fulfill in the case of history as it stands in relation to its interpreter. History always exceeds and surpasses historical understanding; indeed, it
is more than what is understood at any time by the finite mind, "precisely because it is always in the process of creating history, creating
the conditions of self-consciousness."[36] Understanding history as a
meaningful enterprise is to recognize that it should not be "mecha-
nized," according to the perspective of the naturalist (causalist) approach, since it "does not move automatically with the clock hands."[37]

Constructing historical narratives requires that we acknowledge
"the possibility of deliberate efficacious human agency," even if history is made, to some extent, "unconsciously and involuntarily."[38]
There is an element of freedom that allows human historical subjects
to change the course of their actions. Consequently, as the existentialist-
hermeneutic perspective shows, the future is essentially open, it is the
sphere of projection as much as that of foreseeing. A prediction formulated on the basis of knowledge about certain past events may itself
turn into a self-falsifying prophecy if the individuals involved in these
events decide, for instance, to avoid the predicted effects by acting in
opposition to what the prediction foretells. However, to consider "deliberate efficacious human agency" as a source of historical change is
an ontic, not an ontological, step. Accordingly, the concept of freedom (either as freedom-from or as freedom-to) is not an ontological
category. Ontologically, it is the dialectic of openness and closeness of
(our) being that is the source of ontic interplays between indeterminacy and being determined or between freedom and being constrained. Moreover, historical understanding is endangered if human
agency is seen as an invariant substratum underlying historical
change. In fact, human agency and freedom are themselves historical
products. The anti-naturalist and anti-essentialist view of history does
not presuppose any timeless set of features pertaining to human na-

ture, since "whatever possesses a history changes 'nature' as a result of its ongoing history."[39] Persons or selves are "products of their own *praxis,*" they "lack *natures,* have or are only *histories.*"[40]

If we reject essentialism and the belief that we are able to achieve perfect self-consciousness, if we admit that our understanding is finite and situated, we open an ontological perspective on history. This perspective is more fundamental than ontic narratives of historical processes; it allows us, moreover, to thematize the historicity and temporality of both individual and communal being.

HISTORICITY

History, in which we participate, may be characterized — initially and ontically — as "the sum-total of human-made processes in which such human creations as language, law, economy, state, etc., are crystallized," and as "the sum total of those processes in which man harnesses such natural factors as rivers, weather, laws of nature, etc., to his own ends."[41] History understood in this way is what Heidegger calls world-history; but this view is based on an assumption not present in his concept — namely, that "the historical process is the sum total of acts by which it is generated."[42] This assumption distinguishes our conception of history, not only from Heidegger's concept of world-history but also from the Hegelian and Marxian views of history.

According to Hegel and Marx, history is not the total process of the development of humankind — that is, its way of being; on the contrary, for them, history is itself a "historical" product. In Hegel's system, history is a product of the internal logic of the Spirit's drive to self-perfection; history belongs to understanding and not understanding to history.[43] Hegelian historical time is not authentic, it "lacks what is uniquely proper to time: contingency, freedom, exposure to the future," since it is "made safe by eternity, underwritten by reason, regulated by necessity."[44] For Marx, history is the product of prehistory, of earlier developments in human productive powers and social relationships. Both Hegel and Marx accept the idea that there is a given origin for history. Nietzsche and Foucault challenge the quest for primal origins "because it is an attempt to capture the exact essence of things, their purest possibilities, and their carefully protected identities; because this search assumes the existence of immobile forms that precede the external world of accident and succession." For Foucault, the historical beginning of things involves no "inviolable identity of their origin," but only "the dissension of other things"; in a word, the contingent dispersion of events.[45]

To undertake an ontological analysis of history is to consider *historicity* that is an ontological structure of any historical process and—in general—of being in a historical way, of being in history. There are two main ways of constructing the concept of historicity. The first, elaborated by Heidegger, is existentialist; the second is nonindividualist and non-existentialist. This latter perspective has been elaborated, for example, by Gadamer, Foucault, Habermas, and Fackenheim. For Heidegger, the historical way of being (historizing) is neither objective, as an attribute of the world itself, as "a succession of processes, a changing appearance and disappearance of events,"[46] nor subjective, as depicted in reference to an inner stream of experience. It is Dasein's way of being, and history is rooted in its historicity that is "more than a mere temporality of individual existence. It also constitutes 'history,' that is, world-historical process."[47]

History is "that specific historizing of existent Dasein which comes to pass in time."[48] Dasein's historicity is, essentially, its temporality and finitude: "human existence is a temporal happening *(Geschehen)* that is aware of itself as happening and changing, as stretching along between birth and death."[49] This idea has crucial importance for any anti-naturalist and historicist view of history: *human finitude* is an ontological (precisely, existentialist) condition of history. In virtue of this ontological condition, history may be understood as a succession of generations that hand down sociocultural heritage so that its continued inheritance from predecessors takes place.

An infinite subject of being or action—even were it capable of changing itself—could neither historicize nor possess history. As Fackenheim rightly argues, according to the conceptions of Scotus Eriugena, Böhme, Schelling, or Berdyaev, in which God is conceived as a *self-constituting process,* self-constitution is pure self-identification and self-determination and "proceeds from the indifference of sheer possibility of nothingness" to "the differentiation of actuality." God "establishes its own identity throughout this process by returning upon itself," and "actualizes *ex nihilo* the totality of possibilities" because God is absolute self-making.[50] This process stands in opposition to finite, situated existence and differs radically from history because "it must wholly transcend temporality,"—that is, it must be "eternal, or wholly present," given that the absolute returns upon itself. Thus, only an entity "which is a self-making-in-a-situation can be, in its ontological constitution, historical." Fackenheim rightly concludes that the concept of *self-making* distinguishes historicity from temporality, but is nevertheless not sufficient to define history, or to separate it from the quasi-historical being of the absolute.[51] For Fackenheim, it is human situation that distinguishes historicity from eternity. The *human situation* is radically different from both the natural and the historical situation; it is neither an anthropological fact, in the manner of a natural situation, nor part of one's autobiography, in the manner of a historical situation. The human situation is existential: it situates any human agent and "is understood by him as such only when he faces up to it as his own."[52] Seen in this light, the natural situation limits us because we accept it as a necessity; the historical situation limits us but is also open to augmentation and can become part of us. The human situation, on the other hand, is discovered individually and, when recognized *as* human, is understood as universally human. Moreover, the human situation is not produced by us, but given as "the condition of all human producing."[53]

Campbell, following Fackenheim, argues that the concept of the human situation, "in which human being *recognizes* itself as a self-formative process engaged in a finite situation," overcomes natural and historical contingencies, since the concept of the human situation introduces "a special kind of universality which transcends the relativities of particular situations." Campbell emphasizes, of course, that the universality in question is different from that typical of facts "laid out before a detached knower, and describable in an impersonal report."[54] The understanding of the human situation is a personal act of self-reflection and of resolutely taking possession of the situation. However, to ground historicity in the human situation is to establish an existentialist ontological invariant—namely, the universal human situation—understood in purely existential terms as stripped of all social and historical contents.

We believe that an existentialist way of reading the notion of situatedness endangers historicity as much as pointing to invariances jeopardizes it. Even if (cognitive) self-understanding, conceived as a personal act of self-reflection and of taking possession of the human situation, is historically universal (that is, presumed to happen in all human cultures), the historical variability of its content prevents us from considering it as ontologically universal. As Wachterhauser notes, "the 'givenness' of the self is not, in principle, unchanging over time."[55] The self is given through historically contingent languages of self-description.[56] Moreover, it is given through participation in society. It is through both language and social relations "that history often affects our self-identity."[57] Even the broader existential concept, the notion of Dasein's finitude, endangers historicity if it is understood as expressing our universal human condition. Moreover, when finitude is understood individualistically, as one's own historizing, it is not sufficient for the ontological (non-existentialist) grounding of historicity. It is insufficient for stating that the past prior to Dasein's own existence or the future extending beyond its death have genuine existence, independent of us. Contrary to Fackenheim, we hold that historicity cannot simply be constituted by the human situation if it is understood, as Fackenheim does, within the opposition between absolute, nonhistorizing Being and individual, historizing human being.

To avoid existentialist invariants we need to tie historicity to *our* experience of the historical character of our common being, to a historically local experience of historicity. Thus, we perceive our being as historical, which is one of the most fundamental features that being has for-us. We must never forget that historicity is an ontological feature of human being as seen from a particular sociohistorical perspective. This is the view—as Jaspers puts it—that reality "appears to us as historicity";[58] but this does not mean necessarily that it appears as historical to other cultures. To see it as such is itself a historical experience. It is an experience that differs from the experience of time in mythical narrative, in which it is perceived as nonlinear, pendulum-like change.[59] Both experiences of time are historically local; as indeed is my experience of my ultimate individuation by death.[60] To recognize the limitation of both types of experiences, it is necessary to replace the Heideggerian unidirectional dependence of historicity on human temporality (mortality) with the idea of reciprocal dependence between (individual) temporality and (social) historicity. We agree, therefore, with Rotenstreich that "the proper sphere of history is precisely the province of public events, creations, and institutions," such as the state, law or language, and that "the impersonal cannot be conceived as a sum of personals."[61] History is fundamentally *our-world-history*. Let us see, then, whether those who opt for a nonindividualist approach to history can provide a more convincing view of historicity.

The way in which the early work of Foucault, especially the *Archaeology*, mobilizes historicity can be contrasted with Heidegger's attempts at elaborating a concept of historicity free from supra-individual fundamentals. This early work rejects the idea of a constituting subject of historical action or cognition, the subject "which is either transcendental in relation to the field of events or runs in its empty sameness throughout the course of history."[62] The transcendentally conscious subject is a foundation of continuity but not of temporality and change. When the concept of the transcendental subject is abandoned, the idea of continuity in history also disappears.[63] Thus, Foucault's genealogy stresses the jolts and surprises of history, the chance occurrences, and the radical dispersion of events. He is not concerned with individual cognition as a vehicle for expressing signs of inner meaning and intention, but with the discourse of entire periods within which the given individual is anonymously present.[64] He aims to show how discourses, domains of objects, etc., are constituted, how the subject of cognition emerges from discursive formations, and how it is constituted within power-knowledge configurations.[65] Discourse is a sociocultural reality configurating historically and systematically the objects of which it speaks, and beneath discourse there is no "presentifying" substratum, no persistent invariance, no point of termination, "no original or transcendental signified to which all signifiers can ultimately refer."[66] Human thought is constituted through discourses that occur within fields of action, whose structure is defined by rules and social relations more fundamental than the assertions of any individual thinking within this space. Archaeology seeks for what constitutes such discourses, for rules of the formation of objects and concepts, for the selection of theoretical strategies, and for the positioning of the subject;[67] genealogy, on the other hand, articulates these discursive practices in reference to their embedment in the changing nexus of power-knowledge.

   Foucault's approach to history is—in a sense—the opposite of Heidegger's, although both place their work in reference to Nietzsche. Like Nietzsche, both are hostile to the present order and conceive the past-future nexus as a source for critical understanding. But if Heidegger is enthralled with Nietzsche's nostalgia for the past, Foucault is animated by his vision of the past as a liberating force for the future; if Heidegger concentrates on death as structuring our being, temporality, and history, Foucault concentrates on genesis and origins. Both approaches are biased and so are unbalanced and ineffective. As Dasein is not the sole supporter of world-history, since its historizing is not sufficient for understanding the historicity of events as they unfold through time, so historical events, discourses, and networks of power relations cannot exist and belong to history by themselves. Neither human existence nor the being itself of supra-individual structures is the sole source of historicity. Neither existentialist analysis of the phenomenon of death nor the study of beginnings can reveal the structure of history. Rather, it is our self-making that is its source. More precisely, the historicity of our being, which turns being into history, is constituted within the interrelations between the individual and the social. Foucault begins to grasp this point in his late work; but as we have indicated, his conception of historicity remains primitive and unthematized.

Let us now spell out the central ideas of our conception of historicity. First, it is obvious that human being is historical: indeed, both human beings and the sociocultural world are historical; more precisely, their being is sociohistorical. This comes to more, however, than the idea that, "apart from history, man's very being, *qua* being and *qua* human, is deficient."[68] It means that human being is impossible either as being or as human apart from and outside history, an observation that pertains to both individual and collective human being. Historicity is not something that merely supplements the prehistorical, natural being of humankind. Quite the contrary. *Humankind is its history,* and the natural being of humans is realized historically within the movement of history. Hence, no qualitative distinction between the historical and the natural is prior to history and historical narration. Furthermore, it is obvious that, ontologically speaking, the historicity of our being is not simply the endless passage of time. History is not a succession of temporal stages, in reference to which we presuppose—in order to attribute continuity to their succession— that they are states of some atemporal entity. Historicity is an ontological feature of our practice understood as human way of being, and it is equiprimordial with interrelatedness. Indeed, it is an ontological condition of any history understood ontically as a temporal process. Historicity invokes the variability of practice together with all its correlatives; it means also that humans participate in the process of appropriating their heritage and of carrying their achievements forward into the future. Furthermore, history embraces not only human historizing but also the flux of "alienated" products. Moreover, it would not be a genuine process without practice and the scaffolding of social relations, mechanisms, and institutions. It is not only human being that stretches from birth to death; it is also the sociocultural world that stretches from the past, through the present, into the future. Precisely because historicity characterizes practice and its dynamic world, the ontological conditions of history cannot be reduced to the existentialist structures of Dasein. As we argued in chapter 3, the fact that Heidegger recognizes neither human plurality nor an element of self-constitution in equipment leads to his inability to grasp the historicity of the sociocultural worlds that we inherit from our ancestors, dwell in, transform, and leave to our successors. The conditions "under which the notion of a heritage is introduced" cannot be monadic. Heritage is "transmitted from *another* to the *self*, "and the repetition of the past is always a reenactment within a community.[69] In other words, as *temporalizing* is the content of the historical being of humans, so *historizing* is the content of the temporal being of the sociocultural world. Furthermore, the ontological conditions of both individual temporalizing and historicity, on the one hand, and the ontological conditions of supra-individual historizing and the temporality of the sociocultural world, on the other, are realized together through ongoing interrelations between human beings. Seen in this way, history is not simply the possession of the sociocultural world; it is an ontological structure of the being of humankind and its world as constituted in practice. It structures the sphere of interactions between—as Schutz calls them—*consociating* humans and their sociocultural world, between contemporaries, predecessors, successors, and their social worlds.[70] Accordingly, historicity is possessed neither by a human being nor by the sociocultural world in and of itself. It is realized through the dynamic processes of the mutual distantiation, externalization, objectification, and alienation of us and our world. These processes underlie the relation between the present, the past, and the future; they make possible the relations that shape our being in terms of the "here-nowness" of the present, the relations of closing and opening up the past by the present, and the relations of opening and projecting the future. Finally, historicity and our processes of self-constitution are inseparable: they condition and constitute each other. Self-making, with self-reflecting as its component, is the ontological condition of history because—as we have argued—there would be no historical process without self-making. It is also the fact that human being is a process of self-making that entails both its own historicity and the historicity of us constituted as self-makers. It entails "that the scope of human being differs from one historical situation to another; that, to the extent to which it is historically situated, man's very humanity differs from age to age."[71] Again, the distinction between human being and human acting disappears; people effectively constitute themselves through and in their action. "Man is what he becomes and has become; and the processes of becoming which make him distinctively human are historical. But what makes history distinctively historical is human action."[72] We belong to history not merely as a part and product of the historical process; our being is active historizing, the very activity of making history itself.

Accepting the idea that human self-making is inseparable from historicity guards us from a number of misunderstandings: (1) from attempts to conceive human self-making as an absolute creation ex nihilo; (2) from the search for external origins of human history; and (3) from attempts that seek universal principles, patterns, or mechanisms that drive human history. Human self-constitution is always a historically situated self-making that proceeds within given circumstances and is open to contingent change. But the content of our historicity is not independent of us; we make it and this provides another reason why it is factical: it is "the particular mode of historical existence we in part find ourselves in and in part shape in cooperation with others."[73] Hence, our self-making and history are linked together within an ontic-ontological circle: history situates our self-making and is produced by our practice. To situate practice in history conceives it as ontologically different and separate from our self-making; it loses this otherness, however, when it is understood as a human product.[74]

The establishment of historicity as an ontological feature of being equiprimordial with interrelatedness mandates the conclusion that the ontological conditions of history lie in the structures of both individual and social being: in individual situatedness and finitude, on the one hand, and in the plurality of humankind, on the other; in short, in human individuality and plurality. These ontological conditions do not lie beside one another, so to speak; they are linked within an ontological circle and dynamically condition each other. Within this circle the finitude of individual being gives historicity to human plurality: plurality is realized as the succession of human generations; and human plurality situates individuals within their generation, after their ancestors and before their heirs. Because of this ontological circle, human individuality and plurality are not in themselves ahistorical. If being is always already historical, an ontological analysis cannot search for the invariant, universal characteristics of being in the hope of unfolding their ahistorical content. The ontological structures and conditions of being are also historical in the sense that their content is-in-history and changes from age to age.

Human individuality and plurality are the basis of two other ontological circles that connect temporalizing with temporality and historizing with historicity. The historizing of both practice and the sociocultural world — that is, that they exist in history as forms of practice and in particular life-worlds — is the source of the historicity of individual human being, of the fact that it changes from one epoch to another. The temporalizing of a human being, which consists in its stretching between birth and death, is the source of the temporality of practice and the sociocultural world, of the fact that practice is always a complex of subpractices of a particular society and the world is always the life-world of this society. In virtue of these circles the concepts of historicity and temporality join the ontological and ontic levels of our self-understanding within an ontic-ontological circle. Recognition of the circle permits the realization that our ontological analysis is a heuristic search for structures and conditions that we must postulate in order to understand the meaning of what is-in-history. But since both those structures and those conditions are always historically situated, their disclosure does not in itself terminate our self-understanding. This must be followed by the reconstitution of the ontic dimension of our being in a hermeneutic-historical narrative. The following general picture emerges: our relation to history has an interactive structure because our acts of self-constituting are always situated, so that a nonhistorical human situation cannot exist.

As we (empirical subjects) are (individually and collectively) thrown into history and our self-making is always situated by history and within history, so history is always in the state of being made by us. For this reason human historical situatedness cannot mean simply thrownness; it must be composed of both *thrownness* and *transcending*. Their relationship constitutes an ontological structure of our sociohistorical being. To be historically, therefore, is to encounter repeatedly what we were as a prelude to what we will become. It is in this sense that our situatedness is constituted by the effective presence of past-ness in our present-into-the-future-being. The idea of thrownness "not only specifies the limits of sovereign self-possession but also opens up and determines the positive possibilities that we are."[75] Our thrown-ness means that we have to deal immediately with whatever we facti-cally confront here and now. Transcending, or futurity, is our ability to go beyond the historical locality and factual activity of doing this. It means overcoming the limits of this (present) situation toward the fu-ture, toward something that is-not-this-situation; in other words, to-ward something that—from the perspective of a current situation—as yet does not exist. In transcending toward the future, we open up "the limits of sovereign self-possession" to the extent that sovereign self-possession can be changed in the future, and we close "the positive possibilities that we are" to the extent they are either realized or lost. So conceived ontologically, human historical being, individual and collective being-within-history, is constituted by a tension between be-ing thrown into a historical situation and at the same time transcend-ing it.

PRESENT, PAST, AND FUTURE

Objectivist approaches to history maintain that historical and physical time do not differ from one another and that the historical concepts of past, present, and future are reducible to their physical counterparts. Within this framework, the past is only "the temporal direction into which time flows irretrievably and disappears," the present is merely the boundary between what is coming (the future) and what is disap-pearing (the past), and the future is nothing else but "the place of ori-gin of time's flowing away."[76] Advocates of this approach to historical time search for regularities that equally govern past, present, and future occurrences and as such are prior to practice. They ransack the past for examples, presupposing "the implicit atomism, which claims that there is nothing more to a historical development than the repeti-tion and aggregation of isolated events."[77] From our perspective, this stance makes little sense: it is based on the assumption that the past, the present, and the future are ontologically commensurable, so that events taken from any of these frameworks combine together uni-formly. This assumption suppresses their historically relative character and ignores the fundamental fact that they are the past-related-to-this-present, the present-referring-to-that-past-and-future, and the future-related-to-this-present.

Our view also goes beyond the nonhermeneutic historicist view of Margolis, according to which the historical past is "indissolubly incarnated" in the physical past though not reducible to it, because it "is conceptually more complex." Presupposing this, Margolis holds that "what is reversible or alterable in history is confined to what is Intentional, significant, interpretable—without violating the strong constraint: namely, (1) that physical time is irreversible; and (2) that events, once past, are, qua past, unalterable."[78] Margolis claims also that the historical past is "logically emergent" from the physical past. This presupposes, however, the unnecessary and questionable idea that physical time is logically and ontologically prior to history. This presupposition is, of course, a relic of physicalism. Contrary to Margolis, we hold that the relation between nature and history is not one of logical priority, ontological primordiality, and unidirectional determination. So we believe that historical time is not simply determined by physical time, a time conceived to be out-there independently of practice. Moreover, we hold that historical past, present, and future cannot be reduced to physical past, present, and future. In short, we maintain that the temporal modalities of history are correlatives of practice without which they would not exist.

In order to undertake an ontological analysis of the internal structure of historical time, the separation of the past from the present and the future, we will appeal to the idea of the circle of individuality and plurality, which connects "the private time of individual fate" and "the public time of history."[79] Ricoeur himself considers the relation of *contemporaneity* as a structure that mediates between private and public times. The concept of contemporaneity, borrowed from Schutz, allows him to connect both times in so far as they are copresent. Hence, to connect the present with the past, to connect contemporaries to predecessors, Ricoeur refers to the succession of generations, the idea of a chain of memories that turns history into "a we-relationship, extending in continuous fashion from the first days of humanity to the present," and, as we have seen, he develops a concept of historical narrative. The idea of the succession and replacement of generations is particularly important here. It possesses a content that goes far beyond a purely physical or even biological understanding of time. First, it is "the euphemism by which we signify that the living take the place of the dead." In other words, it attributes to history the fundamental meaning that it is our history, the history of mortals. It discloses the fact that the world and practice are relativized to different and successive generations of people. Second, it is linked to our act of intending "a more human Other, whose lack we fill through the figure of our ancestors, the icon of the immemorial, along with that of our successors, the icon of hope."[80] We hope to become other than we are, more human, more perfect, and immortal. Third, the idea of the succession of generations allows us to understand that history is a process during which cultural heritage is transmitted from one generation to another in the form of tradition and that these successive generations are mutually dependent. In short, that the generation of heirs is not merely in passive receipt of the heritage of its ancestors. However, the succession of human generations does not constitute by itself an internal ontological structure of history. It is simply an ontic manifestation of human plurality and finitude. The ontological structure of history is derivative in reference to the sequence of the past, the present, and the future. "The past makes the future possible by constituting the material for consumption, thus letting the future come to past."[81] But past, present, and future do not exist by themselves, since time does not exist by itself. To say this is not to mean, of course, that they are merely our intentional (intellectual) constructs. From the ontological vantage point, the temporal sequence of the past-present-future is a flowing of human practice and, correlated with it, various life-worlds. In order to uncover the ontological conditions of the past, present, and future, as well as the conditions of their sequence, let us begin with an analysis of the phenomenon of the past.

We cannot simply say that a certain sociocultural world is a past world because it no longer exists. Heidegger rightly argues that this would require us to presuppose that its pastness is a quality it possesses by itself or in virtue of objective time. Moreover, we should also refrain from saying, in spite of Heidegger, that a certain world is a past world because it is a correlative of the practice of a generation of people who died. This presupposes that the past has, or acquires, its pastness independently of both the present and us. Yet this is impossible: a past world was not past for the generation of people who created it and died; for them it was a present world. It is a past world only for us, from our perspective, situated as we are in our present. However, it would not be a past world (for us) if it had not been a present world for our ancestors. What this uncovers is an interplay between the objectivity of a past world and its dependence on us. It turns out that both elements: we, with our practice and world, as well as our ancestors, with their practice and world, are necessary for establishing the past. In other words, the existence of a sociohistorical relationship between us and our predecessors is the ontological condition of the past. This socio-historical relationship is a complex ontic interrelation with several spheres of mutual dependence between our subpractice and their subpractice. We can distinguish individual experience of the past—that is, of the pastness of events that have gone by—hermeneutic inter-pretation of past texts—in particular, those treated as manifestations of their own epochs—and practical appropriation of the "material culture" of our ancestors either as part of our own activity or as antiq-uities. In each of these spheres the mutuality of the relation between us and the past can be shown. For instance, Gadamer addresses the objective nature of the past and the impact it makes on us in terms of his concepts of tradition, language, and application. He does not leave unnoticed, however, the dependence of the past on our inter-pretative activity. Next, when Nietzsche analyzes our experience of the past, considering the meaning of the expression "it was," and our atti-tude toward it, he addresses the dependence of the past on us and on our individual experience of historical becoming.

Nietzsche stresses the very special relation we have to "it was." For the child "playing between the fences of the past and the future in blissful blindness" there is not yet a past to disown. This bliss, however, will end; "it will be called out of its forgetfulness. Then it will come to understand the phrase 'it was,' that password with which struggle, suf-fering and boredom approach man to remind him what his existence basically is—a never to be perfected imperfectum."[82] This awareness of something that remains within the relentless temporal chain of extinguishing moments is the emergence in the individual of the idea of historical existence. The individual begins to understand the "it was" in relation to itself and to see the past as past, as a dimension of being that refers to what is no longer and will never be again. It is we, there-fore, who constitute the "it was"—that which has come to be historically—by standing outside becoming, so to speak, and marking off a past from its undifferentiated flow. Yet, to understand "it was" as a past having his-torical meaning means something more; it is to grasp the idea that an-other such "remainder" is yet to come. That such "remainders" come again and again is a salient mark of historical time. When we realize that the "just experienced" is no longer and never to be again we comprehend our existence as "a never to be perfected imperfectum." This sense of the historical Nietzsche vividly contrasts with the unhistorical existence of the animals. They live from moment to moment forgetful of what is gone and unaware of what is yet to come and fail to grasp that both disappear into the flow of time. Our life would be similar if we were immersed fully in the flux of history. We would be like a being who does not possess the power to forget, who is damned to see becoming everywhere: "such a one no longer believes in his own existence, no longer believes in himself; he sees everything flow apart in turbulent particles, and loses himself in the stream of becoming."[83]

From Gadamer's point of view, hermeneutic interpretation must presuppose that the past is not identical with the present or the future, that it has its own way of being, that it is fundamentally independent of contemporary interpreters, and that there is real distance between the past and the present. This distance makes possible historical understanding and enables us to conceive the process of understanding as progress of interpretation.[84] The distance between the present and the past is why certain conceptions of going beyond the horizon of one's understanding toward other horizons (for example, the past) are naive. Naive historicists believe that they can transcend their horizon and enter horizons of bygone historical actors, as if no gap were present between them. Naive objectivists claim that it is possible to transcend every horizon and gain a view from nowhere, to perform the "God trick" and reach a position not limited by any horizon, as if all horizons were unified by an ahistorical human situation or by a universal horizon embracing human understanding and being. Rejecting such naive views, hermeneutic pluralists "argue that although it is true that the cultural backgrounds can never be *entirely* transcendent, this limitation need not entail that the cultural background can never be transcendent *at all.*"[85] Clearly, we are able to fuse different horizons because they are simultaneously closed and opened. This fusion means "rising to a higher universality that overcomes not only our own particularity but also that of the other."[86] It does not mean, however, that the gradual process of fusing horizons recovers one unique and universal horizon of human understanding.[87] The idea that it is possible to fuse horizons, "supposedly existing by themselves," is necessary to avoid radical perspectivalism, which treats historical perspectives as mutually alien (incommensurable) and alienated.[88] To avoid radical relativism requires an acceptance of pluralism and—in fact—of ethnocentrism. If ethnocentrism means that we always start from our own historical and social context, it follows directly from hermeneutics.

If there is a distance, however, that separates the present from the past and the past has its own way of being, we must assume that hermeneutic interpretation not only takes part in the constitution of an ontic link between the present and the past when it fuses horizons. We must also assume that it constitutes an ontological link between them. If, however, this is not presupposed, we must find some form of mediation between the past and the present (and the future) prior to attempts at interpreting the past. Gadamer finds this mediation in the realm of language. Again—as in the case of human finitude and history—it is beyond doubt that the process of historical understanding has a necessary linguistic character; the fusion of horizons requires linguistic articulation.[89] This does not mean, however, that language, even when taken together with the entire web of meanings it conveys, can be accepted as the ultimate, self-standing ontological mediation between the present and the past. It is embedded in practice, and any attempt to extract language from practice leads to its reification and absolutization, so that it becomes a nonhistorical medium within which practice is diachronically entangled. According to us, it is practice that mediates between the present and the past and between any of these and the future.

As the fusion of horizons does not require the assumption of one universal horizon for all historical human self-understanding so, broadly understood, mediation between the present and the past does not require that we presuppose any ahistorical continuance. It is not the sameness of past, present, and future events that guaranties historical continuance and enables a transition from the past to the present and to the future. Forms of mediation occur when they are worked out practically. Mediation between the past and the present is the appropriation and reenactment of the results of the practice of ancestors. Mediation between the present and the future is our transcending into the future that initiates the realization of possibilities we have already worked out. The appropriation of tools, pieces of art, values, gods, ideas, etc., created by our ancestors occurs when we learn how to use those tools, when we appropriate them, when we admire their art, worship their gods, and when we think with the help of their ideas or creatively modify them. So the fusion of different historical horizons occurs if it is worked out practically, in the actual course of interpretation and assimilation of a historical tradition that is effectively being conducted. Mediation and fusion exist as correlatives of practice. "In so far as any of us manifests the human way of being, we are all continually integrating future anticipated possibilities with our own past into our present action."[90] Seen in this way, the pastness of the past does not exist simply and directly by itself, or through the agency of the present effects of past events. The past is historical, and not simply temporal (physical), "only if, and to the extent that, it is capable of present appropriation and re-enactment."[91] The existence of the past world or of a horizon other than our own, "the historical life of a tradition," is a correlative of our (interpretative) practice. Any historical legacy exists presently if, and only if, it is experienced, interpreted, understood, and appropriated; without our active appropriation it no longer exists.[92]

This dependence of the past world and practice on us is, however, only one side of the real interrelation. Our ancestors must, first, have created their world and, second, passed it on to us, toward their future as their legacy. Had they destroyed it there would now be nothing for us to appropriate. Only under these two conditions, namely, our activity and that of our ancestors, can the sociohistorical relationship be established. The mutuality of this relationship is necessitated by the fact that not only must the past be open to our determination, to our interpretation, emulation, and appropriation; we must also be open to the historical tradition that is able to condition us. The process of transferring heritage is also a way of establishing the (historical) universality of the human situation or of the human condition. Such an existential, universal situation exists to the extent that it is established practically, through various universalizing human actions and institutions: from the existentialist idea of the human situation, the ideal of universal human rights, the proliferation of goods or ideas, to political leagues, technological unification, military conquests, and so forth.

It is not only the past that depends upon our present practice, especially upon its hermeneutic interpretation. As both Nietzsche and Gadamer stress, relating ourselves to the past (in hermeneutic understanding) shapes the forms of our being because it prevents the individual from losing "himself in the stream of becoming" and seeing "everything flow apart in turbulent particles."[93] Without historical consciousness our knowledge of the present would have been different and deficient in character. Immersed in immediacy we could not, and would not, recognize its temporal, historical nature. Moreover, without reference to the past the individual could not generate itself anew: "only through the power to use the past for life and to refashion what has happened into history, does man become man."[94] Nietzsche advances three ways in which history belongs to active human life: it pertains to a human being as someone "who acts and strives, as one who preserves and admires, and as one who suffers and is in need of liberation."[95] In parallel to those aspects of human life, three kinds of history can be distinguished: "a monumental, an antiquarian and a critical."[96] The *monumental* type of historical writing freezes the significance of the past into moments of greatness to be emulated. It reduces history to a series of unconnected events without any basis in the historical process itself; moreover, it essentializes history from a supra-historical perspective and thereby attempts to reduce historical continuity to identical patterns of recurring events. It maintains that "the past and the present is one and the same," and, as in religious worldviews, that history embodies unchanging values.[97] Monumental history offers great service in so far as it provides models to be emulated in the quest for significance in life. However, its great danger is that it can lead to the ideal of the supra-historical human being that can turn its gaze away from the contingencies of becoming. If we cannot live unhistorically, if we are always situated within a limited historical horizon, the pretense to supra-historical understanding is a harmful illusion. *Antiquarian* history mummifies the past, turning it into a shrine to be worshipped. It thereby renders history ineffective for the immediate needs of the present and threatens to bury its significance. In the *critical* approach, which Nietzsche incorporates into his genealogies, there is no pretense at recovering the immediacy of the past, and no attempt to reproduce the original production in the manner of Romantic historiographies of empathy. For Nietzsche—and here he is the first great exponent—historical interpretation, as indeed any act of interpretation, must be productive. It involves a critical interaction between the foreknowledge of the interpreter and the entity or event to be interpreted. The past must be critically appropriated if it is to illuminate the present and point to a renewed sense of our possibilities. This type of historical writing approaches history reflectively and diagnostically in the service and enhancement of life: it unmasks history's distortions, errors, wrongful emergences, and harmful turnings to the extent that their traces are effectively embodied in the changing horizon of the present.[98] Critical history judges and condemns. These historical attitudes are, in fact, three interconnected ways of assimilating the past into the drives of life.99 In this regard, critical history is the negative side of monumental history and "constitutes one of its indispensable presuppositions."[100] More importantly, each is a way of belonging to history, and each must balance the other. By treating them as interrelated ways of understanding history, Nietzsche establishes them as an ontological structure of historical being.

If a relationship between a present generation and its ancestors is an ontological condition of the past and contributes to the establishment of the meaning of the present, so a relationship between a present generation and its successors is the ontological condition of the future. Consequently, as appropriation forms the content of our active relation to the past, so *transcending* provides the clue for the relationship between the present and the future. Transcending requires further analysis. The most important questions are the problem of its nature and of its role for human being. Is it, as Campbell seems to suggest, following Hegel, an act of infinite consciousness or an activity that presupposes an infinite consciousness embracing both the present and the future? Undoubtedly, Campbell is not simply taking up Hegel's idea that the human individual (the philosopher, in fact) can surpass all historical limitations, step outside a particular historical horizon, and reach a horizonless situation from which an infinite future becomes visible. Nor does he think with Hegel that this can be done through thinking, in which "I raise myself above all that is finite to the absolute and am infinite consciousness, while at the same time I am finite self-consciousness, indeed to the full extent of my empirical condition."[101] Campbell does follow Hegel, however, when he states that human being is constituted by the struggle between a finite situation and an infinite consciousness. Moreover, he believes that among our different ways of being, in which this struggle is manifested, there are certain privileged and qualitatively distinct ways. These are the ways in which human beings quest for self-understanding that is expressible in words, in "statements, or perhaps poetry."[102] We think, however, that this claim is doubtful.

When self-understanding appears in the form of hermeneutic reflection upon our activity of transcending a given situation, it does not embrace infinity because we are more than we know, and our history is more than we understand at that given moment. In other words, our understanding of our own activity of transcending a present situation is always finite and limited by what we already are without yet fully knowing it. Our self-understanding is always in-history as it "realizes itself in the understanding of a subject matter and does not have the character of a free self-realization."[103] Our self-understanding is existentially and historically authentic when it is consistent with our historicity, when it uncovers our finitude and illuminates our act of transcending as a move from one finite situation to another, from one horizon to another. Thus, for us, contrary to Campbell, infinity does not appear as the feature of consciousness that enables us to transcend locality and reach a (more) universal self-understanding. Infinity is to be understood as a feature of what does not yet exist, of the not yet experienced; in fact, of the future. Thus, a tension constitutive of our being is not a struggle between a finite situation and an infinite consciousness, but a struggle between actually transcending toward the nonexisting, the unknown, the infinite, and recognizing the unavoidability of our thrownness and finitude. This struggle between the actual, practical transcending toward the future and the recognition of the inescapability of thrownness may be viewed as the ontological structure of our historical situatedness. In our activity of transcending, there is not only hermeneutic understanding of thrownness but also, so to speak, operational knowledge actively guiding our projection. This knowledge contains infinity in the form of different (conceptual) invariants, such as the principle of induction, scientific laws, and the very concept of the unknown itself. Seen in this way, infinity does not belong to the future by itself; it is assigned to the future by us. If consciousness connected with our activity of transcending contains infinity, it cannot give us a historical, hermeneutical understanding of that situation. To understand the situation within which it is produced requires not objectifying cognition but self-awareness. So to the extent consciousness involved in the activity of transcending is objectified it cannot be our self-awareness. Self-consciousness is not simply a type of consciousness that refers to a present object. The hermeneutic answer to the question of whether knowing that we know is like knowing an object is in the negative: we cannot be objectively self-conscious. As Gadamer emphasizes, "not all reflection makes what it is directed at into an object."[104] In general, then, a key part of the relationship between our thrownness and transcending is a tension between the nonobjectifying self-awareness of our participation in the particular here-and-now of our sociohistorical being and an objectifying consciousness (that is, operational knowledge) that is involved in the act of transcending toward what does-not-yet-exist.

A last point to note is that transcending toward a future situation requires a change in those who move beyond a present situation. The ontological core of an act of transcending is human self-transcending. Moreover, in an analogous sense we can talk about self-transcending in reference to sociocultural systems. In sociocultural systems self-transcending is realized by the subpractices that are their basis; for example, in the case of science by scientific research.

Elements of the past world and of the future world are—for us—temporal and historical in a more fundamental way than they are for Heidegger. For him, equipment's horizontal temporal structure is composed of being "taken for granted as already a resource," being "applied in present coping," and "being directed toward some outcome."[105] This structure is constituted by Dasein's activity, and a thing lacks any temporal or historical structure independent of the temporal structure of Dasein's activity. We agree with Heidegger that it is our activity that gives temporality and historical localization to events, making them past, present, or future. This is not, however, an individual activity but practice, as performed not only by us and our contemporaries but also by our predecessors and heirs. Consequently, we claim that history and time are neither objective, in the sense that they are independent of humans and human practice, nor constituted by the temporalizing of an individual alone. In particular, both historical and physical times are correlatives of practice. It is not only the concept of physical time but also—and this is an unavoidable conclusion of our ontology—physical time as such that is a correlative of practice: it is a correlative of physical research in the sense that it exists-in-the-research-of-physics, especially as articulated in its various theories. This means, for instance, that to claim that history is irreversible because physical time is irreversible is but one possible view of the connection between them. In fact, it is an example of the objectifying view that brushes aside the connection between physical time and practice. It is equally possible, coherent, and plausible to say that it matters little for the characterization of historical time whether or not physical time is anisotropic, directional and irreversible. What is constitutive for historical time—seen from the hermeneutic perspective—is exactly what is brushed aside by objectivist views of physical time: historical time exists only within the interrelations among successive generations of humans who experience and make their historical being.

Another issue that demands detailed discussion is the problem of the ontic content of the historical being of human and nonhuman entities. This we now examine in connection with the question of individual becoming and persistence.

BECOMING AND PERSISTENCE

Views of how individuals and things exist in time divide into three camps: those of the endurantists and the perdurantists and of those who view entities and persons as a dynamic synthesis of interrelations among processes.

For *endurantists,* like Aristotle, what persists are concrete, self-identical, and countable individuals. For them the historical existence of individual entities does not pose any special difficulty. Notions of becoming and existing through change are based on essences understood as the self-referential ground of the identification of individuals. For substance-ontologies becoming is coming-to-be-this-or-that in virtue of the actualization of prior possibilities, as in the case of Aristotle; the act of God's creation (out of nothing), as in Christian philosophy; or the natural process of emerging from earlier configurations—for example, through evolutionary processes—as in naturalist ontologies. According to the first position, becoming is constituted by a pre-given and unconditioned reality that drives things forward to a final state of fulfillment. Aristotle's interconnected concepts for conceiving change—*dýnamis, enérgeia,* and *entelechéia*—provide a classic example. On this view, the essential self is a fixed entity not open to historizing: if the individual does actualize its possibilities it becomes what it already potentially is. Indeed, the endurantists' fixation on substance-ontology and their unreflective acceptance of its explanatory power constitute a myth of substance akin to the myth of the nonconceptualized given.[106] The second position is—to some extent—a modification of Aristotle's. The third approach finds its home in substance-ontologies as well as in process-ontologies.

The position of endurantists is plagued by a tension between the qualitative differences that change requires and the qualitative nondifference demanded by numerical identity. This they try to neutralize or otherwise undercut. Their chief strategy appeals to the temporal relativization of changing properties; another strategy holds that Leibniz's principle of the identity of indiscernibles quantifies over essential but not over contingent properties of a thing. These strategies pay a price. Contingent properties need to be time-indexed (for example, weighing 180 pounds on August 10, 1996), while at the same time viewed as timelessly part of what a person or thing is. This threatens to invoke the notion of individual essence and incurs the added difficulty of showing what something *is* essentially refers to contingent properties that are tensed or relativized. It seems that the endurantist, in the face of these difficulties, is forced to embrace unanalyzable and individual essences to put together an account of change and self-identity that satisfies Leibniz's principle.[107]

The perdurantists reject substance-ontology and conceive persistence in terms of events, stages, and states. The work of Quine and Lewis best illustrates this approach. According to this view, persistent entities are a series of temporally short stages or different particulars of a four-dimensional space-time manifold. This allows a thing to be conceived simply as "a sum of momentary states of particles, or brief particle moments, scattered over a stretch of time as well as space."[108] Perdurantists avoid Leibniz's principle by denying that what there is before and after change is numerically identical.[109] For them, statements such as "x at t_1" and "x at t_2" refer to different entities, references to which adjudicate a thing's identity across time.

For the endurantist, however, the perdurantist solution misses the real problem: how to account for something's self-identity as it persists all at once at a time with respect to its endurance through successive times? According to the endurantist, no genuine account is available of what comes into being as a whole from the perdurantist perspective.[110] Perdurantists counter that the temporal parts of things do not come into existence ex nihilo; they are merely the result of what occurs in the passage of time. Nevertheless, the perdurantist picture of things dissolved in Quinian fashion into river-stages and cat-processes cannot address the issue of change and permanence in history. In perdurantist ontology there is no real change, only timeless variation;[111] this means that there is no possibility of considering changing wholes, such as those present in history, that change with respect to themselves across time.[112] Yet, to understand historical processes we must make reference to activities that are their own dynamic basis of change and, as such, persist trans-temporally.

As we argued in chapter 1, process-ontology provides a coherent notion of persistence that avoids conceiving a thing's enduring at a time as logically identical to its existing at another. A process is a unitary item, not in virtue of the fixity of a set of essential properties, but because its unity is constituted in its history. Within its historical trajectory it dynamically integrates, consolidates, and constitutes itself providing the temporal patterning of its successive actions. From this standpoint, the features of substantial things need not be invoked: the unity of an item is simply designated by the ways it integrates its processual features as it acts uniformly in relation to other items. Moreover, there are processes that do not have countable "things" as their subjects and which are not the doings of things.113 From our sociohistorical perspective, *becoming* is not a state or an event (whether momentary or extended in time) but a process that is—as Nietzsche emphasizes—destined for noncompletion. Becoming is, in fact, the process of the identification of an individual through appearing-in-history. There are two possible misunderstandings of the idea that entities appear-in-history. First, it may trigger the view that it is merely one possible way of an entity's being, as if an entity could first simply be and then appear in history, as on a stage, and in so doing gain a historical identity. Second, it may foster the belief that an entity that has appeared in history thereby becomes present in history and, as such, is open to changes in an identity it already possesses, to gaining or losing features, functions, activities, and meanings. For us, however, becoming is not merely a happening, a pure coming-into-existence without the acquisition of features other than existence. It is the movement of attainment, of achieving stasis within change, only to change again, and so on. This means that attainment itself is understood dynamically and not as a permanent stage reached when an entity appears-in-history. Appearing-in-history establishes individual identity (in the case of human and non-human entities) that is composed—as we have already clarified—of the element of being constituted and the element of self-constitution. Both elements allow us to understand that the ontological structure of becoming, as appearing-in-history, is an interplay between openness and closeness. It enables us to see the specific character of forness. Appearing-in-history entails that things become this or that, here or there, and then or now, that they start-to-be-for other entities (in particular, for us) to the extent they appear within a certain locality of practice and thereby enter into the relations and assignments that are copresent in that locality. Seen in this light, the identification they gain within a certain place and time is stable but not permanent.

As becoming for substance-ontologies is coming-to-be-this or-that, so *continuance* is the existence of a changeless essence through change that grounds the appearance and disappearance of contingent features. The sameness of essential features provides the ground and guarantee for the identity of an individual in the course of its existence. Since essence or primitive "thisness" secures the continuity of identity, the identity of a individual is its self-identity: a thing is what it is self-identically at each moment that it endures. Process-ontologies and our sociohistorical ontology reject this essentialist picture of coming into being, remaining the same, and persisting cross-temporally; this is why, for them, identification cannot be based on Leibniz's principle. Individual identity and the self-conscious realizing of one's possibilities are not simply a matter of remaining essentially the same through change. But, as Campbell rightly emphasizes,[114] such ontologies need a concept of identity in considering the temporality of human life and that of nonhuman entities.

Nietzsche, endorsing process-ontology, thematizes our experience of becoming. Becoming is a process arrayed with multiple points of emergence in a contingent dispersion of heterogeneous events; it emerges within the self-consuming flow of time. The genesis of each moment is made possible by the annihilation of that which precedes it. Since every moment preceding a given moment is annihilated, the future is that which comes toward us. Seen in this light, the phenomenon of becoming is the future that comes to pass. Moreover, time and becoming intrinsically relate, and because of their mutuality time is not an illusion masking a timeless metaphysical world of permanence. Nietzsche's view of becoming presupposes neither an originating substratum antecedent to the flux of history nor a structuring aim moving toward fulfillment.[115] There are only becomings, diachronic "things" in the making;[116] everything is in flux, individuals and non-individuals alike, and the idea of "steady-state teleology" is rejected. Indeed, this perspective is merely a diachronic version of the metaphysical belief in ultimate, rational harmony.

Becoming experienced as life is a dynamic unity, a never-ceasing flow, an all-embracing reality in which every individual is immersed and dwells. What there is are becomings or processes destined for noncompletion, complexes constituting and reconstituting themselves in relation to one another.[117] Thus, for Nietzsche, an entity—seen as a becoming—is never grasped "in the moment," as a thing-in-itself, a unique "one" strictly unified with itself at different times, a view presupposed by substance-ontologies.[118] Entities are temporally spread out and present as having been this and about to be that. Accordingly, what there is, is not resolved in atomistic fashion into self-sufficient parts capable of existing self-identically at each moment in time;[119] rather, what there is consists in varying sorts of shifting, temporal parts, the identity of which is diffused out into "their contexts or wholes."[120] According to Nietzsche, what is crucial for entities conceived as becomings is that they, like processes and activities, are repeatable occurrents in time and space. Needless to say for us, this is an ontic view of entities as becomings. Understood ontologically, becoming happens not in space and time as what is occurring, but in history, within the context of meanings, as coming to be this or that.

Although becoming and forness, as its content and ontological structure, are common to nonhuman and human entities, there are important differences in the becoming of both types of entities. This can be brought out usefully by considering Nietzsche and Foucault's treatment of the self, self-identification, and self-synthesis. Nietzsche rejects the possibility that pre-given and necessary *archaei* guide the self and unify states of being persisting identically through all changes, and he extends the view of entities as multiple occurrences to human individuals. He also rejects the concept of the subject as the constituting foundation for knowledge and eliminates the idea that persons are identical with an inner mental substance that is the subject of conscious experience. Consciousness arises from the human need to communicate with others and emerges within language.[121] It is "conscious thinking that takes the form of words, which is to say signs of communication, and this fact uncovers the origin of consciousness. In brief, the development of language and the development of consciousness . . . go hand and hand."[122]

That a particular person comes to regard itself as a unique center of action, and sees itself as a conscious "ego-substance," is not a primary conception of the person: it is a perspective derived from self-synthesis-into-unity, a perspective of the ego, the self-referential *I*. It is from this derivative perspective that individualized acts of agency are projected.[123] Thus, a person is for Nietzsche a self-synthesized processual unity who, possessing a high degree of specificity and

uniqueness, actively unifies innumerable drives as it develops its activities. So understood, persons are processes that dynamically fuse
their interrelated activities. Accordingly, the unity of the self's identity emerges from its involvement in acts of self-synthesis that also relate to other processes that are going on in its world. Self-unity is not
something given but something achieved. Nehamas sums up
Nietzsche's position:

> To become what one is . . . is not to reach a specific new state
> and to stop becoming—it is not to reach a state at all. It is to
> identify oneself with all of one's actions, to see that everything
> that one does (what one becomes) is what one is. In the ideal
> case it is also to fit all this into a coherent whole and to want to
> be everything that one is: it is to give style to one's character; to
> be, we might say, becoming.[124]

Nietzsche does not reduce human self-synthesis to self-recognition,
self-imagination, and self-projection. In order to structure becoming
and "objectify" human self-constitution, he introduces (in his late
thought) the conception of will to power. This carries many meanings
and functions in various contexts in his writings. It allows him to say,
among other things, that self-synthesis is differentially achieved:
ineptly by reactive (weak) wills and aptly by active wills. Its core role,
for Nietzsche, is ontological.[125] However, will to power is conceptualized by him in naturalist terms; a move we cannot adopt, even if we
accept Nietzsche's claim that becoming is not just change but self-
development,[126] and his view that the phenomenon of human becoming, understood as self-identification, must be elaborated with the
help of the concept of self-constitution. We prefer his earlier reading
of the concept of self-synthesis, couched as it is in historical and sociocultural terms.

Nietzsche's concept of individual self-synthesis, is close to the
Heideggerian idea of self-understanding, and both bear the same
atemporal, ecstatic overtones. If self-synthesis occurs during each moment of our life, it is as if those moments are not different one from
the other. The temporal structure of self-synthesis, however, is made
manifest when Nietzsche unfolds the distinction between a person as
an already existing synthesis and this person's self-synthesizing activity.
Then the phenomenon of self-overcoming becomes visible. Here the
Nietzschean distinction between self-synthesis and self-overcoming is

paralleled by the Heideggerian distinction between thrownness and projection. Moreover, what is common in both oppositions is an evaluation of the different ways of human self-overcoming and self-projecting: in Heidegger's terms, authentic and inauthentic being; in Nietzsche's, master and slave moralities and ways of living that depend on the strength of an individual will to power. The very idea that ways of life can be evaluated is an element of both views that we wish to reject. It presupposes a primordial, ontological hierarchy of values. For us, self-overcoming and self-synthesis, understood as its component, are purely factual; their evaluation, one undoubtedly made repeatedly by people, is also a fact and not an essential part of the ontology of human becoming. Finally, the concept of self-synthesis has a meaning that is similar to the concept of self-making that we introduced in the first section. Both are forms of self-constitution, although the first concept stresses the elaboration of one's identity, whereas the second emphasizes the collective establishment of sociocultural, historical structures, and systems. Consequently, the concept of self-overcoming may also be applied to supra-individual sociocultural systems, such as science. It is necessary to remember, however, that such systems are not individuals writ large and that the internal structure, both ontic and ontological, of their self-overcoming is not identical with the internal structure of individual self-overcoming.

Foucault's later conception of human self-constitution, in terms of the conceptual constructing of individual whoness, contains a radical version of the idea of the specific character of human self-identification. In switching his focus from the self, seen as constituted within discourse by the power of historical contingency, to the subject's active self-constitution, he concentrates on the "mechanisms" of subjectification, on the procedures that enable the individual to recognize itself as a subject objectified as such-and-such. Lying at the center of acts of self-constitution is the action of thought, the ways "in which human beings 'problematize' what they are, what they do, and the world in which they live."[127] Foucault analyzes the "forms and modalities of the relation to the self . . . by which the individual constitutes and recognizes himself *qua* subject."[128] He tells us that his aim is "to study the constitution of the subject as its own object," to study the procedures by which the subject is led to observe, analyze, decipher, and recognize itself "as a domain of possible knowledge." This is "the history of 'subjectivity,' if by that word is meant the way in which the subject experiences itself in a truth game in which it has relation to it-self."[129] What animates Foucault's thought is the idea of subjectificat-ions of the self by the self that involve indefinite objectifications of the self by the self.

Foucault's history of subjectivity remains intellectual and, in a subtle sense, ahistorical. He rejects the idea of the transcendental sub-ject but not the concept of individual agency. This leads him toward a sociohistorical conception of humans as real individuals acting histori-cally within contingent language games, discourses, and practices, and, as such, being constituted and constituting themselves, through the dynamics of history. However, to concretize the idea of human self-constitution in terms of the activities of self-recognition is to treat self-constitution as an intellectual enterprise. Moreover, understanding self-constituting as something performed by a self, as the constitu-tion of a cognitive, moral, and practical subject objectified as such-and-such, universalizes one particular historical form of constructing human self-understanding. Although he historizes the content of indi-vidual self-constitution, viewing the self as something that becomes such-and-such according to the possibilities of a certain epoch, he his-torizes neither the idea of individual, intellectual self-awareness nor the process of autonomous self-creation. This is a historical phenome-non, however, that emerged in modern Europe. We will return to this issue in the next section.

At this point, concluding observations concerning individual self-constitution are in order. The question is not whether action involves deliberate, self-referential acts of consciousness. Indeed, philosophy, from Descartes through Kant to Husserl, has built self-referentiality into the concept of consciousness. Moreover, this concept is now an integral part of our nonphilosophical view of ourselves. Rather, the question is whether self-referential acts of consciousness exhaust our self-constitution. This is not the case, if persons are viewed as becom-ings, according to which no form of subjectivity is logically prior to personal interaction or independent of collective actions.

For us, human identity is the joint result of what we have done and what we will do; it is constituted by the contingent and multiple ways in which we integrate our traits, habits, and patterns of activities. The unification of the self through its constant involvement in self-synthesis constitutes its identity.[130] It is action that is the core of human self-synthesis. The substance-oriented view focuses on the intentionality of the doer and hence on individual initiation of action. According to so-ciohistorical ontology, on the other hand, action is the drama of actors who copresently interrelate among themselves. From its perspective, the difference between deliberate, self-referential acts of intentionality and ongoing modes of activity involving nondeliberative coping is a matter of degree, specificity, and complexity.[131] Indeed, intentionality is not a consequence of thought but of action, and subjectivity is nei-ther prior to interaction in the public world nor independent of it. The idea that individualized agency and conscious intention are primary takes center stage only if persons are viewed within the framework of substance-ontology. However, if they are viewed as syntheses of innu-merable drives, activities, and processes, and as part of ever more en-compassing and interactive wholes, the intentionality of individual action is a derivative issue. We agree with Foucault that the subject con-stitutes itself in fields of action, both with respect to its own actions and with respect to those of others. The subject does not experience itself simply as a cognitive, practical, or moral subject: it constitutes itself as this particular agent and as its own object within the contingencies of its historical situation. This conception of self-constitution is entirely historical. The historically located individual views the uniqueness of it-self by means of the possibilities for action made available within its sit-uation. There is no prior essence that we have to be, no who that any of us is apart from being at each moment the who we are. Indeed, who we are at each moment is our own free choice, whether or not we actively exercise that freedom. Of course, this does not mean that we can be whomever we want, but we have choices and basic ones at that.

From our perspective, knowledge is not the state or property of self-identical individuals, anymore than action is simply a bodily move-ment caused by individual intention. It is processual and belongs to power-knowledge and thought-action complexes, themselves chang-ing configurations within becoming; they are the "in-between" that re-sults from syntheses of persons dynamically interrelating. What is basic, then, is the interrelation of power and activity driving from the past into the future. This primordial feature of interrelation provides a ba-sis for the connections of knowledge, thought, intention, and action. What makes possible, ontically, our identity is the endorsement of the continuance between us, as we are at the moment, the children we once were, and the older people we will be. We appropriate the past as our past and project the future as our future. In other words, human (historical) identity is neither an objective sameness of essence nor a likeness of consecutive stages, nor even a historical continuity through becoming. It is the process for which remaining oneself, based on the sense of self and the appropriation of one's own past, is crucial. The ontic condition of this process of remaining oneself is the accumulation of experience, as Nietzsche rightly notes. Without accumulated experiences, lodging themselves in the individual or collective memory, we would emerge from past events in uninterrupted continuity; it is in virtue of the fact that experience is accumulated that we do not remain the same through change. Indeed, this is why self-reflective reference to oneself is a source of our whoness and identity. Human identity is always self-identity. But it is never a purely individual, subjective, and isolated self-identity; to recognize oneself as the subject of certain experiences is equally to recognize others as having those same experiences.[132] From this standpoint, the concept of the other (of other persons, not of other minds plus bodies) is an irreducibly primitive feature of the world we inhabit, and the basis from which our ontological investigations of the self begin.

Our remarks do not amount to a rejection of Foucault's idea that the self constitutes itself through forms of objectification understood in terms of a particular moral, practical, or cognitive subject performing its activity in ways appropriate to a particular historical context. Our aim, rather, is to qualify his view in two ways necessary for its appropriation. First, as we have argued, if the identification of a human being does not reduce to self-identification so self-identification does not reduce to conceptual self-definition. Second, as Elias and others show, the notion of individualization understood as a necessary ontic condition of self-definition is itself a historical phenomenon.

THE RISE OF INDIVIDUALISM IN MODERN EUROPE

To complete our ontic-ontological analysis of being and becoming, we turn to the rise of individualism. This phenomenon slowly appeared in history between 1500 and 1800 and comprised the emergence of new modes of self-understanding, new forms for structuring self-consciousness, and new ways of incorporating individual consciousness into the changing *ethos* of society. Moreover, the rise of individualism is connected with the establishment of the distinction between public and private space and with the formation of a cognitive self expressed in the ethos of cooperative inquiry. Its rise reflects the transition from forms of collective and communal sociability in the Middle Ages to forms of privatization in the nineteenth century.

This transition took place across many fronts. Elias shows that a new attitude toward the body, one's own as well as other's, began to emphasize protective zones of privacy; along with this shift came the emergence of codes of modesty directed to the necessity of covering certain bodily parts. The chivalric customs of the warrior knights of the Middle Ages were replaced by rules of bodily conduct and codes of etiquette emphasizing modesty.[133] The sixteenth and seventeenth centuries witnessed the rise of a vast literature on civility and on the discipline and use of the body in matters of dress and social interaction. Fear of breaches in norms of behavior—indeed the mere presence of such norms—molded feelings of shame, embarrassment, pleasure, or displeasure into forms of conformity not possible earlier.

As Elias shows, with advancing civilization, people's lives were increasingly "split between an intimate and a public sphere, between secret and public behavior." This split became a "second nature" now hidden from individual awareness. Public prohibitions supported by social sanctions became reproduced in individuals' self-control. As a result, a new kind of mentality emerged historically from public regulation that radically restructured the human self and raised the threshold of embarrassment and shame. Self-controlled behavior, itself a result of historical contingency, came to be viewed as individual free-will expressing the individual's sense of human dignity and status.[134]

There is always a "continuous correspondence between the social structure and the structure of the personality, of the individual self."[135] Elias makes the point from a sociocultural perspective that Heidegger and Gadamer make from a philosophical standpoint: every human being must understand itself within a cultural world that already encodes specifically possible ways to be human, that provides ways to be what humans can become. Accordingly, there is a continual interchange between sociocultural dynamics and individual and collective responses that forge the trajectory of sociocultural life. People articulate distinctions between the self and the other and between society and nature against particular cultural backgrounds. Historically changing conceptions of the self began to reshape the very notion of what it is to live a life and how best to organize it. More and more the process of living became a matter of expressing one's inner realities and private values. Increased wealth and growth in the network of trade and commerce throughout early modern Europe helped to foster attitudes of sophistication and taste in matters such as decoration, art, literature, furniture, food, drink, entertainment, and the maintenance of private libraries.[136]

Among the factors that influenced the process of the emergence of the modern self are the following: (1) An acceleration in the process of state-formation and the increased presence of the state in the affairs of individual lives.[137] What is important here is a phenomenon Elias calls the "Monopoly Mechanism." This refers to those forms of competition in which more and more sociocultural units are increasingly controlled by hegemonic and centralized sites of authority.[138] (2) The growth of changing forms of sociability, especially the transition from the anonymous forms of the late Middle Ages to the rise of more personal but simultaneously more restricted forms of sociability centering on the family and specific individuals.[139] (3) Acceleration in literacy and in the availability of books. (4) The religious reformation that produced new forms of individualism and new senses of self-responsibility. Let us take these points in turn.

First, the gradual emergence of the modern state, sometimes absolutist but always administrative and bureaucratic, created a precondition for demarcating a private sphere distinct from the public world. This process entailed the encroachment of the nation-state into what was previously a social space open anonymously to communities. The nation-state brought sanctions and constraints in its train and strengthened and regulated social bonds existing between individuals and groups. Indeed, absolutism, as it manifested itself in the centralizing powers of government, reached an ascendancy in the late seventeenth century in both France and England.[140] Elias uncovers a close historical relationship between the emergence of the absolutist state in France, under Richelieu, Mazarin, and Louis XIV and the appearance of various emotional and psychological structurings of the self that consigned to the private sphere what were once public acts. Concomitant with this development was the emergence of court-societies, from the late middle ages into the eighteenth century. They produced new codes of conduct that became more and more constraining and exclusionary and were imitated in various ways by other social strata.[141] Certain modes of uncontrolled physical and public behavior now marked the lower strata and were no longer associated with the privileged part of society

and the body politic. Recognition of the link between the socially appropriate and what was involved in self-discipline marked "the entrance to civility, the entrance that distinguishes the members of the
privileged group from the vulgar, the upper classes from the lower, the
courtly from the rustic."[142] This resulted in the state and the courts in-
stituting new ways of being, inculcating firmer norms of personal discipline, and constituting an ethos of self-conscious constraint designed
to heighten a sense of what was permissible in public as opposed to private life.[143]

There emerged another type of disciplined and exclusionary individual—the capitalist—who took personal initiative to new heights
in the name of risk-taking ventures. The growth of unfettered capital
shaped a wide range of self-making possibilities now open to the individual. It helped to produce the entrepreneurial individual that
defined itself, by the eighteenth century, in terms of property and pos-
sessions. The emergence of capitalism, together with the interiorized
self of Protestantism, helped to produce the idea of the self as an
agent of control that acts to extend its domain and thereby empower
and extend its identity.[144]

Second, the transition from anonymous forms of social life in public spaces (such as the castle court and the village square of the Middle
Ages) proceeded through private gatherings for the purpose of reading secular or biblical literature to the emergence of the immediate
family as the chief focus of the private. In its nuclear role the family became the secure refuge from the uncertainties of the other and the
hostile scrutiny and surveillance of the state.

Third, progress in literacy led eventually to the spread of silent
reading. This opened a vista for private and self-engaged experience
that created a sense of solitary reflection previously present only in
convents and monasteries. As a sense of privacy grew more and more,
it helped to foster an attitude of inner resourcefulness in the face of
the daily contingencies of life.

Fourth, the new religions, while not discouraging collective forms
of parish life, emphasized the place of individual conscience before
God and the pious duty to examine conscience. Moreover, they encouraged personal responsibility for salvation by emphasizing prayer
and solitary meditation. In the Counter-Reformation Catholic church,
personal religious practices were systematically incorporated into the
heart of collective devotions. For example, stressing the confessional

and the duty to make pilgrimages reoriented the faithful toward intimacy with the sacred. In the words of Chartier: "Despite their differences and confrontations, churches on both sides of the Christian divide pursued the same goal: to articulate, within the context of a revitalized Christianity, the disciplines necessary to the faith, coupled with a credo invariably uttered in the first person singular."[145]

Within the domain of religion itself displacements occurred along institutional, social, and individual fronts unleashing forces with momentous consequences. An event of singular importance, in this regard, is the dissolution of the religious hegemony that centered on Rome and spread throughout Europe and the far-flung Americas. The collapse of the Catholic ascendancy during the sixteenth century released tidal forces of uncertainty contributing to the formation of new forms of social integration and conflict between society and individuals. Changing relationships between states, societies, religions, and individuals in the seventeenth century helped to forge a new balance between God, society, and nature. Within these new integrations, the business of understanding and controlling the natural order assumed a more prominent role as the century progressed. Indeed, the story of how science and its forms of organization remained relevant to soteriological concerns while on the way to becoming an autonomous body of knowledge is a decisive event in the consolidation and legitimation of natural knowledge.[146]

These lines of development from which new modes of self-experience arose explain the emergence of *homo clausus,* the closed man. This form of individualization is partly explained by the impact of new patterns of behavior that distanced individuals from the course of natural events and from one another. But for Elias it is also in the early modern period that

> [i]ncreasing differentiation of society and the stronger social pressure for self-restraint, threw people back on their own resources. At the same time, the greater measure of self-restraint in more and more situations . . . was, in self-reflection, experienced by many of them as an invisible wall separating them from other objects and persons. Here lay some of the strongest social and emotional roots of the *homo clausus* feeling and the solipsistic tendencies that went with it. Transcendental philosophy reflected this experience.[147]

What Elias's analysis establishes is that the concept of self-referential consciousness has its origin during the early modern period. For Elias, this form of self-consciousness embodies a tension "which causes the individual to feel that 'inside' himself he is something that exists quite alone, without relations to other people, and that only becomes related 'afterwards' to others 'outside.'"[148] Of course, the concepts of "inner" and "outer" need to be treated with care both historically and philosophically.

The history of the privatization of the self should not be confused with the issue of the constitution of the subject. The latter issue is philosophical and equates the subject with self-referential consciousness and, at the transcendental level, identifies the subject with the "ego" that thinks.[149] The notion of the self-referential self is not new in the seventeenth century and is to be found earlier in the work of Augustine. But Augustine's concept is tied to the theological issue of Divine presence within the self as a source of truth and does not ground the idea of an individualized and separate inner self.[150]

New structures of the inner psyche, Elias argues, are a crucial element that informs the contingent historical environment within which Descartes "meditated" on the inner sources of cognitive certainty. From this meditation emerges the Cogito, the doctrine of certainty, and its relation to the ontological argument, features that became explicit in his ontology of the disembodied ego.[151] In Cartesian ontology, a disembodied mental entity is invented and conceived to be in self-reflective relation to itself and to have indubitable cognitive access to its objects. This is a version of the Platonic idea that the essence of human being is to know essences. But it is also the invention of epistemology as the enterprise which has special knowledge of knowing, and which has immediate access to objective cognitive foundations. Descartes delineates a systematic philosophy of truth and knowledge based on the concept of the self-reflective self. Here is the emergence of the idea that there are ahistorical and neutral standards within history by which to assess what is true and right, and which make possible ahistorical closure in modes of rational discourse. Moreover, Descartes's *Meditations* provide a source for a modern reflexive notion that thoughts, unlike physical things, lack size or position, an idea that is now part of our everyday sense of ourselves. But the idea that a self-reflective self is a part of a broader cultural awareness probably awaited the impact of the psychoanalytic work of Freud.

The seventeenth century also witnessed the emergence of *homines aperti,* who participate in collaborative activities directed by specific programs of action. A case in point is the emergence of cooperative inquiry aimed at producing natural knowledge among English experimentalists such as Boyle and Hooke. The emergence of cooperative knowledge-making is a historical instance of the view we are emphasizing, that knowers are inseparably connected with their projects and cognitive subpractices, and that knowledge is not an inner content which, once analyzed, is "given once for all, and independently of any future experience."[152]

CHAPTER 6

Historicity of Science

THE SHIFT IN PHILOSOPHY OF SCIENCE from synchronic to diachronic views seems irreversible. Nevertheless, many who advocate a historical picture of science have a simplistic understanding of what history of science is.[1] It becomes objectified, subsumed under nomological descriptions, and conceived as existing independently of how it is interpreted. The past of science so objectified is viewed as containing individual episodes that can illustrate nomological models of scientific development. Isolated historical events—seen as instances of theory-change—are extracted from their contexts and made to speak for themselves. Both assumptions, that history of science is an objective passage of events and that historical episodes comprise independent atoms of change, are ontologically as well as epistemologically naive.

It is misleading to see the past of actual science as merely an objective sequence of events separated by temporal distance from contemporary science. It must be viewed in regard to its hermeneutic and narrative connections with modern science, with movements in contemporary history and philosophy of science, and with respect to the life of modern society. As we have seen, Nietzsche, Heidegger, and Gadamer stress that history "is that which is always exerting its influence on the present. The present is circumscribed by a past from which it can never escape, and toward which it is therefore destined to adopt some posture or other."[2] Clearly, we are not part of the past of science; we gaze upon it from within the present horizon and see the past of our science within it. From this perspective, there is continuity and connection between past-science and present-science. This situation cannot be grasped, however, by nomological models of theory-change, precisely because they disregard the internal continuity of science in order to divide it into episodes, epochs, or stages taken to be exemplary and comparable. From this objectivist perspective, identity within the history of science is revealed from the "god's-eye" view and universalized as nonsituated and nonlocal. In rejecting objectivism one is not forced to accept a totalizing relativism, which claims that any and every viewpoint is equal to any other. Understood in this way, relativism and objectivism are twins of one another; indeed, both are "god tricks."[3]

Grasping the historical continuity of science requires hermeneutic understanding rather than subsumption under abstract generalizations. From the hermeneutic perspective the past is continuously and actively appropriated into current patterns of thinking and ways of being. The interrelation between tradition and current thinking in science requires detailed analysis, for which we critically mobilize Gadamerian hermeneutics.

NOMOLOGICAL AND HERMENEUTIC VIEWS OF THE HISTORY OF SCIENCE

Philosophical models of the history of science are typically called models of theory-change because they decompose knowledge or science into units, such as theories or supra-theoretical wholes, that undergo historical transformation. They comprise models of theoretical reduction, theory-change, cognitive progress, or the growth of knowledge, which can be divided into two main groups: epistemological (logicist or rationalist) or naturalist—namely, views that appeal either to psychosociological or to biological (evolutionary) concepts. The models of Kuhn, Laudan, Toulmin, Hull, or Giere represent naturalist models.[4] Epistemological models, which are either cumulative or teleological,[5] reveal some salient tendencies in contemporary philosophy of science: purification, decontextualization, and decomposition. Views regarding theoretical reduction, reconstructing logical relations between consecutive theories, and the inductive models of later logical empiricists represent cumulative models.[6] On the other hand, Popper's conception of the third world and Lakatos's methodology of scientific research programs represent teleological models. According to cumulative models, the growth of scientific knowledge consists in the stable accumulation of facts or theories—for example, theories with higher and higher degree of verisimilitude or theories that solve problems better. These models typically maintain the idea of cumulative retention: that the correct elements of knowledge are carried forward and transformed into the successor framework. According to teleological models, progress consists in getting closer and closer to the truth.[7]

In Popper's model, the history of science is an objective progress, a growth of disembodied scientific knowledge that proceeds through an endless succession of developmental sequences, always composed of a problem, its tentative solution (hypothesis), a falsifying test, and a new problem that emerges from falsification.[8] Each developmental sequence, as well as their global succession, is constituted by the method of science. In particular, the method of science guarantees that false statements are eliminated from science, making its progressive nature secured. As inductivists are unable to take into account the existence of ruptures, breaks, and revolutions that disrupt the steady accumulation of facts and theories, so Popper cannot explain the persistence of falsified theories. To remedy this difficulty, more complicated models of scientific change have been put forward—for instance, by Kuhn and Lakatos.

According to Lakatos, once scientific programs are established in science and systematic research is made possible, the evolution of science proceeds through their inner changes (constituting progressive or degenerating problemshifts) and through sporadic revolutions during which new programs supersede existing ones. However, the history of science is not simply a sequence of research programs; it is an endless competition, during which from time to time degenerating programs are eliminated.[9] Each research program consists in an autonomous developmental sequence, composed of theories and models, and each is governed by common methodological rules allowing for the evaluation of theories and programs as progressive or degenerative in terms of their ability to predict new facts, and with respect to whether they are better than their competitors.[10]

In Kuhn's model, scientific disciplines "undergo development from preparadigm science, to normal science, to revolutionary science, to normal science, and so on."[11] After achieving maturity "the successive transition from one paradigm to another via revolution is the usual developmental pattern."[12] Each mature discipline changes through cyclical developmental sequences that consist of a period of normal science, its crisis, a revolution, and the establishment of a new paradigm. The mechanism that produces each developmental sequence is twofold: rational (methodological) and psychosocial. The first is realized through the values of theoretical preferences, the accuracy of predictions, simplicity, self-consistency, plausibility, and compatibility with other theories.[13] The second comprises subjective factors, such as the previous experience of scientists, their professional initiation, extra-scientific commitments, differences in their personalities, a conversion experience (Gestalt switch), and social factors such as procedures establishing consensus and strategies of persuasion.[14] The heart of the revolutionary turn is, after all, a radical shift of perception that changes the way the world is perceived (but not sensed) by individual researchers.[15] Kuhn's model breaks with the rationalist approach, since it contains a naturalist, psychosociological explanation of change in science: the development of science is not a purely objective and rational process of the growth of knowledge, paradigm succession does not aim at common truth, and the cause of the proliferation of a certain theory is not its verisimilitude but decisions made by convinced scientists that it better solves problems. Kuhn's model views the development of science as a communal enterprise, in which scientists' actual reasons, values, and shared conventions, as well as their individual skills and abilities, play an important role.

There is, however, an essential affinity among these three models; they are theoretical or nomological models of scientific change. They decompose the history of science into units that undergo transformations or replacements and establish a mechanism of change that applies universally and equally to each of those units. In other words, the question of the mechanism of development is posed not in reference to science as a whole but in reference to consecutive historical units of scientific knowledge. Since they contain scenarios that allegedly apply to every case of scientific change and explain how science progresses toward truth, or how the growth in scientific knowledge emerges, or simply how science evolves, they project on science a repeatable, time-independent, nonhistorical pattern of development. As a result, science is not cumulative in respect to its dynamic; what happens in science now, how it operates and what rules or factors direct contemporary scientific activity, does not depend on what has happened in science in previous centuries. What happens now and what happened in the past depend equally on the self-same developmental mechanism. In other words, nomological models "identify what is permanent in the flux of history and thereby make it unhistorical."[16] Nomological models of scientific change are always already infected with the paradox of treating the historical as the product of the nonhistorical, and the nonhistorical as a product of the historical. Kuhn, Lakatos, and their followers certainly regard science as having a historical beginning. Nevertheless, the development of science is, for them, the result of nonhistorical mechanisms. This means that either these mechanisms are independent of the empirical existence of science or that the historical advent of science was simultaneously an act of their establishment.

There are also non-nomological models of scientific development that avoid this paradox. If nomological models "assume that fundamental methodological concepts for the sciences (e.g., explanation, confirmation, and reduction/translation) are in important respects invariant across scientific disciplines and their historical development," in contrast the piecemeal methodological approach of Miller, Shapere, Galison, or Laudan considers all elements of science as relative to domains and developed in the course of research. In this account, methodological bootstrapping and the overlapping periodization of the history of science is presupposed.[17] In Laudan's "reticulated" model, theories, methods, and cognitive values (aims) of science change, though not simultaneously.[18] For him, the history of science is a piecemeal process without any universal developmental mechanism.[19]

Paradoxically, a *hermeneutic* view can approximate nomological conceptions. Heidegger, for instance, pictures the development of science as a sequence of different historically restricted traditions. According to him, changes in the basic concepts of our scientific understanding of being bring about the development of science. "The real 'movement' of the sciences takes place in the revision of these basic concepts, a revision which is more or less radical and lucid with regard to itself. A science's level of development is determined by the extent to which it is *capable* of a crisis in its basic concepts."[20] In Heidegger's picture, scientific development, far from being uniform and continuous, "must be shocked" from time to time "by more or less radical revisions of the existing horizons."[21] A crisis means that a given discipline begins to question its foundations and to replace them with new structures drawn from its own resources. Initially such replacement occurs when a discipline frees itself from commonsense preconceptions. Caputo rightly says:

> Inasmuch as Heidegger holds to a 'horizonal' or holistic theory
> of understanding, he regards progress in a science as possible on

two levels. In the first place, the scientist can continue to fill in
the existing horizon, building up in a continuous way the known
body of information (confirming predictions, refining calculations, etc.). . . . it is possible—and sometimes necessary—that
the horizon itself undergoes revision and that can occur only by
a discontinuous revision or shift of horizons.[22]

Clearly, Caputo provides a Kuhnian reading of Heidegger's position.
He emphasizes that Kuhn's (early) conception of scientific revolutions "is not only congenial to the standpoint of *Being and Time,* but in
fact elucidates, works out and corrects what is only a seminal suggestion in Heidegger."[23] It must be noted, however, that Heidegger is neither a historian nor a philosopher of science. For this reason, he does
not consider issues concerning the empirical circumstances of crises,
the empirical conditions of scientific revolutions, or the place that
information established under an old horizon assumes within a new.
From Heidegger's ontological viewpoint, these are ontic issues to be
considered, for instance, in terms of philosophical models of scientific
change. Some models may, of course, cohere with Heidegger's ontological analysis.

The existence of changes in science does not imply, for Heidegger, progress in understanding. He emphasizes that

it makes no sense whatever to suppose that modern science is
more exact than that of antiquity. Neither can we say that the
Galilean doctrine of free falling bodies is true and that Aristotle's teaching, that light bodies strive upward, is false; for the
Greek understanding of the essence of body and place and of
the relation between the two rests upon a different interpretation of beings and hence conditions a correspondingly different kind of seeing and questioning of natural events. No one
would presume to maintain that Shakespeare's poetry is more
advanced than that of Aeschylus. It is still more impossible to
say that the modern understanding of whatever is, is more correct than that of the Greeks.[24]

Neither Heidegger nor Gadamer, however, addresses the problem of
the historicity of natural science. For them, it remains a cognitive
enterprise alienated from everyday life and governed in its progress by
the method of science. Although they readily admit that science

changes and even progresses, "to see how advances in the natural sciences or in mathematics belong to the moment in history at which they took place" is in their view of secondary interest.[25] Deeply convinced that the natural sciences are constituted by their objectifying method, Heidegger and Gadamer overlook the constant process of change involved in the self-constitution of science, downplay the significance of the role of tradition as an agent of research, and tacitly accept a dubious view of the ontological genesis of science. There is no doubt that recent discussions in the philosophy of science render their position strongly problematic. These discussions invalidate Gadamer's claim that there is "progress of research" which in itself can serve as "the self-evident standard of examination" and his claim that the law of the development of research in natural science or in mathematics is derived "from the law of investigated object."[26] This claim is doubtful, if not entirely meaningless, if this object is not considered to be independent of scientific projection. Hence, the concept of historicity, as it pertains to science, requires elaboration.

HISTORICITY OF SCIENCE

We agree with Margolis that historicity must be applicable to all scientific disciplines: they are historical subpractices, and their methodological competence "has an inescapably historical structure."[27] This is not to deny, however, that peculiarities exist in the development of various scientific disciplines. The issue of the historicity of science can be identified neither with rational progress toward objective truth nor with rational self-correction in science, as performed according to nonhistorical rules and standards. Since we reject the idea that nonhistorical mechanisms govern the history of science, we are led to view scientific research as a spontaneous process, which not only takes place in particular sociohistorical contexts but is part of practice and its history. In particular, science establishes its internal dialectic of rational discourse within practice as that develops within the interrelations of sociohistorical systems. As Margolis notes, models of rational inquiry are themselves "the contingent 'products' of the flux of historical life."[28] On this view, it is not only scientific method that is historical; reference and predication are also historically local and include scientific concepts that "emerge from and are applied to the world through the praxis of historical life."[29] For this reason the developing character of science, its chief ontological feature, cannot be described in purely logical or methodological terms. The logical and method-ological elements of science are not invariant; they change together with other elements of scientific research and are not such as to consti-tute the essence of scientific cognition. Nor can psychological terms describe the historicity of science. A purely psychological approach cannot explain the ability of scientific cognition, which is an open-ended collective activity, to overcome itself. This ability is historical and social from top to bottom. It is the result of historically situated social interactions, the cooperation of scientists, and their interactions with objects of cognition. Nor does it result from individual sensual and mental activities, themselves determined by psychological and socio-logical factors, especially by the existing stock of knowledge and the rules of its production.[30]

Approaching the history of science from the hermeneutic-ontological standpoint shuts the door on any objectifying view-from-nowhere. The basic idea of this approach, cherished also by Nietzsche and Foucault, that the writing of history is guided by an interest in the present has important consequences when applied to the historiogra-phy of science. First, it means that the history of science proceeds from an interest in understanding contemporary science as the product of the past. When approaching the past of science, hermeneutic inter-pretation openly ascribes to it the meaning of the very process that has led to present science. Second, since the hermeneutic approach at-tempts to contribute to our self-understanding, it must embrace the present activity of science and its history as conditioning our situation. It thereby reveals a tension between the objectifying descriptions of the history of science and the awareness that we cannot reconstruct past episodes in an entirely objective way or show who was objectively right in past controversies. If we appeal in such reconstructions to sci-entific knowledge gathered later than an episode under explanation, we should not pretend that we do not know that this later knowledge was accumulated on the basis of that episode.

Let us refer to two examples. Lakatos gives the following rational reconstruction of the research program of Niels Bohr: "Bohr, in 1913, may not have even thought of the possibility of electron spin. . . . Nev-ertheless, the historian, describing with hindsight the Bohrian pro-gramme, should include electron spin in it, since electron spin fits naturally in the original outline of the programme."[31] Is this an objective reconstruction of a historical episode that aims at discovering a real research program? By no means; Lakatos incorporates into Bohr's program an idea that appeared later. Thus, he is right, ironically, when talking about "constructing internal history."[32] Also, Bernstein, in reference to the controversy between Galileo and Bellarmine and the claim that Galileo was right says: "we can now give strong arguments showing why the type of reasons Bellarmine gave were deficient."[33] Is this a demonstration that Bellarmine was objectively wrong? Not in the least. Once again we extract arguments against Bellarmine and his way of reasoning from contemporary physics that purports to embrace the arguments and methodology of his opponent—Galileo. Without access to the Truth[34] or to a different science (produced on a Twin Earth, for instance) we cannot even claim that such embracement does not matter. Of course, Bernstein rightly stresses that there are no atemporal standards of correctness or rationality; we always use a concept of rationality that has been hammered out during the historical development of science.[35] What, then, is the epistemic nature of Lakatos's reconstruction or the objectivist interpretation of the Bellarmine case criticized by Bernstein? Their aim is either to include certain episodes into the past of contemporary science (Lakatos) or to exclude certain episodes (Bernstein).

Moreover, the hermeneutic-ontological approach to the history of science denies historical realism. Margolis characterizes it as a position taking for granted: "(1) that historical narratives may be true or false, (2) that they may represent in a realistic sense what actually obtains in the human world, and (3) that they entail the inherent interpretability of human events."[36] He admits that his historical realism is itself a historical artifact.[37] Nevertheless, it is not for him an arbitrary claim: "the posited independence of the physical world depends on the genuine achievements of empirical science."[38] In many respects Margolis's position is similar to the Strong Programme. Both are based on three assumptions, which together have a superficial similarity to traditional realism: (1) scientific knowledge (and the history of science) should be interpreted realistically; that is, it should be interpreted as assuming that its objects exist; (2) the natural and the cultural worlds are not equivalent ontologically: the natural is genetically and logically prior to the cultural world including all forms of cognition— that is, "the natural world does not change at the behest of our theories";[39] (3) science (including the content of scientific knowledge) is historical, and this implies that it is conditioned by the sociohistorical circumstances of its emergence. There is, however, a fundamental difference between traditional realism and the positions of Margolis and the Strong Programme. Traditional realists accept 1 and 2, but reject 3. Moreover, they claim that the indication of truthfulness, with respect to a scientific theory, suffices for explaining its acceptance. Thus, realists, especially those who separate issues of what exists from those of truth, believe that the acceptance of a scientific theory results from the fact that its content is determined by nonsocial facts that exist objectively, depending neither upon human acts of believing nor upon social context.[40] On the other hand, Margolis and the advocates of the Strong Programme want to harmonize the relativist thesis that beliefs and their content find their primary and sufficient explanation in social reality with the assumption that these beliefs and their content are, as a matter of fact, previously (in both the genetic and logical sense) determined by nonsocial reality. In our view, they fail to realize that to preserve the idea of an objective physical world existing prior to the world of culture and human knowledge is self-destructive of their position. Insofar as this idea is taken for granted, "no cogent and convincing reasons can be delivered against the belief that it is nonsocial reality itself which determines the content of knowledge, and thus, its value, so the acceptance of a certain true proposition does not require any other explanation than the very assertion of its corresponding with reality."[41]

Margolis's project may not be incoherent in combining a radical historicist approach to the history of science with a meta-philosophical realism supporting the belief that the history of science can be objectively represented.[42] What seems incoherent is the tacitly accepted combination of a historicist approach to science and the meta-scientific idea that science is realistic. This idea, as well as the assumption that the natural world is real by itself, are not historicized by Margolis. In short, he believes that it is possible to historize knowledge, reason, and truth-claims, but not truth itself;[43] he formulates the idea of "the meaning of objectivity under historicist condition," but he fails to consider fully the consequences that his position entails.[44]

In order to consider what a historized view of scientific objectivity comes to, we must overcome a presupposition tacitly accepted by Margolis. This states that knowledge, including historical narration, is either representationalist or fictional—that is, it does or it does not take truth-values.[45] This presupposition is present in his discussion of what he calls the canonical view of history (for example, Hempel, Popper, Putnam, or Grünbaum) and in his discussion of the views of Ricoeur and White.[46] Margolis misses the fact that in the ontological-hermeneutic tradition cognition and knowledge are not taken for granted as representational structures. Cognition is disclosed, by Heidegger's existentialist analysis, as constituting (together with other activities) human being, and this is taken to be the fundamental and primordial ontological function of cognition.[47] For Heidegger, the original function of cognition is existentialist rather than epistemic; it is a function inherent in Dasein's self-constitution as self-understanding. Representing the world becomes its secondary and derivative function. Moreover, Heidegger stresses that cognition belongs to the history-of-being and this idea allows him to reveal the fundamentally historical nature of cognition. Contemporary forms of human cognition are the result of historical changes of being itself. During this process representational thinking is created and "gives an indication for the way in which Being itself begins to complete its essence as reality."[48] Different philosophical conceptions—for example, the view that human beings are a unity of mind and body, the idea that being is the system of nature, the conceptualization of the historical nature of being through the concept of the history of humankind, and even reference to objective values—are modern, representational forms of cognition.

Although we reject an existentialist reading of (scientific) cognition, we do consider scientific research as a derivative form of sociocultural being. Consequently, scientific objectivity and scientific forms of representational cognition are not prior and given to science but are constituted and reconstituted during its history. As we argue in the next chapter, objectivity cannot be separated from practice: scientific objectivity results from the fact that science is a socially constituted enterprise. Moreover, science is constituted historically and, therefore, *its historicity is more primordial than its objectivity.*

Historicity characterizes scientific research as a certain subpractice that is realized through interrelations among scientific dialogue, scientific experience (together with reality as experienced by scientists), and the technological involvement of research. Historicity also characterizes the social structures within which scientific research is performed. In short, it refers to science as a sociocultural system. In order to clarify historicity we need to recall an idea introduced in

chapter 5—namely, that self-making is the core of history and therefore constitutes the content of historicity as an ontological feature of
practice. Analogously, the historicity of science means that it is science
itself that is constantly constituted, or, in fact, reconstituted, during its
own history. In other words, the historicity of science means that its
being-in-history is self-making; science is the process of constituting itself. Accordingly, the *continuity* of science may be explained as a property of science that is created during the historical activity of scientists.
Understanding science's historicity as self-making precludes the need
to posit nonhistorical mechanisms of transformation to account for the
internal continuity of science. To talk about science and its history precludes the need to identify continuance in science with cross-historical
sameness, to assume that there is something common and essential to
all instances of scientific change, no matter how vague or general.[49] In
other words, continuity is not sameness and has no need of being attributed to science with the help of a universal model of theory-change
or paradigm replacement.

    That there is no essence common to all instances of scientific re-
search does not mean that there is no evolution or any development
and progress in science. What counts, for instance, as a legitimate scientific theory at one stage in the history of science might differ, even
radically, from what counts as such at another; but "there is often a
chain of developments connecting the two different sets of criteria, a
chain through which a 'rational evolution' can be traced between the
two."[50] Following this line of reasoning, we can say that different epochs
from the history of science may be called "scientific" if they can be
shown to result from their predecessors and be linked through them to
the very beginning of science. This can be established, however, only by
detailed historical studies of change and development in the sciences
and the use of interpretative techniques—genealogical or otherwise—
that are appropriate to the task. The idea that there are "chains" linking different sets of criteria or different epochs requires careful elaboration. Such chains cannot be traced through a purely rational
reconstruction of the rules of cognitive procedures, nor can they be established by hermeneutical interpretation. The existence of historical
continuance may be established only within the ontic-ontological circle.
Moreover, the approach to the history of science that we advocate is
not simply a matter of interpreting past scientific texts. Apart from its
hermeneutic aim, our approach has ontological and ontic aims; this is

why the self-constituted forms of continuance in scientific process are not, for us, merely discursive and epistemic. They embrace all elements and aspects of science understood as a sociocultural system.

Science's self-constitution is, as all forms of human self-making, situated and relative. Its situatedness means that the history of science is determined by inner and outer factors themselves always-in-history. This means, in turn, that their importance, and the ways they influence science, are historically variable. Moreover, the weight of outer factors may be reduced as the inner factors come into prominence. The existence of outside and inside factors that influence the transformation of science manifests itself in the rules governing the activity of scientists. Among these rules we can distinguish outer-controllable rules and (purely) inner-controllable rules. Rules of the first type are established in domains of practice that differ from scientific research—for example, religion, political systems, and education— and are imposed upon science by appropriate social institutions of mediation. Rules of the second type are created and mobilized inside science. The outer-controllable rules show which of the activities performed by scientists are scientific and acceptable from the outside point of view and which are not. The inner-controllable rules show which activities of scientists are scientific and acceptable from the point of view of scientists themselves. Rules of both types can be consistent or discordant with each other, or rules of one type can have no equivalents among those of the other. Of course, both types are open to change.

The relative character of scientific self-constitution means that at every moment of its historical being there exists a particular relationship between what is given from its previous development, or from other human subpractices, and what is new or undergoing creation. There is neither an absolutely given datum of science nor any absolute independence and freedom in science. The content of knowledge and its ideals, cognitive procedures, epistemological rules, types of scientific communities, values, forms of research organization, interactions with other parts of culture, even reality as it is perceived by scientists all change. The distinction between the given and the created is contextual; what is given at a certain time forms a context that directs the activity of scientists and leads either to the respectful contemplation of existing knowledge or to its criticism and perhaps rejection. Consequently, the scientist is essentially in a situation of *indeterminacy,*

like anyone who participates in ongoing practice, because "antecedent
constraints (such as goals, rules, or 'structural conditions')" do not determine it strictly and are often "at variance with one another."[51] Science contains an element of indeterminacy in the sense that the course
of cognitive interactions "cannot be deduced from knowledge of the
individual actors' intentions or interests."[52] Also Latour and Woolgar
emphasize that "a body of practices widely regarded by outsiders as
well organized, logical, and coherent, in fact consists of a disordered
array of observations with which scientists struggle to produce order."[53]
Since scientific research is neither generated by objective, ahistorical
rules nor enforced by social conventions and regulations, indeterminacy is the ontological condition of the necessarily axiological or normative nature of scientific subpractice. As we argued in chapter 4,
participation in scientific research requires personal involvement,
commitment, and responsibility. The indeterminacy of scientific research accounts for the unpredictability of scientific cognition.

By analyzing science from the historicist perspective we can understand how those features of modern scientific knowledge pointed to in
chapter 4—namely, discursiveness, empiricity, the universalist stance,
mathematization, the search for explanatory determinations, and technological involvement—can be regarded as historical processes that together combine the entire process of scientific self-constitution as an
objectifying cognitive subpractice. Scientific self-constitution is like a
tapestry woven together out of these interrelated processes. Each epoch in the history of science is marked by a determinate composition
of these processes and their mutual interactions, and it is conditioned
by their previous development. Most important among these features
is discursiveness, the fact that scientific cognition is composed of concepts and forms of reasoning. This is the core of science seen as an objectifying cognitive subpractice, and the development of science may
be considered as an ongoing transformation of concepts, problems,
and arguments. To an extent, the other three features—namely, the
universalist stance, mathematization, and the search for explanatory
determinations—are internal features of discursivity. None of these processes, however, exhausts the content of scientific development.

From the ontic viewpoint, the crucial aspect of the history of science is the fact that social relations within which scientific research
proceeds constantly transform. Although the concept of the social relations of science is logically subsequent to that of scientific research

and is revealed through studying research, it is ontically prior. Concrete scientific activities are always performed within a given system of social relations that constitute science at a given moment. Their existence is necessary, therefore, for the ongoing functioning of science, and the changes they undergo are necessary for the possibility that science actively participates in social life. Saying that the social relations of science are ontically prior to scientific research does not mean that they explain scientific activity by themselves. They are simply ontically necessary for constituting science as a sociocultural system. Indeed, as we argued in chapter 3, social relations belong to the ontic-ontological circle that embraces fields of action and their dynamics of change. From this perspective, social relations are constituted within interactions of power and knowledge and, in turn, they condition these interactions. The historical transformation of the social relations of science consists of changes in the ways in which scientists interact, modifications of scientific institutions, establishments of networks within science or between science and other subpractices, and changes in the cultural rules that configurate social relations among scientists. These processes depend partially on the totality of social life; they result from the impact exerted on science by other sociocultural systems through the agency of interrelations. We can say that science operates socially in a "functional" way. This does not mean that in every case the "structuration" of the ways science operates will be seen as most profitable from the viewpoint of the needs of the social system as a whole. This may or may not be the case. External influences can obstruct progressive change in science, and influences from different systems may be contradictory, and so forth. Moreover, changes in the social relations of science are conditioned also from within by the activity of scientists themselves.

If contemporary science is characterized as a relatively autonomous and institutionalized domain of practice, carrying its own cognitive authority and interacting with other spheres of practice, so the history of science may be interpreted as the historical process of self-constitution through several empirical phases: first, the emancipation of scientific knowledge from the church, theology, and philosophy, which began in the fifteenth and sixteenth centuries; second, the institutionalization of scientific research that emerged in the seventeenth century and continued beyond; third, the establishment of science as a factor driving the civilizing progress in the eighteenth century; fourth, the dogmatization of scientific knowledge, based on the ideal of scientific objectivity elaborated in the nineteenth century; and fifth, the contemporary process of entering into mutual interactions with other sociocultural subpractices, which place constraints on science. This last phase is reflected by the rapid development of science studies in recent years.

From an ontological point of view, the historicity of science, its activity of self-constitution, can be understood as a particular manifestation of the dynamism that governs human cognition. Ontologically, this dynamism is based on the capacity for self-transcending. However, as we have already pointed out, science is not self-reflective, it does not transcend its particular historical stages through dialectical reflection upon its discourse. Self-transcending in science is the result of dynamic interrelations among scientific dialogue, experience, and the various forms of technological engagement of science. In particular, it is forced by the impact of scientific experience and by the involvement of science in social and technological practices. Experience and technological involvement link scientific discourse to other, already historized subpractices and are the locus of its historicity. As in everyday life, where we experience our finitude and know that "all foresight is limited and all plans uncertain,"[54] so in science, experience that reveals finite and changeable empirical phenomena is the source of historicity. As technological subpractices change the world and the circumstances of humans, so science faces the challenges generated by them. Moreover, the historicity of science means a continuing interplay between the openness and closeness of both scientific research and the world as it exists for science.[55] Scientific research, nature, and society acquire openness, a feature fundamental to all historical beings, as correlatives of practice, when they enter history. Without openness scientific research would not be possible as a historical, developing enterprise, and its objects, both natural and social, would not be accessible to it.

The idea of the interplay between openness and closeness is central to hermeneutic considerations that pertain to our self-consciousness. We are able to understand ourselves, which means that we are open, but our consciousness cannot, as such, become entirely transparent to us. It cannot become self-transparent because it is linguistically patterned. Our acts of thematizing ourselves cannot be at the same time acts of thematizing language, whether as a medium or as a means of thematizing. On the other hand, when we thematize language, when it becomes an object of cognition, we are not "focally aware" either of ourselves as thematizing, speaking, and communicating, or of the conditions of objectification. "Every revelation is also a disguise: every consciousness of an object is made possible by a non-objectivizing consciousness."[56]

The same situation occurs in the case of the relation between the world, as studied by science, and scientific research and knowledge: this relation is also an interplay between openness and closeness. Functioning within science, the world and research operations change their being from readiness-to-hand to presence-at-hand and back again. These changes in their way of being mean that they move from openness, from being visible and thematized, to closeness, to remaining invisible and not thematized. When the world becomes open in a given scientific projection, within which inner-worldly entities are objectively present, the projection itself, composed of the knowledge that presents objects and the research underlying the projection, is simply used but not cognized; it remains invisible, closed, and unthematized. Basic concepts of a certain scientific tradition cannot be fully understandable to the participants of this tradition; at least, not as they arise at its beginning. They are ready-to-hand and they determine the way in which scientists project their objects. As far as they allow scientists to study and understand phenomena they are, like all equipment, transparent. "The interpretative concepts are not, as such, thematic in understanding."[57] Their smooth working must become obstructed to make them visible, to turn the attention of scientists from the subject-matter toward concepts. Only when this is so are the concepts "no longer regarded as transparent—either in reflecting the world or conveying ideas."[58] They attract scientists' attention and demand that an effort of thematization be made. Science "exhibits a rapidly oscillating dialectic between concern with theories and concern with the phenomena they describe primarily because it constantly works at the limits of its equipment's capabilities and aims at extending those limits."[59] Moreover, to the extent to which particular inner-worldly entities or phenomena play an active role in scientific research, are embodied in scientific measuring devices, or are used for cognizing something else, they are handy and remain invisible and not thematized. Any further attempt to uncover within the world what remains invisible requires a modification of the projection itself. At this stage, it is knowledge and scientific research that are objectified and thematized as they undergo critical scrutiny. When they become focused in scientific attention, the world becomes invisible and closed. Furthermore, since the openness of the world is always relativized to a given project—for example, to a theory—and in this sense never absolute but partial and situated, the world is never simply open and objectively present as a whole. Only particular entities, dependencies, or processes can be objectively present when disclosed within a given objectifying projection. But what does it mean that something is disclosed, and how does it relate to truth and historicity?

Heidegger's analysis problematizes the subject/object duality presupposed by the epistemological enterprise of Western thought. As we have seen, this duality holds knowledge to be possible just in case it is a correct representation of an independent reality. Moreover, whether in its idealist or realist version, it divides the world into subjects, whose raison d'être is to know (perceive and represent), and entities, whose characteristic is to be perceived and represented. Furthermore, it assumes that the knowing subject is self-enclosed and aware reflexively of its mental operations and states, each one identifiable in isolation from entities in the external world. Of course, in combination, these assumptions create the classic "problem of knowledge": how is knowledge of an external world possible? How is it that states of an isolated knower are intentionally directed toward things in the outside world?[60]

If for Kant, "the scandal of philosophy" is that it failed to demonstrate the external world's existence, for Heidegger it is "that such proofs are expected and attempted again and again."[61] This proof is a genuine problem only if, as in classical epistemology, we assume tacitly that contact with the world resides primordially in an isolated self, in a spectator who looks out upon the passing show. But to ask who demands proof of the external world is to see at once that it is a being always already situated in the world. "The question of whether there is a world at all and whether its being can be demonstrated, makes no sense at all if it raised by Dasein as being-in-the-world—and who else should ask it?"[62]

What this implies concerning disclosedness and truth emerges if we consider Heidegger's view of how we become aware primordially of something in the world. For him, the standard epistemic example—"That there appears red"—is not paradigmatic. He considers a hammer as it shows itself in hammering; for example, we notice that "The hammer is too heavy."[63] This indicates straightaway that our primordial relation to things is being engaged with them, a stance that contrasts with the situation of the disengaged epistemic looker-on. In the act of hammering, to ask what lies between us and the world is idle. The hammer is too heavy just in case it actively shows that to us. There is no third "entity" (for example, a sense-data) between us and the heaviness of the hammer to which its "too-heaviness" need "correspond." Understood in this manner, the world does not yet appear as an "it." Only in discourse about things, involving the symbolic function of representation, is the world so depicted. Moreover, perception, the situation of being a looker-on, is a rare form of concernful dealing with entities and not a typical mode of being-in-the-world.

These examples have import: the latter points to an understanding of our ways of interacting (or being able to interact) with things that is an irreducible part of our being-in-the-world. Together, however, they indicate that this understanding is itself presupposed by the epistemic enterprise if the latter is to make sense at all. Heidegger shows this in a remarkable way. Given Dasein's existence is being-in-the-world, is lived equiprimordially in the midst of entities, and alone understands its own being and the being of entities, he combines two claims within a transcending movement of thought: the subjective idealist claim that the existence of entities depends on the mind and the realist claim that entities must be *there to be known*, independently of us. Insofar as "idealism emphasizes the fact that being and reality are 'in consciousness,' this expresses the understanding that being cannot be explained by beings." To the extent, however, that consciousness is subjective and posited as the primordial basis for grasping reality, it "constructs the interpretation of reality in a vacuum." Likewise, to the extent realism posits inner-worldly entities it appears to agree with Heidegger. The similarity is illusory, however, since realism "tries to explain reality ontically by real connections of interaction between real things" within the subject/object opposition and thereby turns conceptual opposition into a real and spatialized one.[64] On both views, the real and the knowable are construed epistemically as what is represented within an ontic framework by a knowing subject and being is transformed into reality—that is, into the objectively present there to be interpreted.[65] At this point, Heidegger cuts beneath idealism and realism, as these assumptions commit both to a distinction between interpreted and uninterpreted reality. He is able to do this because the understanding of being is itself an ontological determination of the being of Dasein, and representing is based ontically in our being-in-the-world and conditioned by the ontological structure of being, especially by the temporality of Dasein. So the existence of entities can neither be affirmed nor denied apart from Dasein's innerworldly projections; "If Dasein does not exist . . . it can neither be *said* that beings are, nor that they are not."[66]

So truth and reality are ontological issues in Heidegger's view, and have their constitutive grounding in Dasein's being-in-the-world, in its acts of disclosing, its ability to reveal entities as they are. "Disclosedness (*erschlossenheit*) and discovering belong essentially to the being and potentiality-for-being of Dasein as being-in-the-world. Dasein is concerned with its potentiality-for-being-in-the-world, and this includes circumspectly discovering and taking care of innerworldly beings."[67] Clearly, Heidegger's concern is not to answer the epistemological question: What conditions are necessary for human cognitive experience and awareness of the world? It is ontological: How can things show themselves and be disclosed within the horizon of our situated ways of coming to grips with them? His answer focuses on a central constituent of our disclosedness: the fact that we are thrown into a world and must cope practically in engaged ways with things in that world. This salient fact about us, that we lead our lives in the midst of entities and in immediate interaction with them, means that "The being of Dasein and its disclosedness belong equiprimordially to the discoveredness of innerworldly beings."[68] So, disengaged knowing cannot itself be the pre-given cognitive form of human contact with the world. It is fundamental that we and inner-worldly entities are equiprimordially present as the disclosed background, or "clearing," against which such a stance can be possible. This is the primordial sense in which "Being-true (truth) means to-be-discovering."[69] Essential to worldly involvement is mood or attunement to things as they are. Moods, as our responses to the world, arise from our being-in-the-world, from the inseparable connection between us and the world we inhabit. Given this, there is no irrevocable distinction between inner affective states and an outer indifferent world onto which they are projected, no divide between the rational or the cognitive and the affective or the emotional, the latter accorded no role in disclosing the way the world is. In this sense emotions have a cognitive dimension.

To suppose, therefore, that total disengagement is epistemically prior to acts of engaged understanding of entities is an illusion. Articulating, or disclosing, the constitution of the background against which we act and think is necessary so that we can detach ourselves from it and objectify any of its elements. Digging beneath representations does not reveal irreducible representations, but rather that we already have a grasp of the world as agents in it.[70] From this perspective, truth understood within discourse as asserting agreement with objects is not primitive; its mode of disclosing derives from our being able to reveal features of what is immediately present to us as we engage with things. As we have argued in chapter 4, the traditional concept of truth tacitly assumes that being has already uncovered itself and that its disclosedness is purely discursive and cognitive. From this standpoint truth as agreement is constructed. So a statement is derivatively true in the sense that it "asserts, it shows, it lets beings 'be seen' (*apophansis*) in their discoveredness. The *being true (truth)* of the statement must be understood as *discovering*."[71] This is the level of explicit intentionality and meaning. But for Heidegger "unconcealment" (*alētheia*) is the primordial phenomenon of truth. To Dasein alone among entities is the world disclosed as "there," a disclosure (*erschliessen*) necessary for there to be any entities at all. Only if something is disclosed can we direct our conceptions toward it. So truth as *adaequatio* (correctness) derives from truth as *alētheia* and is captured more primordially. For example, we discover a particular hammer ontically because the action of hammering—that is, its place in the general structure of the world's happenings—is already known to us. Consequently, there is a parallel between an act of "disclosing" and an ontological state of affairs and between "discovering" and an ontic state of affairs. The first pair refer to our familiarity with the world in virtue of how entities stand in relation to us; the second to the objectification of entities made possible only on the basis of disclosure.[72]

Heidegger's radical break with the ideality of truth is important. For him, primordially, truth is neither an abstract cognitive relation that mediates a mental representation and what it represents, nor a logical, universal content of an assertion, nor a primitive property of a proposition. He does not endorse traditional correspondence accounts of truth that construe truth as a relationship between a propositional content that exists by-itself and facts of the world that exist by-themselves (or in virtue of other facts). Primordially, "truth by no means has the structure of an agreement between knowing and the object in the sense of a correspondence of one being (subject) to another (object)."[73] Truth is not what is necessary, timeless, or absolute; it is what, in its primitive presence, Dasein discloses in irreducibly different historical situations. Truth is disclosing and belongs to Dasein, who is "essentially in the truth" in the sense that the disclosing of what is-in-the-world is Dasein's way of being.[74]

From Heidegger's ontological standpoint, disclosure is more than pointing to a bare presence, to a piece of the given. To disclose something, to "unconceal" it *(alētheia)* requires its conceptualization, thus making it thinkable within an interpretative framework, such as scientific research. In the objectifying stance appropriate to scientific research our way of access to entities is projection, observation, measurement, theorizing, and experimentation, each a derivative mode of disclosing comprising assertions which represent entities categorizable as such and so. Only if the condition of disclosing within a particular interpretative framework is satisfied is there a claim to truth, a claim to what *is* the case. Heidegger develops this point in reference to Newton: "The fact that before Newton his laws were neither true nor false cannot mean that the beings which they point out in a discovering way did not previously exist. The laws became true through Newton, through them beings in themselves became accessible for Da-sein."[75] What is disclosed within the Newtonian framework are objects as material entities that were before they are encountered and will continue to be thereafter. Notice, however, that before they are encountered nothing is either true or false about them. But once conceptualized as (Newtonian) objects, they are encountered in a way that, if we are thinking truly, what we think is the case. Of course, the question remains whether Heidegger's way of constituting the objectivity of scientific knowledge, his treatment of the present-at-hand, is sufficient and plausible. We return to this in chapter 7.

For Heidegger, then, a statement's truth depends on our activity. Truth is bound up with existence and "is relative to the being of Dasein."[76] It is we who go to the world in Promethean fashion to deliver it through our acts of understanding. Disclosure is not given: possibilities must be worked through allowing for error and illusion arising from the forestructuring dimensions of our understanding and the corrigibility of our knowing.[77] So any distinction between what we constitute and what is given is not itself given; already we have been at work in the world. It is important to insist on this image. Too often realists unwittingly mobilize a reverse picture. They imply that the world makes true what we think—if we are thinking truly—by affecting our thinking. Certainly our cognitive operations encounter constraint and resistance. This is not, however, tantamount to the claim that sensory elements are truth-makers—for example, that certain of our beliefs result immediately from bits of the world enacting brute pressure upon us. This presupposes that the cognitive meaning of these worldly bits already exists, as if they themselves possessed it.

Furthermore, entities show themselves in our acts of explicit awareness and expressive action; they appear within language and within our collective concerns and practices. We discussed in chapter 3 how entities enter into their own constitution. In regard to truth this means that they are understood as the entities they are only in terms of their situated being. Since we and the entities we understand are in the world, there is no opposition of meaning and being: the being of entities is meaningful as we cope and comport with them. But the being of entities—depending on whether they are discovered, constructed, or encountered—conditions possibilities of establishing ways of access relevant to them. This in turn constrains the manner and outcome of their disclosure as they are.

Thus, disclosing involves acts of self-understanding *(Sichverstehen)*. These do not refer to inner mental happenings but to our coming to know our collective way around in the world or in a worldly activity such as scientific research. What is known in such acts is neither a pure subject nor a pure object but a subject-in-knowing-an-object. So, what is understood by a given self is not an intrapsychic event: rather, it is what lies within a shared world and reveals a balance between creation and discovery.

Heidegger's view regarding the status of entities within situations of disclosure captures the historicity of their appearing-in-history. Before Newton, his laws were not thinkable. This is not the trivial point that Newton's laws were unavailable before the *Principia's* publication in 1687. On the contrary: it points to the historicity of a specific sociohistorical situation in which a gifted individual articulated a powerful set of concepts allowing physical entities to reveal themselves under particular descriptions. In this way, bodies in absolute space and time, unified by gravity's power and able to move in straight lines in the absence of movers, came to appear-in-history. To say Newton's laws were neither true nor false before the entities they describe were encountered is tantamount to saying that in 1687 they became fact and also it became true that they were always fact. This view entails the historically indexed claim that Newton's statements do not possess an objective truth-value independently of *our* means of understanding them and of providing evidence for them. The realist interpretation contrasts sharply with this: on the realist view Newton's laws possess an objective truth-value such that after 1687 we are able to assert that what they state had always been an omnipresent fact.

We have argued that the sociality of scientific activity compels openness to change and underwrites the contingency of scientific practice. This means that new ways of access to the being of entities is never foreclosed and might occur in unthought-of ways. These features of science provide a basic reason for bracketing truth as explanatory. Fixing the truth-claims of scientific practices (for example, in terms of social consensus, or coherentism, or correspondence to a mind-independent world) essentializes their justificatory dimensions. This is incompatible with their unfettered open-endedness and the plurality of knowledge-making strategies that science generates. Moreover, in practice it is notoriously difficult, at a given time, to distinguish truth-indicative criteria from other epistemic criteria embodied in our justificatory procedures.

HISTORICIST-HERMENEUTIC VIEW OF CONCEPT-FORMATION

Hermeneutic ontology, being itself an awareness of historical continuity and discontinuity,[78] allows us to problematize concept-formation with the help of the Heideggerian notion of the (scientific) projection of ways of being and the Gadamerian concepts of horizon and tradition.

A paradigm may be understood—in accordance with Heidegger—as a tradition of scientific research that establishes a certain horizon, within which a particular case of scientific objectifying projection takes place. The horizon of a research tradition is not, however, only a framework within which the projection of nature or of society proceeds. As a perspective within which scientists work it is simultaneously—in accordance with Gadamer—a structure from within which scientists encounter other traditions and attempt to understand them.

Scientific concepts always exist within a particular tradition and are connected into a hermeneutic circle, within which meanings are conveyed. This circle contains—as Dreyfus points out—not only explicit or implicit assumptions but also activities.[79] The process of concept-formation proceeds within a given tradition and its language, which is a medium of understanding embedded within scientific research. However, neither a tradition nor a concept is a present, stable, and well-defined structure relativized only to an equally stationary language of science and its semantic rules. Every tradition and every concept is open. Thus, concepts exist in a dynamical way; they are becomings and not eternal structures of reality or language. When established, they are nothing more than temporal crystallizations of the process of the elaboration and transformation of meanings, which is relativized simultaneously to the projection of the objects of science and to the understanding of other traditions through their absorption or interpretation. From this perspective, to specify the "nature" of an object is to relate the history of its concept and to delineate the creative fusion of its descriptions over time.

Since neither language nor scientific research is an object of scientific cognition, concept-formation cannot be objectified and described in external theoretical terms in a way that allows one to compare two systems of scientific concepts as objects. In particular, the process of concept-formation cannot be described from outside as the objectively stateable incommensurability of certain theories or paradigms. From the hermeneutic perspective, the thesis of incommensurability is only a critical device that makes possible the disclosure of the prejudices of the received view of scientific theories.[80] Studying any process of concept-formation is an event of participatory understanding that does not "bring the object of its knowledge before itself in an objectivating fashion," but requires the recognition that in understanding, as in speaking, "what one says opens up a variety of possibilities of reformulation."[81] Understanding the phenomenon of concept-formation also demands a recognition that "a word multiplies itself not only uniformly, as logic would have it, but also creatively, according to a varying context, in what might be called 'the living metaphoric of language.'"[82] Concept-formation is, in this light, a linguistic activity, although not enclosed within language.

The process of concept-formation is always shaped by the hermeneutic circle: it starts from foremeanings, because we always understand in light of our anticipatory prejudgments and prejudices, and their content develops until meanings are established and combined into a new conceptual whole. Foremeanings of a given tradition are either concepts inherited from an older tradition or preconcepts elaborated in opposition to it. Other traditions, not necessarily scientific, constitute a context within which concepts of a given tradition are understood. Commonsense forestructures allow the comparability of theories at the early stage of the development of science, mathematical structures—at the mature stage. So from the historical-hermeneutic perspective, the process of concept-formation cannot be separated from the process of the appropriation of existing concepts. In science the appropriation of past theories, ideas, rules, and procedures happens according to two main patterns for establishing the continuance of (scientific) knowledge. The first is absorption, more typical of the natural sciences; the second is hermeneutic interpretation, more characteristic of the human sciences. In the case of natural science, past ideas become either invalidated and abandoned or absorbed through transformation into a new form compatible with a current tradition through restating and incorporating them into contemporary theories. "The past of a science hardly continues to belong to the science as such. Whatever is not eliminated is assimilated—often in theories that remain only marginally related to the original concepts."[83] Absorption means—as Kuhn puts it—that science "destroys its past"; or—as Gadamer puts it—that "the natural scientist writes the history of his subject in terms of the present state of knowledge."[84] While the natural sciences take their past into their possession, in the humanities appropriation of the past has a form of hermeneutic understanding that "lets itself be *addressed* by tradition."[85] Also, the work of the historian of science requires hermeneutic interpretation; even in the natural sciences—or particularly within them. For a historian "the same scientific documents really are texts precisely because they require interpretation to the extent that the interpreter is not the intended reader so that the distance that exists between him and the original reader must be bridged."[86]

The separation between absorption and hermeneutic understanding should not be automatically identified, however, with the borderline between the natural and human sciences. It is rather a difference between interpreters, who act as addressees of a text, and interpreters, whose understanding is permeated by the awareness that they are not addressees of an interpreted text. The problem of whether the two types of scientific disciplines are radically separated by the presence or absence of hermeneutic interpretation is itself a matter of interpretation. Nowadays not only hermeneutics but also postempiricist philosophy of science claims that interpretation appears in both the natural and human sciences; "we never encounter the natural world unmediated by previous understanding."[87] It is true that the human sciences have to be interpretative, because what they are trying to understand is "a field of meaningful human action, which was constituted by a shared understanding of what would count as intelligible and significant action within that context." Is it not, however, also true that "a field of natural scientific research is a work world in which the objects of research are understood as equipment, and they make sense only through the ways we deal with them in research and development"? From this perspective, the objects of the natural sciences are as meaningful and socially constituted as human action itself, and "the distinction between those areas of inquiry in which our self-understanding is at issue and areas in which it is not" is drawn by our self-understanding.[88] From the perspective of hermeneutic self-understanding, "interpretation, characterized by the hermeneutic circle, and the entanglement of understanding with self-understanding, does not demarcate a particular domain of activity or inquiry (or its objects), but is a general characteristic of how things become manifest."[89] The reason for the universal presence of interpretation is our historicity that affects also our linguisticality: "all understanding is an interpretative act, that is, a finite grasp of something from a relative perspective and never a complete or otherwise final vision of things."[90]

For hermeneutic consciousness, which "infuses research with a spirit of self-reflection,"[91] the past does not disappear; it remains as a web of traditions. Hermeneutic understanding "links the historically past *as such* with the time of the interpreter. What happened then *(das Damalige)* cannot be stripped of its 'then-ness' *(Damaligkeit)* and in such a way construed as a contemporary possibility; for in that case its 'then-ness' would be missing."[92] Using Pannenberg's phrase, we can say that what happens in the course of the absorption of previous ideas (wherever it occurs, in the natural sciences or in humanities) is exactly the process of stripping them of their "then-ness." Hence, absorption is a degenerated form of conversation, because it is a unidirectional communication, during which the historicity of a text is ignored and the interpreter acts as an addressee of the text entitled to appropriate it. The distance between an author of the text and the interpreter becomes blurred, and there is no possibility of their mutual influence through criticism, clarification, or persuasion. Hermeneutic understanding is also a degenerate form of conversation, because it contains a gap, a historical distance, that turns a direct dialogue into an interpretation of a text, and this interpretation is always a dialectical play of reading *from* an interpreted text and its tradition and reading *into* it prejudgments of an interpreter's own tradition. Pannenberg rightly stresses that "interpretation in the proper sense can hardly be designated as a 'conversation with the text,' as it is by Gadamer."[93] Gadamer does not say they are identical but minimizes the difference between them, proclaiming that "the dialectic of question and answer that we demonstrated makes understanding appear to be a reciprocal relationship of the same kind as conversation."[94]

Both the absorption of old ideas or theories into a new theory and the hermeneutic interpretation of past ideas involve more than just the purely linguistic translation of terms and statements. Their merit is the tension between strangeness or alienness and familiarity. The concepts of familiarity and strangeness are not, of course, objective, external concepts but subject-related concepts: ideas are neither familiar or strange by themselves nor for themselves; they are familiar or alien to us. Moreover, it is not simply the case that we do not understand an alien system. It is the difference in our attitude toward something strange as opposed to our attitude toward something familiar. "Whatever is known has always seemed systematic, proven, applicable, and evident to the knower. Every alien system of knowledge has likewise seemed contradictory, unproven, inapplicable, fanciful, or mystical."[95] There are at least four spheres in a given discipline that participate in the transition from an old to a new body of knowledge. First, there is the conceptual subpractice, in the course of which old concepts become absorbed into new conceptual frameworks or are hermeneutically appropriated for expanding one's horizon. Second, there is the empirical subpractice of a given discipline that conveys empirical problems, data, measuring devices, etc., and intervenes in conceptual changes. Third, sometimes the practice of the historian of science, mediating between the contemporary horizon of a given discipline and its history is necessary for understanding the past of this discipline. Fourth, social practice that is external in reference to science, with its routines and habits (for example, technology), may also play the mediatory role.

The understanding of a distant historical tradition requires intermediary agents when "the tradition that supports both the transmitted text and its interpreter has become fragile and fissured."[96] Contemporary physicists working, for instance, in the field of general relativity communicate directly with their colleagues working in the same field or, for instance, in quantum mechanics. Here the interaction of traditions is direct and takes the form of dialogue, even if the main forms of communication are preprints, papers, and e-mail letters, and even if their dialogue requires the elaboration of "trading languages" operating within "local trading zones" of a discipline.[97] But when they address previous traditions—for example, Newtonian mechanics, Galilean kinematics, or Aristotelian physics—preserved in the works of their creators, they need additional mediation to prepare them for understanding the message of past traditions; here they have to appeal to the works of the historians of physics, at least in order to understand that part of the content of past traditions not preserved in contemporary theories.

THE ONTIC-ONTOLOGICAL GENEALOGY OF SCIENCE

We have already noted that Heidegger substitutes scientific cognition for science, and scientific cognition, in its turn, he replaces by Dasein's theoretical attitude toward entities. Objectifying thematization is simply one of the various forms of individual being;[98] it is not, however, a straightforward instantiation of everyday taking care of entities within-the-world. Dasein cannot discover objects (nature) within its everyday modes of being. Science is not part of everyday life; Heidegger excludes it "from the structures of meaning and significance which this practice displays."[99] Were this the case, science could not question, and thereby successfully transcend, commonsensical knowledge of the world that arises from human concern for mundane things. Scientific research requires a different mode of being: "We first arrive at science as research when and only when truth has been transformed into the certainty of representation."[100]

However, the derivation of science from the more primordial deeds of Dasein is analyzed neither by Heidegger (nor by Gadamer, for that matter) in a satisfactory way. Heidegger merely alludes to the temporality of existence as a necessary condition for the theoretical attitude. The only part he elaborates in detail is the analysis of the process that begins when a tool ceases to operate smoothly, when it is unsuitable, or when it is missing—in short, the moment that its ontological status begins to change. When equipment becomes unhandy it is "conspicious"; its way of being does not change, but we become aware of its presence and its unreadiness for use. "Once the tool loses its functionality, the unity of the referential *Gestalt* dissolves and we are left with a lifeless 'object' which manifests itself as a potential subject of predicates."[101] In such a situation, we are forced to concentrate our attention on equipment, which was previously transparent, either to fix or to improve it. Dasein moves gradually from concernful dealing with equipment to circumspective deliberation, in which it holds back from any manipulation or utilization and just looks, as it encounters entities "in their mere outward appearance."[102]

Circumspective deliberation, "looking-at," is the attitude typical of science, philosophy, or contemplation: it makes entities present.[103] The change in the way of being of an entity from handiness to objective presence means it undergoes decontextualization, or—in fact—recontextualization.[104] It is recontextualized because it does not slip out of all referential systems, as if it were a thing-in-itself that exists by-itself and for-itself. It remains related to Dasein as the subject that undertakes scientific research. Moreover, its recontextualization is the transformation of its being-within-the-world, into objective presence, or, to speak more precisely, into being-objectively-present-within-the-world, and reveals a new dimension of the way of being of a given entity. When this happens, "nature, as what 'stirs and strives,' what overcomes us, entrances us as landscape, remains hidden. The botanist's plants are not the flowers of the hedgerow; the river's 'source' ascertained by the geographer is not the 'source in the ground.'"[105] The transition to objectifying thematization also changes our way of being. When being is reduced to objective presence and the real ontological nature of being cannot be understood, Dasein is hidden and "the isolated subject is all that remains, and becomes the basis for being joined together with a 'world.'"[106] This new way of its being "can be independently developed" and, "as scientific knowledge, can take over the guidance for being-in-the-world."[107]

There are three essential differences between our primordial attitude toward things and the scientific (or, at least, physical) attitude.

First, science changes the quotidian way of determining the location of things into abstract coordinates of space and time, without any relation to Dasein, which itself also becomes a point in space-time.[108] Consequently, there is a fundamental difference between everyday language and the language of science. "Science aims to produce assertions stripped of all indexicality." Its statements are deprived of any relativization to an observer or speaker; in this way they are not contextually situated.[109] Second, objectifying thematization deprives things studied by scientists of the significance they have within-the-world, in everyday life.[110] They lose their practical intelligibility; "the nature studied by physics is abstracted from the everyday world, i.e., does not fit into the referential whole and connect up with our purpose and for-the-sake-of-whichs."[111] Third, objectifying thematization "has the character of depriving the world of its worldhood";[112] science cannot grasp the worldliness of entities—that is, the fact that they are-within-the-world. The issue referring to worldliness is "*understanding*, not *explanation*—making sense of how things *are*, not explaining how they *work*."[113] The natural sciences cannot grasp the belonging of their objects to the world, their entanglement into a web of meanings that is prior to objectification: the scientist views entities as determinate and isolable and not as enmeshed within the network of equipment.[114] "Heidegger contends that nature can explain only why the available works; it cannot make intelligible availableness as a way of being because nature cannot explain worldliness."[115] Moreover, scientific cognition cannot understand the phenomenon of the "world," the worldhood of the world. The term world designates here an existential-ontological concept that refers to the world-of-Dasein and not to the world understood as a universe of objects.[116]

Heidegger is convinced that to understand the genesis of science ontologically, it is sufficient to ask which conditions are existentially necessary for the possibility of Dasein's existing in the way of scientific research.[117] However, he does not show how existence or its modes of temporality necessitate both the theoretical attitude and the transition from everyday coping with things to objectifying thematization. It seems that Heidegger's analysis is too narrow and one-sidedly existentialist to allow for an understanding of science and its historical emergence. It "results in a negatively defined science derived from the interpretations, interests and instrumentalities of everyday life."[118] Moreover, if the theoretical attitude is in *opposition* to Dasein's average attitude in everyday life, ontological analysis cannot show that it simply *emerges* from that attitude.

Richardson, minimizing the opposition between everyday life and science, sees Heidegger's problem as the question: "What tendencies in our everydayness might incline us to encounter entities explicitly and in their own right?" He answers by pointing toward "deficiencies in concern" as well as to curiosity. Curiosity arises "when a mechanism at the heart of these concernful dealings—our desevering tendency—is released from its subordination to some particular project, and seeks to improve control outside that project's limited scope." Although Richardson is dissatisfied with Heidegger's explanation, he agrees that the theoretical attitude is rooted in the desevering tendency that permeates everyday concern: "It is the persisting operation of this desevering, on occasions when we put aside our efforts on behalf of our more immediate involvements, that makes us attempt to overcome the implicitness and context-dependence of our everyday discovery of equipment."[119] His solution, therefore, remains individualistic and certainly not sufficient if science is more than a form of Dasein's being, if it is a sociocultural system in which scientific research proceeds.

What Heidegger and Richardson miss in their account of scientific objectivity, and its emergence, is the role of practice as the ontological condition of the web of objective concepts, and the role that a community plays as its bearer.[120] Objectification does not emerge from the nature of human being; it stems rather from the collective and practical nature of science. An entity becomes an object only with the help of objectifying means (concepts, procedures, etc.), which cannot be rooted exclusively in the existential structures of Dasein, even from Heidegger's viewpoint. Thus, the practical activity of historical communities of philosophers and scientists constitutes the condition of the possibility of a scientific, objectifying form of thematization. In order to explain objectifying cognition, it is necessary to consider the cognitive community as well as the ongoing dialectic relationship between an individual subject and this community.

The Heideggerian derivation of scientific theorizing from Dasein's everyday attitude toward entities has another drawback. Heidegger does not consider the order of our ways of being and of encountering entities as historical. Everyday coping with tools is prior to philosophizing and scientific cognition, not only ontically but ontologically. That Dasein is a user of equipment is not, for Heidegger, a historical event; it is prior to human history, although it is involved in Dasein's temporality.[121] However, even if it is typical of our historical situation that we encounter entities for the first time in everyday life, that does not make it an ontological necessity. On the contrary: it is a historically contingent situation. Is it not true that contemporary scientists encounter many entities earlier than they are recognized by laymen? Is it not the case that entities become objects of science first and only later enter into everyday human activities through scientific technologies? If these questions can be answered positively, as we believe they should, the primordiality of everyday life over science is as historical as the primordiality of science over everyday life. Moreover, even if we agree with Heidegger that the scientific attitude and acts of objectification arise from human everyday practice, we cannot ignore their ties to philosophy and theology which are historically prior domains of culture. "All attempts to explain the rise of a scientific type of knowledge which do not take into account the armed monopolization of the European fund of knowledge by the medieval Church and do not link it to the partial break-up of this monopoly, are bound to fail."[122] Let us develop this point more broadly.

The fragmentation of the church monopoly culminated during the Reformation movement. The Reformation irrevocably divested the church of its overarching role as the mediator between the secular and the divine realms; this historical fact had far-reaching consequences for the organization, hierarchization, and structuring of knowledge, consequences that reached well into later centuries. First, it revitalized, within the Catholic Church itself, the effort of the clerical and intellectual elites to integrate profane knowledge into the traditional doctrine of the revelatory status of ultimate truth.[123] The tension between revealed and rational truth was negotiated in various ways within the Reformation Church. The Council of Trent (the third period, 1562–63) stressed the role of education in reforming the church and recommended a restoration of the Thomistic synthesis of secular and divine knowledge in relation to the goals of salvation. Although consensus existed within the church concerning the adoption of Thomism, conflicting interpretations emerged from within the Dominican and Jesuit orders. These conflicts significantly affected the Church's conception of the organization and hierarchization of knowledge. This state of affairs influenced Galileo's attempts to legitimize his conception of natural knowledge within the existing cultural environment.[124]

Second, medieval thought concerning nature was guided by theological concerns, and suffered theological interference demanding that secular thought contribute to the concerns of soteriological knowledge, to knowledge of God and the redemptive role of Christ in achieving salvation. Although in the seventeenth century scientific knowledge acquired autonomy and legitimacy, it still had to claim "to contribute to knowledge of ultimate concerns."[125] However, the manner in which secular learning was related to soteriological knowledge changed radically: it moved from the task of producing illustrations of the operation of God's providence in the world to supplying soteriological knowledge based on natural, philosophical foundations.

Derham's lectures (1711–12) serve as an example of this stage in the "naturalization" of theology. Derham created a physico-theology: he attempted to defend the fundamentals of Christianity with the help of Newtonian experimental philosophy. Heyd observes:

> Here was a new synthetic organization of knowledge which somewhat resembled—and perhaps was partly influenced by—the late sixteenth-century attempts to construct a 'Physica Sacra' and reconcile Scripture with physics. There were two crucial differences, however. Not only was the new Physico-Theology based on experimental science rather than Aristotelian physics, but more importantly, science was no longer subservient to the external supervision of theologians. In fact, it now tended to dictate the nature of the new synthesis. The theologians now relied on the work of the scientists who, for their part, were free to pursue their research.[126]

What is evident in the wake of the Reformation crisis is the "secularization" of natural knowledge and the emancipation of scientific research. The new forms of cognition and knowledge were profoundly different in their ways of treating the world. "Common to most ancient and medieval epistemologies was their receptive character: whether we gain knowledge by abstraction from sense impressions, or by illumination, or again by introspection, knowledge or truth is found, not constructed. Implicitly or explicitly, most 'new sciences' of the seventeenth century assumed a constructive theory of knowledge."[127] According to this view, knowledge is acquired in action, not in contemplation. Moreover, instead of contemplating essences, substantial forms or occult qualities, the new sciences presuppose a hiatus between God and a world created and endowed with Divine order. The new knowledge was restricted to the created world, which it treated as existing in-itself externally to our forms of conceptualization. This presupposition allowed natural philosophers to claim that the laws of nature, freely imposed by God—as voluntarist theology teaches—must be discovered and not inferred from the intelligible nature of things.[128]

Anyone familiar with the seventeenth century cannot fail to notice the cultural importance of the distinction between first and second causes. As the century progressed, the notion that human reason can comprehend the secondary causes of natural regularities became a deeply embedded cultural presupposition. Consequently, scientific research in the seventeenth and eighteenth centuries directed itself mainly toward the search for secondary causes, which were thought to mediate, like gravitation, between God and phenomenal reality.

Seventeenth-century secular thinking established yet another division: the division of self and society from nature. Reality became divided into everything human, social and normative, on the one hand, and everything nonhuman, nonsocial, and factual, on the other. As the idea of two irreconcilable "realities" became culturally embedded, human alienation in the face of an indifferent natural world became a datum of educated common sense. The divide that emerges between society and nature, mind and matter is a profoundly sociohistorical difference: it is a configuration inscribed within a society and embodies the historicity of that time and place; it is an amalgam of historical accidents that bears historically specific social, cultural, religious, and institutional conditions.

It is not sufficient, however, to consider science simply as knowledge that grows out of metaphysical and theological knowledge or in opposition to it. Given that science is a particular domain of practice, its ontic genealogy must reveal its development from earlier forms of practice. In fact, among students of the Renaissance, the Reformation, and the Enlightenment it is commonplace to see the emergence of early modern science as a multifarious historical phenomenon embracing social, cultural, educational, religious, cognitive, political, and institutional dimensions.[129] Mobile networks of relationships between states, societies, religions, and individuals in the seventeenth century forged a new balance between God, society, and nature within the emerging cultures of the early modern period. Within these new integrations, the understanding and control of the natural order assumed a more prominent role as the century progressed.

These remarks support our criticism of the Heideggerian view of the genesis of science. If the radical historicist perspective compels us to claim that there are no human or social situations that are strictly universal, scientific practice and the objectifying character of scientific cognition cannot be grounded in existentialist structures, whether these are viewed from an individual or a social perspective. It cannot be denied, moreover, that *to be an objectifying cognitive practice, science must establish itself as an objectifying social activity.* The conditions of scientific objectivity are located within science and are elaborated there, and, in turn, they render the activity of science objective. We have briefly analyzed the causalist way of thinking as an instantiation of objectifying cognition in chapter 3; and we will discuss objectification more broadly in the next chapter. Here, let us summarize our general considerations. If a human being is what it becomes and has become, and if its becoming is engendered historically, this also applies to science and to the processes that engender its appearing-in-history. This means, in turn, that new elements of scientific development continually emerge out of existing practice as scientific activity proceeds. After all, invariances are not historical necessities, processes are not historically inevitable, origins do not predetermine outcomes, and organizing categories do not capture essences.

SCIENTIFIC EXPERIENCE AND ITS HISTORICAL VARIABILITY

Traditional empiricism is responsible for the identification of the objective (real) with the perceived. From this perspective, all knowledge derives meaning from perceptual experience, and all meaningful statements are judged true or false, acceptable or unacceptable, by reference to it. Neopositivists, of course, reject the mentalist overtones of concepts like "idea" or "impression"; in general, they treat scientific experience and knowledge in a purely linguistic way. They believe that this way of understanding it not only eliminates psychologism but also disposes of nonessential, pragmatic aspects of scientific observation and reveals the objective basis of science.[130] The linguistic purification of scientific experience has an important philosophical consequence: it eliminates the metaphysical from philosophy of science. However, if "any intelligible talk of the objects of science, the external world, or anything else is reduced to talk about possible experiences,"[131] scientists must either embrace solipsism or the world must be treated as a phenomenal correlative of our sensory experience, which, in turn, is directly, and without any mediation, connected with observational terms and statements.[132] It is well known that neopositivism assumes that there is a borderline between theoretical and observational languages and held—initially—the ideal that the meaning of theoretic terms is supplied by reduction through "introductive chains" to observational terms. Both languages and the borderline between them are decontextualized—that is, viewed as universal and independent of any psychological or sociological circumstances. Moreover, neopositivism considers sensory experience as fundamental, not only for the meaning and acceptability of single theories but also for theory comparison. It is imperative, therefore, that theories presuppose a stock of observation statements common to all science.[133]

However, the empiricist program failed. "In verifying a theory we move in a circle from hypothesis to data, and data to hypothesis, without even encountering any bare facts which could call our whole theory into question."[134] Logical empiricists had to admit that reductive sentences only supply theoretical terms with indirect and partial interpretation; the empirical meaning of theoretical terms is accompanied by independent, nonobservational meaning. Moreover, as Putnam argues, a difference between entities referred to by theoretical and observational terms (if it exists at all) is not captured by the neopositivist version of the observational/theoretical distinction, because theoretical terms do refer to observable entities, observation statements frequently contain theoretical terms, and observational terms cannot be understood as referring only to observable things.[135]

After Quine, philosophers rejected more overtly the belief in the purity and the privileged position of evidence[136] and accepted that "things are not labelled 'evidence' in nature, but are evidences only to the extent to which they are accepted as such by us as observers."[137] They rejected the idea of a direct and coercive determination of scientific knowledge by empirical facts, admitted that the language of natural science is irreducibly metaphorical and inexact, and began to regard meanings as determined by theories and their relations.[138] The traditional empiricist picture of scientific observation as the passive reception of stimuli and the recording of empirical data has been replaced by the view that it is active because always theory-laden. Not only in rationalist conceptions but also in constructive empiricism, experience is seen as a creative social enterprise, containing decisions and communal agreement; and what is taken to be observable is always observable-to-us.[139] Radical anti-empiricists claim that every proposition of science is theoretical. Consequently, theories do not clash with facts; only theories clash: "an interpretative theory to provide the facts and the explanatory theory to explain them."[140] Moreover, if experience is no more basic than theory, and if every theory does not need to be tested independently, then theorizing may proceed by itself: theoretical disciplines are relatively autonomous and solve problems generated by their own practice—for instance, those that arise from mathematical difficulties.[141] Feyerabend also emphasizes that experimental evidence "does not consist of facts pure and simple, but of facts analyzed, modeled, and manufactured according to some theory."[142] If facts are not independent of interpretation, each and every entity to which science refers is a projection of scientific theories. Theories are how the facts themselves are seen: facts and entities are "tied to the theory, the ideology, the culture that postulates and projects them."[143] Thus, only by referring to a theory or an ideology and, in general, to a historical tradition, can we identify experience and determine what experience tells us. "Experience taken by itself is mute." If it is not interpreted within a certain language, "described in some ordinary idiom," it does not tell us anything.[144] Feyerabend goes further. He does not stop at replacing the empiricist version of the myth of the Given with a rationalist version, which treats theories (commonsensical or scientific) as given. He criticizes both myths, concluding that reason cannot be viewed as independent of practice; "reason and practice are not two different kinds of entities that can exist separately but *parts of a single dialectical process.*"[145]

Feyerabend's position does not, of course, deny the empirical nature of science. Instead, he reopens *the problem of scientific experience.* His criticism of both views of observation (as pure and as theory-laden) allows us to ask (as has been done many times in the history of philosophy) the question: "What is scientific experience?" However, apart from a few exceptions, an answer to this question is not obvious in postempirical philosophy of science. Among the exceptions are Quine's naturalized epistemology, Hesse's network model, Kuhn's analysis of the procedures of learning scientific taxonomies, and Mc-

Dowell's antinaturalist view of experience.[146] The first three refer to the process of learning and are largely naturalist in scope, whereas McDowell considers the nonreducible and active role of conceptualizing in experience, attacking the myth of the nonconceptual given.

A reconsideration of the problem of scientific experience from our perspective should begin with the recognition that scientific experience is a social and historical phenomenon, a phenomenon with a historical beginning that has transformed together with the entire activity of science. This means that the problem of scientific experience is open both as a philosophical (ontological) issue and as a historical and empirical (ontic) issue.

Ontologically, experience is part and parcel of our-being-in-the-world; it is an event, "over which no one has control and which is not even determined by the particular weight of this or that observation."[147] As an event of our being, it is reducible neither to subjective acts of perception nor to sets of observational sentences. As a constitutive element of practice, experience is intersubjective, always linguistic, and, thus, conceptual. It occurs among and between people; it presupposes social relations and interactions and enters into relations with other elements of practice. In particular, when it becomes part of cognitive subpractices it gains autonomy from everyday life experience and enters into mutual relationship with other elements of those subpractices. In the case of science, it enters a relationship with theorizing that is a particular form of conceptual understanding. As Feyerabend, Kuhn, and McDowell stress, experience and theorizing do not exist separately; in science they interact in a dialectical way. Conceptual understanding is also an event of our-being-in-the-world. Only within science it opposes experiencing; in everyday life they form a whole composed—as we have argued—of the synthesizing experience of being, elaborating communal self-understanding, and cognizing the world.

The separation of experiencing and theorizing is a historical product of the replacement of participation, as the human way of being, with determination; this occurred during the transition of Western societies from a religious way of being to the scientific form of being. During this transition different forms of synthesizing our experience of being, elaborating communal self-understanding, and cognizing the world emerged and gave birth to scientific understanding of being; human beings were transformed into abstract objects present—in an objectified way—in an objective, meaningless world. As a result of this change, the experience of a participant in the world became an objectified experience of an external observer, reduced to the receptivity of sensual stimulations taken to mediate between the observer and meaningless facts. The world became deprived of the intelligibility that it had within direct dialogue, in which we deciphered symbols and messages and uncovered their hidden meanings. So now, the only way left for making the world intelligible to us is to *think about it* as something different from the mind, as something external and presumed to be there for the sake of description and explanation. Statements like Galileo's, that science reads the book of nature written in the language of mathematics, are seen as metaphorical. In overcoming the model of participation, the objectifying view of scientific experience disconnected it from the logos of the world and from the intentional movement of understanding that directs us toward things. The world, the subject, and experience became three separate realities; and although they are connected with one another by ontic, causal links and by epistemic, semantic relations, they no longer form an authentic whole. The subject and the object remain opposed, and experience is no more than a mediation between them. This is why this picture remains in conceptual difficulty: experience as such can be explained neither as a phenomenon belonging to the world nor as a phenomenon belonging to the subject. Explained naturalistically it loses its *epistemic* meaning and function, and accounted for in subjective terms it loses its *referential* character. Moreover, attempts at conceptualizing experience as prior in reference to both subjects and objects (if there are any) hold out little promise.

An objectifying view of experience (especially any form of naturalized epistemology) becomes part of science, in particular, natural science, when scientific research turns into a totalizing form of cognition. This totalizing thrust of science is the source of the fact that the objectifying view of experience "takes no account of the inner historicity of experience."[148] It presents experience as a natural or abstract (epistemological) phenomenon totally deprived of historical dimension. Nor can it account for the cumulative scope of experience. From an objectifying, scientific perspective, no human activity is a periodic recovery of mythical time, of eternity; this is why the *universality* of experience in science cannot be guaranteed by the identity of repeatable rites performed in mythical time and directed by the same rituals. Since any act of experience is located within a particular moment and place, any absolute identity of repetition is unthinkable. Thus, the universality of (scientific) experience requires the *cumulation of experience* and, hence, a methodological justification of its *repeatability*. However, when the ritual of the repetition of experience is replaced by cumulation, the very nature of experience changes. Experience must acquire variation since one act of experiencing must be distinguishable from another to be accumulable; but the variability of experience cannot be understood simply as differentiation in virtue of the spatio-temporal localization of acts of experience, as recognizable from outside in another act of experience. On the contrary, it requires *internal* variability. The objectifying approach to experience conceives this internal variability as the separation of two groups of experiences (sensations): experiences that are identical, because they are caused by the same stimuli, and experiences that are different, because they are caused by different stimuli. The internal variability of experience cannot itself be experienced, it can only be conceptualized. This analysis is not meant to state that cumulative experience fails to happen in prescientific everyday life. Certainly it does. It is, however, only in professionalized cognition (first in philosophy and then in science) that, together with the cumulation of experience, its recognition and justification are, and must be, elaborated to support the ideal of the universal validity of scientific cognition and knowledge. The requirement of the internal variability of experience leads to a complication in epistemological (methodological) norms justifying experience. For instance, from the perspective of objectifying conceptualization, in order to have the force of universal validity, acts of experience must be repeated and reconfirmed in other places and times; only in this way can internal variability be established. Endless duplication is—from this perspective—a criterion of the universality of scientific, empirical knowledge. The ideal of duplication provides experience and its objectifying theory with another impulse to abolish "its history and thus itself."[149]

In contrast to this, however, we must recognize that acquiring internal variability experience becomes dynamic, *unlimitedly open* toward what is different and new, and *negative* in the sense pointed out by Hegel: it contains the constant possibility of realizing that something is not what we supposed it to be. And again, these features are not apparent from the perspective of objectifying accounts that tend to view experience in terms of its result—namely, generalized knowledge. As Gadamer notes, those who adopt such a view ignore "the fact that experience is a process" which "cannot be described simply as the unbroken generation of typical universals." The inductivist view of experience as the unbroken accumulation of knowledge not only misses the fact that "this generation takes place as false generalizations are continually refuted by experience and what was regarded as typical is shown not to be so,"[150] it is also unable to understand modification in the way of being of experience. Experience construed in terms of generalized knowledge changes its way of being: it becomes a frozen product of its own history. Not only does its history disappear, it too disappears and gives way to the illusion that nothing important exists in science except linguistic structures, observational and theoretical terms, predicates, sentences, and so forth.

A genuine philosophical conception of science cannot, however, relegate experience, understood as a historical process, to oblivion. Moreover, it should realize that openness and negativity are constitutive features of scientific experience understood as what is never fully repeatable. It is exactly the openness and negative character of experience that account for its role in the constitution of scientific creativity and objectivity. Without these features, discovering unknown objects, properties, and even uncharted regions of reality would not be possible. We return to this issue in the next chapter.

From an ontic and social perspective, scientific experience and theorizing are practical and collective. Scientific experience involves the quasi-orchestrated activity of many people who must have special competencies, both manual and mental, to perform such elementary operations "as measuring, pippeting, storing of the sera, washing of the vessels, etc." The practical and social nature of scientific experience and theorizing reveals itself in their developmental dialectics that involve several stages: "(1) vague visual perception and inadequate initial observation; (2) an irrational, concept-forming, and style-converting state of experience; (3) developed, reproducible, and stylized visual perception of form."[151] From an ontic and *historical* point of view, scientific experience is a process that began to emerge in the seventeenth century as an effect of scientific self-liberation from existing everyday life experience and its world. Science emancipated itself from common sense, empirical data used in the practice of everyday life, and from the Aristotelian codification of both the data and the nature of (everyday) experience.[152]

Galileo's research demonstrates the role of the interrelation between experiencing and theorizing in the establishment of an autonomous form of scientific experience. He replaced everyday empirical data, which had been theoretically structured by Aristotelian philosophy of nature, by data produced in thought experiments directed by theoretical premises and hypotheses. In his research, thought-experiments began to play a new role. For Aristotle and his followers, idealized thought experiments served as purely speculative (philosophical) instruments. They were used to argue that ideal phenomena, limiting cases such as motion in the void or a weightless body, were physically impossible. Aristotle and Aristotelians "saw no mediation—either in principle or in practice—between the factual and counterfactual conditions of the same 'body' (or, as we would say, the same phenomenon). Aristotle, and with him medieval physics, saw both as incommensurable." Galileo rejects the assumption that real and ideal phenomena are incommensurable: limiting cases, as considered in thought-experiments, "explain nature even while they do not describe it."[153] The importance of Galileo's use of thought-experiments is not exhausted, however, in the appropriation of a new device for scientific argumentation. It was an element in the process of discrediting the value of everyday empirical evidence for science, a process that began with the critical effort of medieval thinkers and Copernicus. Everyday experience was deprived of its right to create true descriptive knowledge concerning the world. It was discarded as misleading, and the claims of everyday life to offer true descriptions of the world were unmasked as groundless. Appearances known to everyone were no longer taken for granted, since they did not necessarily allow us to know reality as it ought to be known.[154] Knowing reality as it ought to be known, perceiving it objectively and understanding the nature of what is perceived, required—for Galileo—bringing "thought and representation together."[155] This, in turn, required infusing representations, images of the perceived, with intelligibility—through mathematics—and returning them to the perceived through real experiments.[156] The critical move against prescientific experience did not annihilate the everyday way of perceiving phenomena;

> we can mentally liberate ourselves from the evidence of our senses, and because we can do this we can see things from the rational viewpoint of the Copernican theory. But we cannot try

to supersede or refute natural appearances by viewing things through the 'eyes' of scientific understanding. This is pointless not only because what we see with our eyes has genuine reality for us, but also because the truth that science states is itself relative to a particular world orientation and cannot at all claim to be the whole.[157]

This critical move was an element of the liberation of scientific cognition from existing practice and all forms of authority functioning within it. Consequently, scientific knowledge ceased to be the "natural" expansion of human experience regarding the world. Instead, it becomes "an independent setup, indeed an attack upon nature, which it subjects to a new but only partial mastery."[158] The limited character of scientific projections was not, however, recognized. Scientific cognition was proclaimed to be better than both everyday understanding and understanding elaborated by other forms of cognition (for example, philosophy and religion), because it alone had at its disposal the integrated experimental-theoretical machinery for manufacturing true knowledge. Radical scientific realism became indispensable as a means of enabling scientists to justify the claims to truth of newborn scientific cognition. It was necessary for the idea of the truthfulness of the theoretical-experimental picture of the world to become infused into the entirety of human practice.

The establishment of scientific realism required the elimination of existing natural interpretations and ideals of perception "so closely connected with observations that it needs a special effort to realize their existence and to determine their content."[159] Galileo replaced natural interpretations codified by Aristotelian dynamics with a new interpretation that included a narrower picture (to be precise, a "kinematic" theory of locomotion and free fall). This was accompanied by certain metaphysical assumptions, for instance the principle of the relativity of motion.[160]

> Galileo was suggesting that the relevant descriptions which need to be included in an observation report are those motions which the observed body has which are not also shared by the observer. Motions which are shared by both the observer and the object being observed need not be reported since they are, on the suggested programme of redescription, both imperceptible and irrelevant.[161]

It was not simply the new theory of motion, however, that supplied Galileo with justification for his redescription of experience. This came from "the collective experience of a considerable number of competent inquirers spread through the three centuries before his own day. These men were perfectly familiar with the details of Aristotelian physics. They had frequently established the fact that the Aristotelian conception of movement was inadequate."[162] In undertaking telescopic observations, Galileo violated the traditional ideal of naked-eye observation and the belief that terrestrial phenomena are incommensurable with celestial. In this case, it was Copernican theory that justified using the telescope, because telescopic phenomena, for instance, the variation in the brightness of Mars and Venus observed with a telescope, agreed with it.[163] Galileo did not, of course, create scientific experience single-handedly. He carried out concrete experiments, both real (rather few, in fact) and imaginary, improved measuring and observational instruments, changed existing philosophical ideas, and inferred observational predictions from theoretical hypotheses. Nevertheless, what happened was the beginning of a new form of scientific experience and a reality that corresponded to it. From Galileo onward, telescopic investigations suggested the limitations of the unaided human senses and promised to reveal more detailed marvels as technology improved.

As our brief analysis shows, there are two crucial factors in the creation of scientific experience: the swift development of measuring and experimental devices and the rise of cooperative experimentation. The first factor has been well recognized and analyzed: "Before 1590 the instrumental armory of the physical sciences consisted solely of devices for astronomical observation. The next hundred years witnessed the rapid introduction and exploitation of telescopes, microscopes, thermometers, barometers, air pumps, electric charge detectors, and numerous other new experimental devices. . . . In less than a century physical science became instrumental."[164]

The second factor is less obvious though no less important. The core feature of experience, as it arises within the new experimental context of the seventeenth-century science, is not the cognitive activity of anonymous individual observers but the development of cooperative inquiry. Shapin, in accounting for Boyle's laboratory culture, makes it clear that the key issue of collective experimentation became the practical management of trust and testimony within local settings of experimentation.[165] The socialization of scientific experience required the elaboration of techniques for the practical management of factual testimony and for new epistemic stances toward traditional knowledge and its institutional forms of organization. The spirit of Bacon's natural philosophy took shape against this background, and the modern outlook is encapsulated in his dictum that truth is "the daughter of time, not of authority." The impact of this newly formed scientific experience on the life of societies of the time is well known. In short, early modern Europe had to absorb marvels revealed in telescopes or microscopes and discovered by the exploration of distant geographical worlds. They outran the traditional schemes of classification and raised doubts regarding their criteria of plausibility.[166]

The fact that scientific experience and its life-world are different from everyday experience of reality was recognized for instance by Descartes and later by Kant. For them, sense experience became an instrument of theoretical reason. Their efforts were limited, therefore, to the establishment of the autonomy of scientific experience construed in a purely intellectual way. According to Descartes, in the case of scientific cognition, understanding turns toward the external senses or decides that "the senses must be banished and the imagination as far as possible divested of every distinct impression."[167] Understanding renders a subject "who philosophies in an orderly way," a critical realist, one who "will never assert that the object has passed complete and without any alteration from the external world to his senses, and from his senses to his imagination, unless he has some previous ground for believing this." Finally, understanding makes sense perception into a systematic procedure when the aim is either to discover the real nature of something—for example, the nature of the magnet[168]—or to test theoretical hypotheses. Kant, aiming at reforming and restoring metaphysics, elaborated—in fact—the justification of nonmetaphysical knowledge. This is valid knowledge about nature, for instance, Newtonian mechanics, although it is empirical, hypothetical, inconclusive, and only relatively universal. He justified modern natural science, where a revolution in thinking had taken place, through the codification of the principles of (pure) reason that science projects on nature.[169] Thus, according to Kant, reason not only controls sense experience, it makes possible receptivity and the world of our sensitivity. Experience, together with its subject and object, is possible only under certain conditions. "The principles of possible experience are . . . at the same time universal laws of nature"; and nature is "the totality of all objects of experience."[170] This means that for Kant cognition is constructive in a deeper sense than it is for a moderate realist, as represented by Giere. A moderate realist admits that "Nature does not reveal to us directly *how* best to represent her,"[171] but at the same time tacitly assumes that nature reveals to us directly *what* to represent—that, in other words, nature reveals itself to us.

That nature reveals itself to us was an element of the view of experience that appeared in nineteenth-century science together with different automatic, self-recording instruments and the accumulation of their results; for example, photographs. Scientists praised them as producing better observations than human observers:

> machines offered freedom from will—from the willful interventions that had come to be seen as the most dangerous aspects of subjectivity. If the machine was ignorant of theory and incapable of judgment, so much the better, for theory and judgment were the first steps down the primrose path to intervention. In its very failings, the machine seemed to embody the negative ideal of noninterventionist objectivity, with its morality of restraint and prohibition.[172]

However, collecting machine-recorded data and the ideal of mechanical objectivity did not abate the role of manipulation and intervention. Because of modern technology, scientific experience transformed itself into manipulative experimentation mediated by complicated equipment. Even the very concept of data changed radically: "in current research what empiricists would call data turn out to be numbers or graphs produced on computer terminals or printouts. There are no longer any counters to read."[173] As certain postempiricist philosophers of science admit, modern experimentation has a life of its own in the sense that it does not need to be informed directly by theories or involved in testing them.[174] Experimentation is not an activity of discovering phenomena, publicly discernible events or processes that occur regularly under definite circumstances, but a process of producing, refining, and stabilizing them. What constitutes the core of scientific experiments is intervening, manipulating and creatively using (unobservable) entities.[175] The technological saturation of scientific experimentation makes it more similar to a process of production than to passive observation. Laboratories, and the phenomena constructed in them by scientists, do not simply mediate between knowledge and a pre given objective reality. Measuring devices, machines, and the industrial complexes used by scientists are not just "man-made extensions of senses."[176] The aim of technology is not simply to provide "the connection between our evolved sensory capacities and the world of science."[177] Technology becomes a necessary condition for scientific research and its world. It allows scientists to access phenomena not present in the pretechnological world, the world of the naked eye, or even that of enhanced experience. In a real sense, therefore, the role of technology in scientific research is parallel to that of the forms of intuition in Kant's theory: the empirical world of (modern) science exists as its correlative.

Ontically, this new situation in scientific research can only be understood with the help of a holistic historical view of the complex composed of theories, realities, and technologies. Feyerabend expresses this holistic and constructivist picture as follows: "Scientists, being equipped with a complex organism and embedded in constantly changing physical and social surroundings, used ideas and actions (and, much later, equipment up to and including industrial complexes such as CERN) to *manufacture*, first, metaphysical atoms, then, crude physical atoms, and, finally, complex systems of elementary particles out of a material that did not contain these elements but could be shaped into them."[178] Theories, realities, and technologies are related reciprocally and are integrated into historical complexes conditioned by technical and social subpractices. Technologies mediate between theories and realities by reifying theoretical entities and idealizing realities; but the relation between instruments and realities is not direct. It is mediated, in turn, by theories that are sources of justification for instruments and for measuring results.[179] Finally, the relation between technologies and theories is mediated by realities, because without their intervention the application of technology could not serve the purpose of empirical testing. In general, as Kuhn emphasizes in commenting on Wise's study, each of these three elements is constitutive for the other two, and each is necessary for producing knowledge.[180]

Hacking claims that scientists (experimenters) need not treat the (theoretical) entities they investigate realistically until they begin to use them to manipulate other entities.[181] Unfortunately, his realist view is based on an idealization of science. He suggests the following alternative: either unnecessary (superfluous or even unjustified) realism in reference to theoretical entities or necessary (and justified) realism in reference to entities used by scientists for experimentation. This alternative presupposes that research begins with a theory, as if nothing precedes theorizing (with the exception of earlier experiments and theories). Hacking misses a phenomenon discussed by Latour and Woolgar—namely, the deconstruction of reality already existing outside science.[182] The deconstruction of reality out-there, transforming elements of the life-world into objects of science, into, in particular, theoretical entities, precedes theorizing and experimenting. Pasteur translated (transferred) the anthrax disease from a farm to his laboratory.[183] This deconstruction is, of course, the recontextualization of entities; this is why they are never things-in-themselves but always already correlatives of this or that sphere of practice.

It is useful to juxtapose Hacking's views on experimental entities with Heelan's.[184] For Heelan, theoretical quantities become known in virtue of how we respond to appropriate empirical procedures, such as the use of measuring instruments. There are two possible responses. We can theoretically infer the value of the quantity, provided we know the relevant theory. Or, we can "read" the value of the quantity directly from the measuring instrument if we are skillful in its use. In this case the use of the instrument serves what Heelan designates as a "readable technology." He illustrates this with a simple example. In "reading" a thermometer we do not proceed from a statement about "the position of the mercury on a scale to infer a conclusion about the temperature of the room by a deductive argument based on thermodynamics: of course, one could, but then one is not 'reading' the thermometer."[185] In other words, the use of instruments in science, such as the thermometer, come to embody theory in the process of standardization; as such they can "serve a public often quite ignorant of the mathematical model on which it is based."[186]

Heelan has effectively eliminated any principled distinction between theoretical and observational entities. If Hacking views experimental entities as having a life of their own independent of high theory, in Heelan's view the theoretical and experimental life of entities converges in the very process of making themselves visible and thus part of the "furniture of the earth." Hence, the practices of science, its instrumentation, and its disclosing equipment can deliver information about phenomena just as directly as the unaided senses do.

Accordingly, Heelan overcomes the distinction between what is postulated or inferred scientifically and what is real, since for him observability "is the criterion of all physical things, including theoretical states and entities." In this sense, both perceived and technologically observed entities are naturalized "in the scientists' world through the use of a readable technology."[187]

CHAPTER 7

Science and Objectivity

At least from the time of Kant scientific cognition has been seen as a special form of thematizing that makes possible the discovery of (empirical) objects.[1] Less clear, however, is what this means and how objects are discovered in science. The job of clarifying scientific objectification was straightforward when philosophers were able to refer to rational foundations, such as synthetic a priori propositions or to the unalterable data of perceptual experience. Now the situation is more complex. In the light of critical attacks on foundationalism nothing in science seems stable and self-sufficient.

Present difficulties with the objective have their source in the distinction between the contexts of discovery and justification. A form of the genetic fallacy, it presupposes that origins are irrelevant to evaluating what emerges. If philosophy is the methodological justification of science, the historical, the psychological, and the social become philosophically irrelevant. In our view, however, it is vain to separate what is whole: the social, the historical, and the cognitive are united within human practice, itself the ultimate source of the objective. From this perspective, it is necessary to mobilize the ontological circle that connects the whole and its parts. In what follows, the central concepts we analyze are scientific *objectivity* and *creativity*.

OBJECTIVITY AS THE PRODUCT OF REASON

Although current debates on science assume "that objectivity is and has been a monolithic and immutable concept, at least since the seventeenth century," objectivity has its own history.[2] It can be understood in two ways: as pointing to the nonsubjective, to the detached, impartial, and disinterested nature of scientific cognition, or to the fact that science presents its subject matter in terms of the objective, the physically real. According to the first view, scientific cognition cannot be realized from any particular or idiosyncratic perspective: it must be characterized by aperspectival objectivity that appears within the opposition individual/public, the opposition between individual, subjective opinions and acts of believing and shared, communal statements and acts of argumentation. Aperspectival understanding construes scientific cognition as a view from nowhere, as (re)presenting its subject matter (ideally) in a mechanistic rather than an interpretive way. The second view takes science to have access to a shared object of cognition. On this view, objectivity constitutes itself within the subject/object opposition, and scientific cognition is construed as representing an independent object.[3] Both concepts of objectivity are intertwined in discussions of the rationality of science and are usually viewed as constituting an epistemic unity. This reflects the tacit conviction that rational cognition is the sole source of truth and can only be achieved through procedures that are themselves rational and cognitive. We now concentrate on the first issue—namely, on the idea that scientific cognition is not subjective but objective.

The idea that genuine knowledge *(epistémē)* is objectively valid for all (rational) human beings, or even for all subjects of cognition, is central to Greek philosophy. However, its modern formulations reveal more clearly what is necessary for securing this type of objectivity. As Machamer observes, the problem, "How could ideas and practices that were first-person based ever become universal and objective?" gradually emerged in seventeenth-century consciousness. This problem presupposes individualist epistemology: knowledge is acquired individually, in the course of thinking and experiencing, but is nevertheless universalizable and certain. On this view, knowledge belongs to all individuals in principle and, when justified, reveals what is necessary in states of affairs. To secure the universalist ideal of knowledge, seventeenth-century philosophy articulated the method of science and adopted the entrepreneurial model of I. This view characterizes "the elitist genius who sees more clearly, more deeply, and more correctly what is true and what needs to be done." Each great philosopher of the period "wanted power, to be the leader, to design *the* new system of science, knowledge, and method. Each wanted to command all other natural philosophers in a collective endeavor, and wanted acknowledgment by them as a leader." Paradoxically, although the entrepreneurial conception emphasizes the role of the intellectual leader, it does not presuppose a subjectivist ideal of cognition. The ideal of the elitist genius was combined with the ideal of objective knowledge, guaranteed by the correct method of science, and with the idea of mathematical-experimental cognition that accompanies it.[4] Scientific method, correctly applied, allows any cognitive subject to know things, not only as they are but also as they must be physically, on the basis of the principle of causality, or metaphysically, in virtue of their defining nature. Accordingly, the subject's idiosyncrasies must be abstracted away; for objective knowledge can only be established by cognitive abilities common to all people and shaped by the universal method of science (as established single-handedly by the intellectual leader). On this ideal, "every scientist has the obligation to remain in the background" and "to withdraw his own individuality."[5] From this emerged the ideal of cognitive egalitarianism, the claim that "anybody that acquires knowledge" may be regarded as an equal among equals.[6]

Appealing to the subjective (human cognitive abilities) and the objective (methodological) communal conditions of knowledge failed, however, to solve the problem. First, empiricists and rationalists disagreed on what comprises the common cognitive equipment of human beings. Moreover, philosophers of the period articulated methodologies that possessed little overlap in aims and means. Second, the entrepreneurial model of I refers to speculation and experience (treated in abstract fashion), but not to activities such as measuring, calibrating instruments, making drawings, and taking pictures. Third, the seventeenth-century conception of objective knowledge based the attempt to establish necessary connections among things on the ideal of epistemic certainty. While this conception subordinated personal cognitive processes to the methodological authority of science, it did not depersonalize scientific cognition. For the most part, objective grounds and subjective motives were insufficiently contrasted. This process occurred later, when Husserl and Frege, for instance, criticized psychologism, and when philosophers began to excise the context of discovery from the philosophy of science.

The nineteenth century elaborated new forms of scientific objectivity: aperspectival objectivity, constituting itself in terms of the view from nowhere, and mechanical objectivity that "forbids judgement and interpretation in reporting and picturing scientific results."[7]

These involved the imposition of constraints on individual activity in order to establish depersonalized scientific procedures. The ideal of aperspectival objectivity appeared in philosophy, especially in epistemology, aesthetics, and moral philosophy, in the eighteenth century, but only

> in the middle decades of the nineteenth century was aperspectival objectivity imported and naturalized into the ethos of the natural sciences, as a result of a reorganization of scientific life that multiplied professional contacts at every level, from the international commission to the well-staffed laboratory. Aperspectival objectivity became a scientific value when science came to consist in large part of communications that crossed boundaries of nationality, training and skill. Indeed, the essence of aperspectival objectivity is communicability, narrowing the range of genuine knowledge to coincide with that of public knowledge.[8]

Thus, within nineteenth-century scientific research aperspectival objectivity became the ideal of "eliminating individual (or occasionally group, as in the case of national styles or anthropomorphism) idiosyncrasies."[9] It contributed to establishing the idea of depersonalized and collective scientific research. We return to this issue in the next section.

In contemporary philosophy of science, as we mentioned, objectivity is contrasted with the psychological, with mental processes structured by emotions and interests, with subjectivity, and with perspectival cognition and idiosyncratic arbitrariness, as expressed by the nineteenth-century ideal.[10] Since the psychological and the subjective are equated with the irrational (or, at best, the arational), objectivity is identified with rationality. However, the abandonment of traditional epistemological individualism removed the ultimate model of absolute objectivity (rationality) that had been ascribed to God. If there are only finite subjects of cognition, objectivity must be understood as human *intersubjectivity*. For instance, Popper admits that "objectivity is closely bound up with the *social aspect of scientific method,* with the fact that science and scientific objectivity do not result from the attempts of individual scientists to be 'objective,' but from the cooperation of many scientists. Scientific objectivity can be described as the intersubjectivity of scientific method."[11] This socialized view of objectivity permits a normative view of the sources and guarantees of objectivity: philosophy of science should establish (or rather, discover) the rules of scientific rationality (the method of science) that constrain individually executed scientific procedures.[12]

For philosophy of science the concept of scientific *rationality* takes its significance from two mutually connected sources: Cartesian anxiety and the threat of relativism. The discovery that scientific research involves disagreement, or even lack of communication (as between incommensurable paradigms or language games), creates anxiety as to whether this disagreement can in principle be solved rationally.[13] If it cannot, science is threatened by nihilism and defeatism that can lead—according to critics of nihilism—to historico-cultural relativism.[14] And should such a disaster eventuate, the philosophy of science would be in danger of becoming nothing more than sociology or history of science. This explains, of course, why the search for the rational core of science is so crucial for many philosophers.

Expositions of the rationality of science usually refer to scientific beliefs or to the actions of scientists. In the first case, rationality is equated with conforming to logical canons.[15] In the second, it is a standard applied to the pursuit of goals—in particular, to decisions concerning the acceptance or rejection of theories, hypotheses, or research orientations, as well as to preferences for certain forms of representation or experimental techniques. As Laudan reminds us, in these accounts of rationality two components are present: the methodological and the axiological.[16] Axiology sets the aims or values for scientists and is normative; but methodology may be a descriptive enterprise. When it is combined, however, with axiology, as is usually the case in discussions of rationality, it becomes normative. These components take various forms. In logical empiricism the axiological ingredient was embodied in the criterion of demarcation that clearly delineated scientific knowledge from other forms: epistemic value was attributed to well-confirmed scientific theories. Logical empiricists offered a theory of the logical and probabilistic structure of confirmation, designed to explicate and justify theories, explanations, and acts of appraising the credibility of hypotheses on rational grounds.[17] In postempiricist philosophy of science, positions on rationality stretch from epistemological conceptions, according to which "scientific rationality" is a normative and methodological concept, as in the work of Popper, Lakatos, Newton-Smith, or Siegel, to instrumentalist views, usually descriptive and naturalist, such as those articulated by Quine. Epistemological views are dogmatic in outlook. Dogmatists lay down the a priori aims of scientific activity and its methodological rules and seek to delineate standards of what is rational. They minimize the significance of the factual and historical components of science and consider scientists as fully rational and cognizing automata. Their concept of the method of science, or the idea of rational rules for evaluating or choosing theories, reflects an attempt to purify research, to essentialize it, and thus to find normative foundations that plausibly justify the "well played game of science." For them, the axiological ingredient of scientific rationality is the aim of science couched in terms of truth or verisimilitude. Methodologists who follow Popper view increasing verisimilitude as the intuitive aim of science that justifies its method; they accept it as the ultimate, although not the only, epistemic value.[18]

Less dogmatic adherents of an epistemic understanding of scientific rationality—for example, normative rationalists or "temperate rationalists," such as Newton-Smith, Toulmin, Siegel, or Meynell—seek a middle way between dogmatic and relativist views of scientific method and rationality. According to them: "if we are to understand the type of rationality that is involved in theory-choice we must recognize that theory-choice is a judgmental activity requiring imagination, interpretation, the weighing of alternatives, and application of criteria that are essentially open. But such judgments also need to be supported by reasons (reasons which themselves change and vary in the course of scientific development)."[19] Like the dogmatists, they believe that the rationality of inquiry (and beliefs) depends (at least partly) on the aim of science: an action is rational if it satisfies the aim of science or at least one of its subgoals.[20] They also believe (Toulmin is an exception) that the aim of science must be established by epistemological considerations and not merely by studying real science.[21] The normative, axiological component of their view is complicated in the sense that, apart from the ultimate aim of science, they consider a hierarchy of subgoals. Although they believe that science has permanent aims, they admit that historical change affects the methodological rules that guide it.[22] While dogmatists look for permanent standards of rationality, whose universal character they presuppose as self-evident, middle-way adherents (normative rationalists) seek standards of rationality that allow them to treat science as an enterprise having permanent rationality. Unlike dogmatists, who declare themselves immutabilists and cannot allow methodological revolutions in the standards of science, adherents of the middle way recognize change, either gradual or evolutionary, in scientific methodology.[23] In their view, standards of scientific rationality are neither absolute nor purely local. There is another important difference between them and dogmatists: as the dogmatists see the rationality of science, together with its method, as self-sustaining, so temperate rationalists seek ontological or epistemological grounds for the axiological component of scientific rationality. They point to invariant features of human cognition: they appeal to constitutive characteristics of science itself, or root rationality in the dictates of reason linked to human survival, or in common problems facing people in all cultures, or in transsocial mental faculties involved in cognitive self-transcendence.[24]

The methodological approach to scientific rationality can be dismissed, as is done by advocates of naturalism, without, however, eliminating an epistemological perspective on the normative features of the scientific enterprise. Laudan attempts to naturalize the concept of rationality while maintaining the normative character of methodology. Scientific rationality in his view "is agent-and context-specific." He rejects the need for an external axiology or methodology; both may be considered as empirical without eliminating their "normative consequences."[25] The choice of which means (actions) are conducive to stated ends is an empirical issue; but also cognitive goals along with their means should be established empirically, in accordance with values implicit in scientific research.[26] For him, these goals should not be utopian. Accepting a methodological rule demands considering whether empirical matters are such that our cognitive ends will be furthered by following this rule rather than another. Methodological rules are not categorical but hypothetical imperatives "whose antecedent is a statement about aims or goals, and whose consequent is the elliptical expression of the mandated action."[27] Accordingly, knowing that an action is conducive to an end provides reasons for engaging in that action, and acting in accordance with reason is normative and, consequently, rational.

Laudan's attempt at naturalizing both the aims of science and scientific method is criticized by advocates of "epistemic rationality."[28] They believe that the axiological components of rationality can (and must) be established by pure epistemological reasoning alone capable of underwriting its universal standards.[29] Accordingly, the rules of scientific research are objective in the sense that they are prior to scientific activity and discovered "by logical analysis and reconstruction of the rationale of the scientific search for knowledge."[30] Their objectivity indicates the ahistorical nature of their content or, at least, of the way they determine the research of (mature) science. Hence, epistemic rationalists emphasize that disputes referring not to the "instrumental efficacy of alternative rules," but to the "epistemic standing" of those rules, cannot be resolved empirically;[31] and, moreover, that without establishing the rationality of ends epistemically, naturalist methodology cannot warrant the rationality of accepting rules that lead to those ends.[32] Laudan is not moved by this type of criticism. He rightly rejects the distinction between epistemic and instrumental rationality, pointing out that the justification of both rules and aims appeals to reasons and is always already instrumental, that is, relative to certain cognitive ends.[33]

Finally, those who advocate the naturalist turn, such as Feyerabend, Quine, Giere, or Hacking, replace epistemological conceptions of rationality with an instrumentalist outlook and abandon normative questions regarding rationality. They do not believe that there is an objective, noncontextualized, and transhistorical justification for the aims, rules, or standards of science. They claim, moreover, that we can at best only describe the methods and standards of rationality used by scientists as they change historically. Taking a stand in the history of science, they agree that scientific rationality is practical and embraces a wide spectrum of local rationalities, typical of different intellectual traditions.[34] This view leads certain philosophers to drive the standards of the scientific enterprise "onto the relative and transitory authority of some local reference-group."[35] They stress the pluralism of traditions and values.[36] In this respect, Feyerabend's view is particularly insightful. Opposing Popper's way of objectifying the methodological component of science, he subsumes it under the idealist conception of the relation between reason and practice. The opposition to idealism is naturalism, according to which reason is completely determined by practice: "Reason receives both its content and its authority from practice. It describes the way in which practice works and formulates its underlying principles." Following Wittgenstein, Feyerabend claims that "standards or rules are not independent of the material on which they act" and, accordingly, treats them as two sides of (scientific) activity: "I regard every action and every piece of research both as a potential instance of the application of rules and as a test case."[37] As acting may violate existing rules and change them, so perseverance in following rules may change action. Methodological rules cannot be objectified by separating them from real action and treating them as prior. They are embedded in scientific traditions, in value judgments made by scientists, and in their concerted actions.[38] This is a "dialectical," holistic, and relativist picture of the relation between reason and practice, unacceptable to partisans of the epistemological position. As judged from the normative rationalist perspective, Feyerabend, Quine, and Rorty treat epistemic standards as arbitrary, subjective commitments and not as matters of reasoning. If they doubt "that the activities, theories, and hypotheses of science are fundamentally justified in terms of reasons and evidence," they doubt that science is rational.[39] Standards and norms that are only described, and not justified, are—from the epistemological perspective—nonrational "dogmas." From our perspective, however, the dialectical view is the acceptable position. It rejects the justificationist ideals of traditional epistemology and embraces the idea that standards of scientific pursuit are constituted by the historical and communal activity of scientists, beyond which they do not require further explanation or justification.

SOCIAL PRODUCTION OF OBJECTIVITY IN SCIENCE

For epistemological rationalists, as indeed with seventeenth-century philosophers, scientific objectivity is connected with the ideal of constraining, disciplining, and depersonalizing individual cognitive activity, and not with the communal content of research. This communal content not only contains shared knowledge but also unified procedures performed by actors who use uniformly calibrated instruments and interact within institutionalized social relations. Moreover, it contributes to the establishment of scientific objectivity. As the history of science shows, in the seventeenth and eighteenth centuries "the real communicative bonds were friendships (or enmities) between individual scientists, nourished by lifelong correspondences"; later networks of formal relations and institutions were imposed on personal bonds and contributed to the establishment of a new form of scientific objectivity.[40] The inability of rationalist models to account for communal objectivity (intersubjectivity) is clearly visible from the perspective of the social studies of science.

Recognition of the nondiscursive dimensions of scientific cognition, realized mainly in the sociology of science and in social studies in the history of science, "helped shift the terms of the rationality debate from the domains of reason, cognition, and logic to those of practice, writing, and instrumentation."[41] To the extent sociologists and historians recognize the situatedness of scientific research and reject essentialist belief in universal rules and values, they can consider objectivity as intersubjectivity, as a matter of scientific communication and participation, and not simply as a matter of the objective background of individual beliefs and cognitive acts.[42] They can then show that isolated laboratory settings are among the real sites for the production of scientific knowledge, and that their situatedness renders the ideal of the universal method of science dubious. Scientific inquiry, in its laboratory setting, is an interplay between the place where experimental events are produced, the means by which they are produced, the protocols of experimentation, and the discursive strategies used to report the process of experimentation. Each of these elements conditions the significance of what is produced.[43]

The view of science as a set of local settings that may form an interconnected system (in spite of their relative isolation and idiosyncrasies) only recently entered the conception of scientific knowledge. This realization affected, of course, the concept of scientific objectivity and the way it is studied. As Rouse notes: "The claim is not that scientific knowledge has no universality, but rather that what universality it has is an achievement always rooted in local know-how within the specially constructed laboratory setting."[44] So from the sociological and historical perspectives, as well as from the hermeneutic viewpoint, the very ideal of objectivity is contextualized: "ideas as to what constitutes an objective judgment or rational decision are themselves ideas of a particular tradition. What we count as an objective, unconditioned determination of a given object-domain is itself conditioned by our tradition or, more precisely, by what in our tradition counts as objective and unconditioned."[45] Arguing that the contraposition of reason and tradition makes no sense, hermeneutics shows that the belief in scientific method, as a foundation of objectivity securing full self-control based on full self-knowledge, leads to attempts at eliminating the historical situatedness of human cognition by expanding its local intersubjectivity into universal objectivity. From the hermeneutic and sociological perspectives, rationality, too, is unavoidably social and "essentially dialogic."[46] Standards that function in scientific dialogue, among members of research teams, co-authors, or between authors and referees, far from being abstract, take an indexical, localized form.[47] These standards allow scientists to decide "what counts as competently construed data, what methods are reliable, which citations are controversial and which are not, what claims require argument, which alternative interpretations must be foreclosed, and so forth."[48] Moreover, as Shapin and Rouse emphasize, "the adequacy of data and procedures is often assessed on the basis of the credibility of the investigator."[49]

Stressing the situatedness of research, sociologists query the idea that one of the objectives secured by the rational rules of science is the *replicability* of experiments. As Collins notes, replicability—"the touchstone of common sense philosophy of science"—is seen as a condition of the public recognition of scientific results. It is rarely practiced, however, and usually occurs "only when the existence of some phenomenon is cast into doubt."[50] In a laboratory there are "skills researchers gradually build up in developing and adopting the equipment to locally determined needs and purposes." Such research skills "do not immediately transfer outside their physical setting and local community; one must first learn one's way about in the new context and adapt one's prior knowledge to it."[51] Consequently, "when scientists do try to replicate others' results, they often modify the original equipment and procedures to suit their own programmatic interests."[52] Frequently experiments cannot be reproduced. Fleck states, for instance, that "the first experiments by Wassermann are irreproducible," although they are "of enormous heuristic value."[53] Replicability is not of itself an objective feature of scientific experiments, or a possibility guaranteed by scientific method itself; *if necessary it must be worked out.* Paradoxically, when it is worked out, "when experiments become certain, precise, and reproducible at any time," then "they no longer are necessary for research purposes proper but function only for demonstration or ad hoc determinations."[54]

The unification and universality of scientific research is secured only if the locality of knowledge and its production (together with laboratory tools, research strategies, and skills) is actually transcended by social relationships that link different settings, and through certain scientific phenomena, processes, and procedures carried by those relationships, such as migration (proliferation), standardization, and adaptation. The *migration* of locally produced scientific results is, first, a spatial migration and, second, a migration between theoretical contexts, problems, and ranges of application. "Conceptual objects are regularly transferred to instances beyond their original range of application, and displaced into contexts which differ from their estab lished situation. Moreover, they are extended to problems clearly distinct from those they have previously been used to solve."[55] Furthermore, scientific results and tools migrate not only within science to other laboratories in which they are used, but also outside science, into other subdomains of practice. Proliferation, far from being a simple act of transferring an element of science from one context to another, is a complex, difficult, and creative process. This is made clear, for instance, from Collins's study of the TEA laser. As he reports, no one to whom he spoke "succeeded in building a TEA laser using written sources (including preprints and internal reports) as the sole source of information." The laboratories he studied "actually learned to build working models of TEA lasers by contact with a source laboratory either by personal visits and telephone calls or by transfer of personnel."[56]

Using results elaborated in another laboratory requires *standardization,* "the process of adapting the work of the vanguard of research into material usable in further investigation."[57] Seen in this light, the need for standardization arises from the locality of research and is necessary for adopting and using a result in a different context and for extending the use of a tool "to a wider class of problems and objects, unanticipated at the time of its first devising."[58] As processes of standardization proceed references to "the particular local contingencies" are removed from products or from tools of laboratory work undergoing standardization, but—as Rouse emphasizes—"only by making these references indefinite." That is, in the course of standardization, problems, tools, or research procedures "are transformed into versions that can be extended beyond their original domain, though not without some loss of content and clarity."[59] In giving an example of standardization Rouse appeals—quite obviously—to Fleck's account of establishing the Wassermann reaction as a scientific fact through the diffusion of the practical experience underlying the application of Wassermann's method. In so far as it strips scientific ideas, tools, and procedures away from results of idiosyncrasies, standardization is an undoubted element in the transition from what is under creation, to the given, to what already exists and is routinized and objectified.

Standardization does not entail, however, the establishment of smooth mechanical unification. As Rouse emphasizes, even in the case of long-standardized procedures, "each laboratory develops its own variations on the standardized schema." Scientists deliberately modify and alternate standardized procedures "because they often believe they can do it 'better,' because they must adapt to local conditions, and because they do not want their work to appear superfluous."[60]

Standardization of scientific tools and procedures should be distinguished from the objectification of the subject matter of research, from presenting results and artifacts as objects, even if—as Ravetz argues —standardization is a necessary condition for the objectification of scientific facts and for making them available to scientists working in other local settings. It is by confusing standardization with objectification that Rouse insists on replacing Heidegger's concept of decontextualization with that of standardization. Standardization is not the same phenomenon as Heidegger's decontextualization. Contrary to what Rouse says,[61] for Heidegger the concept of decontextualization applies to objects and not to acts of theorizing. For him, it is entities, primordially within-the-world, that become decontextualized, and not theoretical activity. Moreover, Rouse disregards a central function Heidegger attributes to decontextualization. For Heidegger, the idea of decontextualization is an existentialist expression of his epistemological view that the subject/object dichotomy provides the foundation of scientific (theoretical) cognition. In this light, decontextualization appears as the ontological crux of the transition from an everyday life attitude toward entities to a scientific attitude. On the other hand, Rouse's view, which states that appropriate access to scientific objects requires practical manipulations demanding unification of local practices, refers to a phenomenon quite different from decontextualization. The concept of standardization is an ontic notion that refers to phenomena already present in science, and, moreover, it may indicate a certain historically limited way in which science functions, for example, as a system of unified local settings.

Four reasons make the replacement of the concept of decontextualization with the concept of standardization impossible. First, the concept of standardization belongs to the ontic description of scientific research, whereas, the concept of decontextualization belongs to its ontology. Second, the latter refers to the relation between science and everyday life practice. Third, decontextualization affects what is ready-to-hand, and alters its way of being into presence-at-hand, whereas standardization affects the way of being of those elements of scientific research that are manufactured in a local setting and transmitted to other laboratories. It is a necessary condition for the possibility of their operating in a unified way in different local circumstances. This is why it is "akin to the transformation of a tool originally designed for a highly specific task within a particular context into a more general-purpose item of equipment."[62] What is subject to standardization is not, however, a thing or a tool (something that Dasein takes care of) during its transfer from everyday practice to science, but a procedure, a piece of equipment, or even a problem, or an interpretation that already functions in science, although only in a certain local setting. Fourth, in the case of everyday practical involvement with equipment standardization is not necessary; in different local settings—for example, in different isolated cultures—the same needs of humans may be assisted by various tools. There are no internal factors acting to force their cross-cultural unification. Processes of standardizing everyday equipment that have occurred in certain societies are historical events whose conditions lie—among others—in the impact of science upon technology. This does not imply that standardization is a necessary phenomenon. It is as necessary as the ideal of universal scientific cognition and knowledge. And this ideal is as necessary as other ideals that function in research, none of which is in fact really necessary. Moreover, standardization exists to the extent it is worked out; it depends, therefore, on practical needs, strategies, and activities, such as the external (technological) need for unified scientific results, the strategies necessary for aligning local scientific settings, and the activity of standardization itself.

THE NORMATIVE ASPECT OF OBJECTIVITY

Unfortunately, the sociological perspective does not parallel the philosophical standpoint on rationality and objectivity. It only reveals social processes constituting the factual objectivity (intersubjectivity) of scientific knowledge, tools, procedures, or skills as traced through their proliferation, standardization, and adaptation. So normative issues of scientific values and rationality remain untouched. However, when "we concentrate on the nature and role of community in sci-

entific inquiry, on the ways in which rationality is essentially dialogic
and intersubjective, then we must not only clarify the descriptive
aspects of actual scientific communities but their normative dimensions as well."[63] Without considering these issues, we cannot arrive at
conclusions germane to scientific creativity and its interrelation with
objectivity.

There are two ways of approaching the normative aspect of scientific objectivity: external, from the perspective of (normative) philosophy of science, and internal, by presupposing that its normative
aspect appears only from the perspective of participants in scientific
research. We discussed the first approach when considering epistemological rationalism. The internal approach does not refer to universal
scientific methodology but rather to *values* that regulate research. In
studying the normative aspect of objectivity from this perspective we
need to refer to hermeneutics that "can contribute to the debate
about rationality by developing a concept of rationality derived from
the communicative and discursive competence of speakers of natural
language." We must go beyond hermeneutics, however, since it "considers participants in conversations to be rational only to the extent
that they are committed to a search for truth which makes them adopt
certain attitudes toward each other."[64]

Although there is nothing new in considering values in science, a
reconsideration of the constitutive role they play has been undertaken
recently by postempiricist philosophers of science. They reject the reason given by logical empiricism for excluding values from science—
namely, that the notion of "objective value" denotes an impossible state
of affairs. In fact, Popper anticipated the importance of values in considering the role of decision making in science.[65] It is in the work of
Rudner, Kuhn, McMullin, and Laudan that a strong claim for the
value-ladeness of scientific research is made.[66] In their view, acts of
theory-choice and theory-appraisal are value judgments referring to
epistemic values as objective properties of scientific knowledge that
are, for particular reasons, desirable guides to its outcome.[67] The
epistemic values deployed in theory-choice are—according to Kuhn—
"permanent attributes of science."[68] Appealing to values, these authors
stress that criteria of assessment or choice act not as rules dictating
choice but as standards that constitute it.[69] Kuhn asks what standards allow the evaluation of the epistemic goodness of a theory, and he establishes a (tentative) list of values that, "whatever their initial source," are

"fixed once and for all, unaffected by their participation in transitions from one theory to another," although in part they are learned from scientists' experience and evolve with it.[70] Following Kuhn, McMullin discusses five characteristics of a good scientific theory: predictive accuracy, internal coherence, external consistency, unifying power, and explanatory fertility.[71] Empirical adequacy (the accuracy of predictions) and explanatory power are "goals of the scientific enterprise itself," and other values are means to these ends.[72]

For epistemological rationalists, the appeal to value judgments means introducing nonrational dogmas into science. Some Popperians are convinced that philosophy of science without nonrational dogmas is possible and articulated by Popper's conception of unlimited criticism that overrules any need for nonrational commitment.[73] They miss the fact that according to Popper the notion of objective truth is an important regulative standard, excluded from the domain of bold criticism.[74] To proclaim this is to adopt faith in reason. It is a dogmatic commitment, a stipulated principle neither rationally argued for nor empirically tested. It allows Popper to attribute universality and sociohistorical insensitivity to the method of science; moreover, he postulates it independently of what empirical studies show concerning the methodological rules followed by scientists.

Acceptance of nonrational commitment seems the only way to avoid the following dilemma: either cognitive paralysis (caused by an infinite regress of justification) or the threat of a vicious circle in reasoning. However, for normative rationalists this is the worst possible solution. If it is necessary to admit that certain values are chosen according to a nonrational option and must be so chosen, any rationalist program fails as both a descriptive and normative project. It seems, however, that rationalists cannot avoid this unfortunate choice. They are put in this position by their acceptance of the fact/value dichotomy, for which there are, according to them, good reasons. The separation of values from facts drives them toward regulative principles. If one accepts the opposition between fact and value or reason and cause, one cannot derive value judgments from factual statements. One cannot infer from a description of scientists who follow rules for producing, evaluating, or replacing theories that, for instance, truth is an ultimate (even if unachievable) aim of science, or that accumulation of knowledge is necessary, or that science is able to transcend local situations and produce knowledge that is more and more truthlike.[75] Equally unacceptable are transitions from value judgments to descriptive theses, for example, from the verisimilitude thesis to realism. What is even worse for a rationalist is that this difficulty cannot be solved by dismissing the fact/value dichotomy. Its dismissal means treating values as facts!

Help may come from the recognition that the detachment of values from facts is a historical, cultural event.[76] Both dilemmas, endless justification or a vicious circle and description or evaluation, are unnecessary. No factual description is void of values: they enter the very process of describing and are sometimes constitutive of what is described.[77] Values are not established independently of facts. Even the most abstract axiological stipulations are related to pictures of the world, human action, history, etc., and some are simply empirical statements.[78] The interrelatedness of describing and evaluating simply indicates that our attitudes toward the world (including the cognitive) involve values. The rejection of the fact/value dichotomy should not mean—as it does for McMullin—the adoption of the view that values simply compose a type of fact—namely, subjective aims or desired objective features of things. This view obscures the particularity of the ontological status of values. In particular, epistemic values function within practice in a different way than objects or psychological states: they do not condition research, they regulate it.

However, if we take into account that principles regulating practice take different forms, such as, customs, technical rules, ethical norms, etc., we should compare epistemic norms with other types of regulation by considering the results of their violation. As Kolakowski argues, the violation of a *technical rule* produces merely the expectancy of failure; the violation of a *social custom* results in the feeling of incorrectness; the infraction of the *law* leads to the fear of punishment; but only the transgression of *moral norms, religious commandments,* or *taboos,* produces the phenomenon of guilt.[79] The experience of guilt is an existential act, "an act of questioning one's own status in the cosmic order," or "a feeling of awe in the face of one's own action which has disturbed the world-harmony." It belongs to the realm of the Sacred, in which knowledge, the feeling of participating in ultimate reality, and moral commitment join together in a single act.[80] We believe that scientific epistemic principles stretch from technical rules and habits to those whose status is similar to ethical norms. The commitment to truth seems closer to adhering to ethical norms than to other types of principles. This does not mean that the experience of a scientist who realizes the falsity of a projected theory or violates another epistemic norm does not differ at all from the moral or religious experience of guilt. However, the unity of knowledge, the commitment to create it, and communion with the world involved in participating in scientific research differs only to a degree from the convergence of understanding and believing in acts of religious or mythical participation in the sacred. If we do not admit their similarity, we cannot easily recognize that the norm commanding commitment to truth is fundamentally different—that is, stronger and more obliging—than a simple technical rule belonging to scientific protocols.

This approach to values allows us to see that they ought to be considered as elements embodied in the attitude of the participant in research. As we mentioned in chapter 4, Polanyi shows that emotional involvement of scientists in research is a source of the normative dimension of scientific activity. Emotional involvement of scientists transforms the *subjective* into the *personal*. A personal act (of scientific cognition) "is neither an arbitrary act nor a passive experience, but a responsible act claiming universal validity."[81] Scientists' recognition of the universality (objectivity) of knowledge as an epistemic value is a subjective act; but their commitment to objectivity, the fact that they undertake responsibility for its achievement, transforms subjective acts into personal acts, acts guided by passions and immersed in communal meanings and values.[82] Polanyi's categories of the personal and the person do not refer to the subjective pole of the subject/object opposition. The personal transcends this dichotomy allowing us to realize how the objective and the subjective are linked: they come into being in the course of responsible (cognitive) action based on commitment.[83] Moreover, there is a close, dialectical, relation between the personal and the universal: "the personal comes into existence by asserting universal intent, and the universal is constituted by being accepted as the impersonal term of this personal commitment."[84] Analogously, as Siegel emphasizes, scientists' commitment to empirical procedures and evidence is necessary for the existence of the empirical aspect of science.[85]

The recognition of the transformation of the subjective into the personal and the introduction of the ethical dimension of scientific activity reveal one more aspect of the ontological interrelation between individual participation in scientific research and this research itself. If the value of scientific statements is guaranteed neither by logic nor by the objective method of science, we need a concept of the knower—Polanyi argues—which does not bankrupt the notion of personal *responsibility* for statements made.[86] From Polanyi's perspective, there are no epistemic values attributed to scientific knowledge a priori, simply in virtue of its being scientific; nor are such values justified by methodological rules. They reveal themselves as *indispensable correlatives* of personal cognitive activity: they both condition personal acts and are simultaneously constituted by them:

> We attribute absoluteness to our standards, because . . . we rely on them in the ultimate resort, even while recognizing that they are actually neither part of ourselves nor made by ourselves, but external to ourselves. Yet this reliance can take place only in some momentary circumstance . . . and our standards will be granted absoluteness within this historical context. So I could properly profess that the scientific values upheld by the tradition of modern science are eternal, even though I feared that they might soon be lost for ever.[87]

Epistemic values (for example, the universality of knowledge or its intersubjective validity) do not exist for Polanyi without personal acts that acknowledge them as goals and without accepting responsibility for their achievement. For Polanyi, the relation between the personal freedom of scientists and their compliance with rules is dialectical and marked by the paradox of dedication, which he formulates in a Kantian manner: "a person asserts his rational independence by obeying the dictates of his own conscience, that is of obligations laid down for himself by himself. . . . Any devotion entails an act of self-compulsion."[88] His approach to rules, standards, and values permits an understanding of the dialectical nature of the dependencies between ourselves, values, and the actions we perform, together with their rules and consequences, without appealing to any external axiological authority.

Following methodological rules and standards constitutes scientists as scientists. Their actions gain meaning and can be evaluated; their results, such as scientific knowledge, appear as if they were the embodiment of objective, absolute, and coercive rules and standards. On the other hand, science as tradition and as an organized system of authority does not exist without scientists' commitment. It is the responsible participation of scientists that maintains science as tradition; their obedience to tradition confirms science as an authority.[89] Individual acts of obeying methodological rules and values constitute and objectify them as obligatory and universal.

That Polanyi sees the essence of scientific activity in axiological and existentialist terms does not mean that he adopts an existentialist view of how we constitute values in an absolute axiological vacuum, as elaborated, for example, by Nietzsche or Sartre. According to Polanyi there is "no existential choice comprising the whole world and claiming responsibility for it. Such a choice would leave neither a center to which it could be responsible, nor a criterion by which it could be judged."[90] Polanyi rejects nihilism, not only because an axiological stipulation cannot comprise the whole world, but also because nihilism neglects our social nature, our being-together. Even a most innovative scientist works within a community and refers to existing results of research (knowledge) and its standards. It is science as a whole, "as mediated by thousands of fellow scientists," that "he accepts unchallenged."[91] Although Polanyi's view is predominantly psychologistic, he executes a first step beyond the traditional absolutist/nihilist dichotomy when he grounds values in the relation between individual scientists and collectively produced science. Both sides of the absolutist/nihilist dichotomy are based on a first person assumption that a primordial relation (whether ontological, existentialist, or epistemological) links a single human being and the World (God included). Consequently, values and norms may be rooted either in the Individual or in the World. The first choice leads to nihilism, the second to absolutism. If the individual is placed within social communities and, moreover, values are considered as intersubjective, nonabsolute, cultural, and historical becomings, this picture is significantly changed: it becomes possible to avoid nihilism without embracing absolutism and still be able to locate the source of values in the world. However, two further steps should be taken. The first is made by Shapin in his consideration of trust that we analyzed in chapter 4; the second involves connecting ethical values, embedded in scientific research, with epistemic values and with the network of social institutions.

For Shapin, it is trust, and not, as for Polanyi, personal emotional and normative commitment that is fundamental for forging science.[92] It is fundamental in two senses. First, the phenomenon of trust, located within direct interactions among scientists, is the social framework

within which individual *obligation* and *responsibility* appear and which simultaneously conditions acts of entrustment and the trajectory of their
consequences. Second, the moral nature of interactions between scientists is dialectically connected with their cognitive nature. Entrustment
is built upon a shared view of the world, and a world-known-in-
common "is built up through acts of trust."[93] Shapin's analysis of the social nature of the moral features of scientific cognition needs development by taking into account historical changes in scientific research.
The normative aspect of research evolved, as the web of social relations
in science changed from informal friendships to formal connections
ramified through networks of social institutions and—what is most
important—networks of recording instruments. Intersubjective forms
of objectivity, guaranteed by trustworthiness based on the social and
epistemic position of the entrepreneurial cognizer, underwent modification due to the employment of instruments able to record and de-
pict phenomena in an automatic, mechanical way. Their employment
was dialectically connected with the ideal of eliminating interpretation
and intervention, the ideal of "*self*-surveillance, a form of self-control at
once moral and natural-philosophical."[94] This new, mechanical objectivity, entailed heroic self-discipline, which needed "the honesty and
self-restraint required to foreswear judgment, interpretation, and even
the testimony of one's own senses" and "the taut concentration required for precise observation and measurement, endlessly repeated
around the clock."[95] As the fabric of scientific research transformed
into networks of formalized organizational structures, institutions, systems of recognition, rewarding, and so on, the locus of authority
shifted from individuals to social structures. The standards of rational
acceptability became social and now "are embodied in the institutions
within which research is referred for funding or publication, cited and
reviewed, and reformulated for inclusion in standard texts and reference works."[96] Thus, authority is disseminated in faculty and editorial
boards, committees who decide about grants, rewards, and so forth.
Taking into account trust and the fact that social institutions have become the locus of authority, we can conclude that the source of
epistemic values is neither an absolutely solitary individual nor an all
embracing Absolute, but scientific research as it constitutes itself in history. This way of overcoming the absolutist/nihilist dichotomy does
not satisfy those who emphasize another understanding of scientific
objectivity, namely, that it refers to an independent physical world.

OBJECTIVITY AS THE PRODUCT OF REALITY

The second way of understanding scientific objectivity—namely, that (scientific) knowledge presents its subject matter as *objective*—is typically discussed under the rubric of realism. However, realism is only one of three answers to an epistemological problem that refuses to go away: "what grounds provide the warrant for the relationship between the object of study and statements made about those objects?"[97] The first is reflective, and according to it, "(at least some) real-world entities enjoy an existence independent of their description," so that "the description of an item becomes an ancillary to the reality of the item itself"; the second is constitutive and claims that accounts "*are* the reality, there is no reality beyond the constructs we imply when we talk of 'reality'"; and the third is mediative, since it sees theories, interpretations, and experimental results as underdetermined by the facts of the natural world, so they "can be thought of as products of the social, cultural and historical circumstances which *intervene between* reality and the produced account."[98] The first position is realist and adopted by many philosophers and by the Strong Programme.[99] The second does not involve commitment to the independent existence of any reality. However, to the extent it decontextualizes and absolutizes discourse it is unacceptable. The third position is the basis for the constructivist program in sociology of science.[100] It may also be found in the work of social historians of science, such as Foucault, Bourdieu, Ringer, and Althusser, in pragmatism, and in praxist versions of Marxism. We concentrate here on the first answer, on realism construed as a philosophical doctrine that applies to science as well as to the realistic attitude of scientists themselves. The former task is particularly difficult, however, since no issue is more fraught with dissension. There are as many versions of realism, antirealism, and irrealism as there are philosophers who discuss the issue.[101]

To introduce order into this cornucopia of views the only way forward is to delineate them with respect to those positions they oppose, which we will collect under the general rubric of antirealism. First, there is ontological realism, which opposes solipsism, understood as the view that nothing exists except selves and their ideas; second, there is semantic realism, which opposes instrumentalism in reference to theories—that is, the view represented by logical empiricists or by Quine's linguistic instrumentalism. Third, semantic realism is typically part of epistemological realism, which opposes a pragmatic approach to theory-choice. Finally, there is Hacking's manipulative realism that opposes idealism as a view that the real is what is theoretically posed.[102]

Ontological (metaphysical) realism maintains—as Devitt notes—a commitment to the existence of types of objects and a commitment to their independence from mental states and beliefs. On this view, items "of most common-sense, and scientific, physical types objectively exist independently of the mental."[103] For ontological realists—such as Popper, Newton-Smith, or Tuomela—the world is real and totally independent of us; it is ready-made in the sense that "it consists of a fixed set of objects which are independent of the human mind."[104] Ontological realism is usually a presupposition of semantic and epistemological realism, though this is not necessary, as the views of Bhaskar or Tuomela demonstrate: "Things exist and act independently of our descriptions, but we can know them only under particular descriptions."[105]

Semantic realism, as applied to science by many contemporary philosophers, is based on the idea that science searches for truth and that its achievement is possible.[106] Semantic realists "speak of the relation between language and the world" and "seek referents for theoretical terms, employ correspondence theories of truth, or make claims for relations of approximate resemblance between theoretical language and some aspects of the world."[107] Thus, semantic realism is realism about truth or about reference;[108] this means that it applies to theories or to (scientific) terms and concepts. According to the realist approach to (scientific) language, "we must be able to form concepts that can serve as descriptions of items that exist apart from our cognitive relations to those items."[109] We are able to form such concepts because of an objective relation between linguistic terms and nonlinguistic entities. The core of the realist approach to scientific theories is the idea that "the sentences of a theory are true or false in virtue of how the world is independently of ourselves."[110] Truth is, therefore, reducible neither to justified assertion nor to empirical adequacy. Realism about truth, on this view, is the core of scientific realism, and we will refer to it in this way. For scientific realism, "science is *about* something; something *out there*, 'external' and (largely) independent of us."[111]

Scientific realism leads to an epistemological thesis that links the acceptance of a theory or the choice between rival theories with truth value. It states that truthfulness is the reason for accepting a theory or that "we sometimes have good reason for concluding of two rival theories that one is more approximately true than another,"[112] or—still more cautiously—that "it is in principle possible to have good reasons for thinking that one of a pair of rival theories is more likely to be more approximately true than the other."[113] Because of this thesis "scientific realism is more than an ontological commitment to theoretical entities as 'methodologically or epistemologically legitimate'"; it "offers an *explanation* for that legitimacy."[114]

Radical semantic realists about reference believe that there are objective referential links between language and reality determining the truth value of sentences. Antirealists about reference deny this. For them, these relations are culture-laden; in particular, they are impregnated with language that is part of culture: "inputs upon which our knowledge is based are conceptually contaminated."[115] Antirealists emphasize that signs "do not intrinsically correspond to objects,"[116] refer to the fact that there is no "one-to-one correspondence between stimuli and sensations,"[117] and reject the causal theory of reference, even if they had accepted it earlier.[118] They either consider reference as related to a certain conceptual system or taxonomic structures, or they accept Quine's thesis that reference is indeterminate unless it is relativized to a chosen translation manual.[119] Moreover, they adopt "a theory of meaning that explicates sentence meaning in terms of warranted assertibility or verificationist conditions."[120]

Let us now comment critically on these versions of realism. At least two elements of ontological realism, which beg for clarification, are not usually considered directly by realists (Sellars is a notable exception). First, an existence claim may be more or less specific and refer to "something," to "common-sense observable objects" or to "scientific non-observable entities." The position that results from the acceptance of the existence of commonsense observable objects (and the rejection of commitment to scientific nonobservables) is, in fact, a completely different sort of realism than one that results from the commitment to scientific nonobservable entities. Second, the independence claim is also largely unclear and provokes a more troublesome issue of whether strong realism and radical antirealism are possible. If (weak) realism claims—according to Devitt—that something "objectively exists independently of the mental,"[121] and "the mental" is understood in individual terms as "my mind" or even as referring to the joint cognitive capacities of participants in a community, antirealism does not exist. Even Bishop Berkeley fails as an example. Any antirealism that claims nothing exists independently of my mind is nonsense. Also an antirealism that denies the exposition of the "independence of existence claim" offered by Brown—namely, that "something exists independently of our attempts to know it"[122]—does not exist and is epistemologically impossible. Even sociological constructivists accept the notion of independence (when they talk about the resistance of facts); otherwise they would not be able to separate conceptually attempts to know something from acts of knowing it and from the intentional object known. In general, to deny the independence of (cognized) reality from an individual mind or from individual cognitive activity makes it impossible to entertain the idea of the intersubjective nature of cognition.

If, on the other hand, realism is taken to claim—as critically characterized by Putnam—that "the world consists of some fixed totality of mind-independent objects"[123] or to embrace "the general question of the possibility of access in knowledge to objects that are independent of any practices of thought and action through which knowledge of them may be achieved,"[124] it is an impossible epistemological position. If the world is mind-independent, both our thoughts and our inventions as their correlatives fail to belong to it and so cannot interact with objects or be influenced by them. At least, they cannot in a worldly way.[125] If objects are supposed to be independent of our practices of thinking and cognizing, it follows that they cannot be thought of. If there were such objects, we could only be silent "about" them. They could not in any sense begin to exist cognitively for us, since what is "thinkable" must lie within the boundary of the conceptual.[126] This difficulty cannot be remedied by an appeal to a realm of absolutely noncognitive subpractices, in which the cognitively independent world is accessed by us. Regardless of what the result of such noncognitive subpractices were, it would not be knowledge—that is, a meaningful cognition of this independent world. Moreover, the realist position is not helped even by the presupposition that our sense-contents are determined by an independent reality. We must still be able to conceptualize and recognize that the knowledge accrued refers, in fact, to the world that realists treat as determining us.

In general, it seems to us that an ontological antirealism, which rejects the existence of any entity independent of the individual mind and its activity (or even from any historically local collective cognitive activity) is not a reasonable epistemological position. Nor is an ontological realism that believes in a reality completely independent of the entirety of human cognition (even that involved in everyday life). As extreme solipsism tries to ignore the social character of cognition, so extreme realism tries to ignore the fact that cognition is an inescapable aspect of our being-in-the-world.

Furthermore, if we consider both positions as referring only to science they also lead to unacceptable views. If an extreme antirealism claims that there is no reality independent of scientific cognition, it ignores all other forms of practice—in particular, those historically prior to scientific research. Their reality is independent of scientific cognition; it may, of course, be shaped by it, but it cannot be created by it, under the pain of inconsistency. If, in turn, extreme (scientific) realism claims that reality presents itself to us within the activities of science as it is in-itself, uncontaminated by any forms of thought, it opposes science to all other forms of practice and treats it as superior. Both views run counter not only to sociological and historical portrayals of science but also to the mainstream of postempiricist philosophy of science.

The most controversial element of scientific realism is its belief that the truth of a theory consists in "its corresponding to the world as it is in itself."[127] This view is usually accompanied by the belief that "scientific procedures are capable of leading us to truth."[128] The realist belief that theory corresponds to reality in itself, in virtue of its referential terms, obliterates all forms of mediation between reality and a scientific theory or, indeed, the totality of such theories. For instance, scientific realists either ignore foremeanings present in any act of understanding or they believe that foremeanings disappear as soon as the genetic process is over. They also ignore historical changes in the conceptual contexts within which objects of science are studied. Only on the basis of these assumptions can realists hold that "we are able to arrive at a unique theory in every domain."[129] These beliefs are the reason why underdetermination of theories by data and their incommensurability cause troubles for realism.[130] They are also the source of the problem of access, which realists need to solve on the pain of becoming agnostics. They face this problem even though they are not naive realists. That is, even if they do not take truth-correspondence as a sort of picturing relation[131] but "highlight the metaphorical nature of scientific language and scientific models and theories . . . as partial, tentative representations of what there is."[132]

The problem of access appears in two forms. If realism is understood as a rational reconstruction of the realist position of scientists, the problem of access becomes an internal scientific question of how the world is in order to determine what is true.[133] Without any direct access to the physical world, the evidence for the truth of a scientific theory comes only from its empirical testing; what convinces us about the empirical adequacy of a theory, or of its explanatory power, is also supposed to convince us about its truthfulness. Thus, the realist can appeal only to "the coherence of the explanations we give and the accuracy of the predictions they enable us to make."[134] The possibility cannot be excluded, however, that at bottom the realist's rational reconstruction of realism remains a sort of hopeful thinking, rationalizing the fact that scientists accept a theory because they *hope* it is true and consequently *put faith* in its verisimilitude. If, on the other hand, realism is understood as an external, realistically interpreted philosophical theory of scientific knowledge, the problem of access becomes a question of how scientific theories and the world correspond. Is such a relation possible? Realism on the part of philosophers of science is a "second order" realism.[135] Its problem is that without extrascientific means to assess the correspondence relation or, at least, the value of (scientific) knowledge itself, such as intellectual intuition or the means offered by Kant's transcendental critique, the realist cannot answer this question. The philosopher's theory of scientific realism remains, at best, a hypothesis.

The problem of access may be avoided by accepting the *relational* nature of truth by admitting that the truth of a theory consists "in its fitting the world as the world presents itself to some observer or observers."[136] This leads either to a weaker form of realism, such as Putnam's views on internal realism, or simply to antirealism about truth, to an instrumentalist or a pragmatic view of theory-choice. As Putnam sustains the realist view that truth is not reducible to rational acceptability, that it is an idealization of justification,[137] so antirealists emphasize that "truth value is an epistemic feature of sentences that cannot transcend our recognition of truth value."[138] And since antirealists doubt whether scientific procedures can secure anything more than empirical adequacy, they are satisfied with rational acceptability or with "the instrumental reliability (in the broad sense) of a theoretical story"; they treat theories as if they were true and rely on them "as useful guides for whatever practical and theoretical jobs may arise."[139] For instance, according to van Fraassen's constructive empiricism, scientific activity is more a construction than a discovery, and its aim are theories that are empirically adequate. Such theories explain phenomena, and they must give a correct and adequate account of what is observable. Therefore, saving phenomena by a theory is a condition of its acceptability.[140]

In the light of these considerations, in our view, truth even when construed as *adaequatio* (correctness) does not presuppose a world of mind-independent facts to which assertions correspond. On this view, an assertion is true if facts make it true, if it is understood in reference to mind-independent states of affairs, and guaranteed if it correctly represents them. Given that the necessary comparison cannot be made, there is no way of knowing whether such conditions are satisfied. If, however, understanding an assertion is knowing its conditions of right assertibility, truth links with the epistemic notion of warrant: an assertion is true in virtue of states of affairs we conceptualize and in terms of our cognitive ability to know when they are optimally satisfied. On this view, our grip on objects is through language and the practices of referring, and objects and reference emerge through these practices. Thus, ontological commitment is based on judgment relative to context and purpose and does not presuppose a mind-independent world of "self-identifying objects"[141] (intrinsic essences) whose inherent "meaning-relations" fix reference. To embrace this view, in our opinion, goes beyond the realism/antirealism antithesis and preserves the primacy of truth as disclosure.

The idea that philosophical considerations of realism are nothing but the reconstruction of realism as it functions within science faces its own problem. The realist attitude is attributed to scientists, not only by adherents of realism in postempirical philosophy of science but also by Husserl, Heidegger, Gadamer, and by sociologists of science.[142] However, the manner in which this attribution is made differs. In postempiricist reconstructions of scientific realism it is done in a depersonalized way: it is scientific knowledge, or the objects of science, that are to be interpreted realistically. In phenomenology, hermeneutics, and sociology, realism is part and parcel of the attitude of the participants in science, who "constantly raise questions as to whether a particular statement 'actually' relates to something 'out there,' or whether it is a mere figment of the imagination, or an artifact of the procedures employed."[143] From this perspective, it is not the rational, argumentative, semantic, and epistemological realist account of scientific theories (with its concept of truth as an objective feature of theories) that governs scientists, but truth as a *value* belonging to personal commitment and to the undertaking of responsibility. This means that commitment is a relation within which something can be believed to be true.[144] Moreover, those who see realism as an attitude of participants in scientific research usually emphasize that the attitude toward science of a sociologist, a historian, or a philosopher should be that of a relativist, not a realist. Relativism means here, first and foremost, *contextualism* (relationism).

A particular case of the view that scientists are realists is found in the phenomenological philosophy of science of Heelan, Crease, and Ihde, all of whom accept a realist attitude toward science. They locate sensory perception at the heart of science and hold that a primordial hermeneutic understanding underlies all linguistic, perceptual, and scientific activity. Certainly, they do not reject conceptual and instrumental mediation between perceptual and observational acts and reality; nor do they fail to articulate the fact that theories, as well as observers, are embodied in the technology of scientific instrumentation.[145] Heelan puts their position in a nutshell:

> theoretical states and entities are or become directly *perceivable*. . . because the measuring process can be or become a 'readable technology,' a new form of embodiment for the scientific observer. In this view, the term 'observation' no longer means *unaided* perception. It implies that theoretical states and entities are real and belong to . . . 'the furniture of the earth,' *because (and to the extent that) they are perceivable in the perceiver's new embodiment*. It also implies that the nature and aim of scientific explanation is to make manifest the processes and structures of the real, the real now being taken as what is or can be given in some World.[146]

Through the use of various sorts of instrumental embodiments in "readable technologies" we learn "to 'read' like a 'text' those scientific and technological artifacts of material culture which give us direct perceptual access to classes of things and objects that are not otherwise objects of perception."[147]

Heelan's position chimes in many ways with the views of analytic philosophy regarding realism, but there are important differences characterized by what he calls "horizontal realism" or "hermenuetical realism."[148] For him reality is not "*totally unrelated* to human life and culture";[149] on the contrary, given his view that what is perceived is a readable text, Heelan assumes that "plural veridical realistic perspectives are possible consonant with the plurality of different horizons of perception within a World: there is not just one empirical basis of fact but a plurality of empirical bases."[150] Each horizon has its own descriptive language and its own appropriate context. Moreover, horizons are "plural, each linked to its own individual context, each connoting its own individual embodiment (or Body structure) and cognitive praxis, and each denoting objectively a manifold of possible perceptual profiles."[151] Thus, Heelan's realism admits of plural, incompatible, and empirically descriptive frameworks, some of which are complimentary and others divergent. This is in general accord with the ontic pluralism we advocate with regard to scientific objectification.

Consonant with this approach, Crease stresses the action-oriented thrust of experimental inquiry: experimentation "is not merely a *praxis*—an application of some skill or technique—but a *poiesis;* a bringing forth of phenomena."[152] Crease understands making entities manifest in experimental settings according to Heelan's stance on realism—namely, that "scientific entities are phenomena accessible to perception via *readable technologies*."[153] He mobilizes theatrical imagery. Experiments are performances—that is, unique events that take place on nature's stage for the purpose of allowing something to become "visible";[154] theory is a script that directs successive experimental performances; "read noetically an experimental script is something to be performed; read noematically, it describes the object appearing in the performance," that is, a scientific entity.[155] This does not preclude the fact that descriptions of an experimental entity can change together with its perceptual profiles and the noetic-noematic correlation within which it is brought forth. Ihde stresses the embodiment of science in its technics and the transformative and mediating role that instrumentation plays in science. Thus, he emphasizes a "phenomenological realism," the view that an instrument displays itself as "simultaneously the condition of the possibility of certain types of knowledge and yet as a non-neutral transformation of what is known. In its use the instrument has the 'phenomenologi-

cal' capacity to bring into presence that which was previously un-
detected and even invisible, but precisely in this *difference* it also
transforms the way in which the phenomenon may appear."[156] For these
writers phenomena are passively given and can only be accessed cognitively by means of instrumental and interpretive extensions of human intentionalities.

Our historical picture of science employs the full and dynamic resources of the ontic-ontological circle. Thus, from our perspective,
horizontal or hermeneutic realism is one among many ontic interpretations of science that can be articulated and which can be incorporated into our larger interpretative framework. It is, of course, valuable
that these writers locate genuine hermeneutic and phenomenological
elements at the heart of scientific experimentation, observation, and
the use of instrumentation. Inspired by principles of Continental philosophy their work throws considerable light on scientific cognition
within the subject/object antithesis. Nevertheless, their epistemic approach to interpretative understanding, directed as it is to perceptual
and observational contexts, tends to obscure the importance of the
ontological character of human understanding and the ways in which
objectified knowledge and the ontic claims of science stem from it. Indeed, their views on forehaving, foresight, and foreconceptions are informed more by the hermeneutics of the textual tradition than by a
fundamental ontology of human being.

CONTEXTUALIST VIEWS OF SCIENTIFIC OBJECTIVITY

Contextualists reject semantic realism in both its parts without, however, accepting instrumentalism. They undertake the more radical
move of surpassing both scientific realism and instrumentalism by
locating science within a broader nonscientific context. They locate it
within the social world and among other cultural activities, in the manner of sociologists or pragmatists; or they anchor it in the life-world and
the structure of everyday practical involvements, in the manner of Husserl, Heidegger, Schutz, and ethnomethodology; or they relativize it to
tradition and its modes of understanding, as should any hermeneutic
approach to the natural sciences. All forms of contextualism agree that
"the pursuit of truth and reality must have a practical aspect" since
"truth and reality are related to human agents, rather than being mere
objects of speculative thought—in other words, truth and reality are

meaningful *for* the existence of human agents."[157] In particular, from the contextualist perspective, understanding and truth are necessarily partial and incomplete. "There can in principle be no final, absolute interpretation of reality which escapes revision in view of the open horizon of future projects. Indeed, because world-horizons are projections of sedimented presuppositions, they essentially remain relative to concrete historical circumstances."[158]

To adopt contextualism is not to accept a solipsistic version of ontological antirealism; rather, it is to accept a constrained understanding of the notion of objectivity.[159] A qualified ontological position is the choice of those who admit the existence of incommensurability in science. They accept the relational nature of reality. Feyerabend stresses that the world "is not directly given to us, we have to catch it through the medium of traditions";[160] and he states that "we certainly cannot assume that two incommensurable theories deal with one and the same objective state of affairs." Therefore, "unless we want to assume that they deal with nothing at all we must admit that they deal with different worlds and that the change (from one world to another) has been brought about by a switch from one theory to another."[161] Also, Kuhn, when considering the essence of scientific revolutions, says: "Examining the record of past research from the vantage of contemporary historiography, the historian of science may be tempted to exclaim that when paradigms change, the world itself changes with them." Consequently, "after a revolution scientists are responding to a different world."[162] The world of scientists is not "fixed once and for all by the nature of the environment, on the one hand, and of science, on the other. Rather, it is determined jointly by the environment and the particular normal-scientific tradition that the student has been trained to pursue."[163] What Kuhn aims at is more than just an acceptance of the (critical) realist claim that the scientist's interpretation of observations changes with a paradigm. While answering the question of whether there is "any legitimate sense in which we can say that they pursued their research in different worlds?" he states that we must learn to make sense at least of a statement that "though the world does not change with a change of paradigm, the scientists afterward work in a different world."[164]

Kuhn and Feyerabend are not isolated in their effort to make sense of the idea that the subjective (for example, scientific theories) and the objective (things, processes, or phenomena) mutually influence each other. Also hermeneutics recognizes that "those who are brought up in a particular linguistic and cultural tradition see the world in a different way from those who belong to other traditions." Furthermore, it admits that not only knowledge and the way of perceiving the world change, but—in a sense—the world also changes. Various worldviews refer to *their worlds* and "the historical 'worlds' that succeed one another in the course of history are different from one another and from the world of today."[165] The same relativization of worlds to worldviews works in science. "Neither the biological nor the physical universe can, in fact deny its concrete existential relativity. In this, physics and biology have the same ontological horizon, which it is impossible for them, as science, to transcend." The world of physics, for example, is not a world of being-in-itself, "where all relativity to Dasein has been surpassed and where knowledge can be called an absolute science."[166]

However, having introduced plurality of worlds Gadamer adds, almost instantly, that "it is always a human—that is, verbally constituted—world that presents itself to us," and stresses that since different historical worlds are verbally constituted "every such world is of itself always open to every possible insight and hence to every expansion of its own world picture, and is accordingly available to others." In other words, historical worlds are not really separated ontically from one another. Moreover, "the world can exist without man and perhaps will do so. This is part of the meaning in which every human, linguistically constituted view of the world lives. In every worldview the existence of the world-in-itself is intended. It is the whole to which linguistically schematized experience refers."[167]

Realism is thus an attitude connected with every worldview. It does not mean, however, that in considering this multiplicity of worldviews we may refer them to one common World; they "are not relative in the sense that one could oppose them to the 'world in itself,' as if the right view from some possible position outside the human, linguistic world could discover it in its being-in-itself." This plurality of historical worlds makes the expression "world-in-itself" problematic, and, moreover, the reason "for the continuing expansion of our own world picture is not given by a 'world in itself' that lies beyond all language." The World "is not different from the views in which it presents itself."[168] Gadamer here rejects any retreat to the traditional metaphysical position.

> The concept of the *life-world* is the antithesis of all objectivism. It is an essentially historical concept, which does not refer to a universe of being, to an 'existent world.' . . . the *life-world* means something else, namely the whole in which we live as historical creatures. . . . the life-world is always at the same time a communal world that involves being with other people as well. It is a world of persons, and in the natural attitude the validity of this personal world is always assumed.[169]

The world-as-intended-by-a-worldview cannot become a subject matter of an objectivist ontological theory, as any such theory attempts to treat itself as ontologically neutral: it overlooks both the projection on which it is based and the cultural tradition to which it belongs. Consequently, it attempts to remove from the world any relativization to us.

For Heidegger, too, realism is an attitude present in certain forms of cognition, in particular, in science. He emphasizes that only humans can make intelligible the way of being of each domain of entities, including that of natural things; without humans there would be no categories to articulate true judgments and no truth.[170] "Beings are discovered only *when* Dasein *is,* and only *as long as* Dasein *is* are they disclosed."[171] Heidegger, as we have seen, can question and reject both realism and the idea of the universal rationality of science on the basis of his idea that projection is simultaneously an act of disclosing. He questions, as we have also seen, the traditional distinction between the objective and the subjective, as well as the ideal of pure theoretical cognition that is transparent to itself and free from any practical components. Like many postempiricist philosophers of science, he believes that the opposition usually accepted by modernist thinkers, *either objectivism or relativism,* is misleading; we do not face such a dilemma because the whole framework of both positions has been called into question.[172] As Caputo emphasizes

> Heidegger holds *both* that it is impossible to gain access to bare and uninterpreted facts of the matter—which is to reject any notion of objectivism or absolutism, as if we could jump out of our skins and make some absolute contact with things—*and* that our hermeneutic constructions, when they are well-formed, do capture something about the world—which is to provide for the objectivity of knowledge.[173]

This view does not obviate the need, however, for science to spell out the warrant for its claims to knowledge (whether it takes a realist stance or not) and to distinguish itself from pseudo-science. After all, forms of cognition that constitute themselves within the subject/object nexus incur this responsibility. Philosophers have concentrated on the question what, in addition to truth, turns mere true belief into knowledge. Responses range from foundationalist and internalist views of epistemic dutifulness and the right to be sure, to coherentist and reliabilist claims as to what confers the necessary warrant for objective knowledge.[174] Of course, it is incumbent upon the philosopher who probates scientific knowledge to assess its inherent evidential structures of support. And this philosophical task differs with respect to each ontic framework, as each has its own practice-sensitive criteria of correctness or truth, whether scientific, religious, aesthetic, or metaphysical.

According to Heidegger, the scientific claim to objectivity stems from science's way of cognizing, its objectification, and does not have universal validity. Science does not have, for instance, "a special access to ultimate reality."[175] It tells us its own, particular truth about reality thematized in a scientific way. Moreover, when Heidegger uses the concept of objectivity in reference to science, his goal is not normative; he aims neither at the approval of science as it is nor at its criticism and improvement. He aims at understanding, at revealing the character of the scientific way of cognizing, as distinguished from everyday cognition or philosophy. The objectifying character of the scientific attitude, its search for objective truth, is authentic because science "has its origin in authentic existence."[176] Heidegger does not think, like Nietzsche, that a disinterested pursuit of truth is "a self-deceptive activity arising from disguised psychological motivation."[177] In his search for an existentialist understanding of science, Heidegger "(1) spells out what everyday scientific practices take for granted, namely that there is a nature in itself, and that science can give us a better and better explanation of how that nature works, and (2) seeks to show that this self-understanding of modern science is both internally coherent and compatible with the ontological implications of our everyday practices."[178]

The Heideggerian view is different, therefore, from the standpoint of postmodernists, such as literary theorists, some social scientists, or feminists who are ready to reject even the Heideggerian version of objectivism, because they believe that scientific theories are simply interpretative texts, or that scientific truths are the product of social practices, or that science is a bastion of male domination. "All these groups would like to believe that natural science is just one more interpretative practice that has somehow conned our culture into thinking that it alone has access to the real."[179] For Heidegger, science is not merely a text, or a social practice, or a form of domination. Science has its place within ontology: scientific research is one of Dasein's ways of being and—simultaneously—a way of being of entities that reveal themselves as objects. When we encounter entities in acts of their objectifying thematization, we understand being as occurrentness, and so we understand that the way of being of objectively present entities is such that they would have been and will continue to be even if Dasein (a human being) had never been.[180] Heidegger puts the point in this way:

> only as long as Da-sein *is*, that is, as long as there is the ontic possibility of an understanding of being, 'is there' [*gibt es*] being. If Da-sein does not exist, there 'is' no 'independence' either, nor 'is' there an 'in itself.' Such matters are then neither comprehensible nor incomprehensible. Innerworldly beings, too, can neither be discovered, nor can they lie in concealment. *Then* it can neither be said that beings are, nor that they are not. *Now,* as long as there is an understanding of being and thus an understanding of objective presence, we can say that *then* beings will still continue to be.[181]

The claim that entities existed before us and continue to be even if we cease to exist can be articulated within a discourse that understands being as objective presence (as, for instance, Newtonian and post-Newtonian science), and it is valid only within such discourse. From the perspective of Heidegger's ontology, it is clear that the key thesis of realism, namely, that the "external world is objectively present in a real way,"[182] establishes a separation between a cognizer, who states it, and the world as cognized. His ontology, on the other hand, shows that Dasein is never a worldless cognizer, but is always already in-the-world. Consequently, the ontological perspective is not a worldview that presupposes realism; on the contrary, it reveals the ontological conditions of any worldview as well as the realist attitude that accompanies it. Indeed, as we argued in chapter 6 in regard to truth and disclosedness, Heidegger transforms the very basis of the epistemological enterprise and the traditional dialogue between idealism and realism.

We cannot, however, adopt Heidegger's position without modification. We have argued throughout the book for the necessity of escaping the individualistic implications of his view, emphasizing the ways in which cognition is a collaborative enterprise involved in historical practice. With this in mind, let us return to sociological constructivism. For constructivists, there is no objective world, independent of practice and language; in order to see things "as things we need to interact with them and with other members of society through them";[183] in order to demonstrate the existence of an object, fact, or event we must appeal to a certain form of representation.[184] Objects are constructed as being out there. The world is always already "a cultural object"; it is "the outcome of a process of inquiry"; and "realist ontology is a *post hoc* justification of existing institutional arrangements."[185] Although the organization of our perceptions of the world and communication with one another presuppose realism, the natural world is, in fact, constituted by scientific research and its discoveries.[186]

However, as the analyses of Latour and Fleck show, the scientific construction of facts can be understood only on the basis of the presupposed existence of something that offers resistance. Although we think that Fleck's conception requires modification, we want to protect and mobilize its core. Fleck rejects the realist belief in facts existing there to be interpreted theoretically by scientists. He studies the process of the emergence of a fact within a definite thought style and states that a fact is "a signal of resistance in the chaotic initial thinking, then a definite thought constraint, and finally a form to be directly perceived."[187] Latour also appeals to the idea of resistance. "The real is not one thing among others but rather gradients of resistance."[188] In other words, the emergence of a fact is a process constituting the real, the objective, that which opposes free, arbitrary thinking.[189] The out-there-ness of facts is produced, and facts are correlatives of that production. Thus, for Fleck and Latour, facts are not absolutely objective; their objectivity is social and historical, it is relativized to scientific communities and to their research. Facts do not exist by themselves; any fact exists in relation to a thought collective: it must "be expressed in the style of the thought collective," perceived as "a thought constraint," and "directly experienced" by the members of a particular

thought collective.[190] Moreover, facts "are never completely independent of each other. They occur either as more or less connected mix-
tures of separate signals, or as a system of knowledge obeying its own
laws. Every change and every discovery has an effect on a terrain that
is virtually limitless. It is characteristic of advanced knowledge, ma-
tured into a coherent system, that each new fact harmoniously—
though ever so slightly—changes all earlier facts. Here every discovery
is actually a recreation of the whole world as construed by a thought
collective."[191] In a holistic way, Fleck links particular and whole: they
are given together, when one changes, so too does the other. From
this holistic picture, it is impossible to separate precisely the discovering
of something in the world-in-itself from its construction. Their difference may be stated only in reference to the facts-as-known-by-the-
thought-collective and the world-as-known-by-the-thought-collective.
The *discovery* of a fact, previously unknown to a thought collective,
changes (only) the world-as-it-*was*-known-by-the-thought-collective; it
does not change the world-in-itself, because the discovery does not
add anything to it; the fact was there before the discovery. The change
of the thought collective's world means that a new world, a world-as-it-
is-known-by-the-thought-collective, is constructed by the collective. So
what actually happens is the transition from the world-as-it-was-known-
by-the-thought-collective to the world-as-it-is-known-by-the-thought-
collective, from the collective's-world-without-the-discovered-fact to
the collective's-world-with-the-discovered-fact. Thus, discovery, at best, is
a means of transition from one world of the given thought collective
to another. This analysis would neither surprise Fleck nor be unwel-
comed by him. It simply uncovers the reasoning underlying his constructivist and holistic view of the construction of facts.

Fleck's way of conceptualizing the objectivity of scientific facts
generates the following question. Is the objectivity of a given fact
purely local and relativized to a given thought community, or is it
arranged into a hierarchy of less and less local objectivities according to the number of communities experiencing this fact? Does the
migration of a given fact to other local settings give it a higher degree
of objectivity? May we, then, define facts of the highest level of objectivity (as objective as possible) by appealing to what is objective for
all thought communities? We would then be in agreement with
Nietzsche (and Gadamer perhaps) who states: "the *more* affects we allow to speak about a thing, the *more* eyes, various eyes we are able to

use for the same thing, the more complete will be our 'concept' of the thing, our 'objectivity.'"[192] We believe, however, that this is not a promising way to proceed. It is not a belief or an attitude or an activity common to any group of scientists that constitutes the objectivity of scientific objects. For us, reality and objectivity, as they are correlated with the subject/object dichotomy, are constituted by practice.

Fleck's appeal to resistance in thinking is, therefore, as insufficient for understanding the constitution of the objective as the belief that the understanding of the constitution of scientific objectivity arises from the study of science itself and its relation to the World. Together with Heidegger and Gadamer we contend that the constitution of scientific objectivity cannot be accounted for without reference to the context of human (manipulative) practical activity. Moreover, we agree with Grondin that hermeneutics "successfully dismisses the naive objectivism that would evaluate the truth of knowledge claims or a theory while bracketing the context of the *praxis* that makes them meaningful."[193] Although Gadamer himself applies this conviction only to the historical and human disciplines, we apply it to the natural sciences.

The reference to human practice allows us to avoid a danger to which contextualism is exposed—namely, absolute relativism. It is—as Latour convincingly argues—a mirror image of universalism as expressed in realism or rationalism. What universalists see as the opposing perspective, and fiercely criticize, is—in fact—absolute relativism. What is common to absolute relativism and universalism is the belief "that the reference to some absolute yardstick is essential to their dispute." Consequently, both claim that "if no common, unique and transcendental measuring instrument exists, then all languages are untranslatable, all intimate emotions incommunicable, all rites equally respectable, all paradigms incommensurable." They differ only in the conclusions they draw from this shared thesis. Moreover, this "yardstick" of commensuration from both perspectives is seen as given. Universalists and absolute relativists tend to forget that "measuring instruments have to be set up." It is only relative relativism, *relationism*, that realizes that "all measures, in hard and soft science alike, are also measuring measures, and they construct a commensurability that did not exist before their own calibration." So, relationists look for yardsticks for constituting commensurability, they try to establish relations, they "point out what measurements and what chains serve to create asymmetries and equalities, hierarchies and differences."[194]

The relationists' search for measures of both similarities and differences has its own difficulty—namely, the need to establish a common space in which "all collectives, producers of natures and societies, find themselves equally immersed."[195] Latour, like Gadamer, rejects the idea that this common space is in nature, or in society, language, God, or Being: "Naturalization, socialization, discursivization, divinization, ontologization—all these '-izations' are equally implausible. None of them forms a common basis on which collectives, thus rendered comparable, might repose." His own solution abandons the modernist perspective in which we remain obsessed with the problem of where we are, "with the construction of one immanence [*immanere*: to reside in] or the deconstruction of another." He believes that we can traverse this perspective, as well as its obsession with homogeneity, and undertake an understanding of hybrid entities.[196]

The rejection of foundationalism in favor of contextualism need not lead to an emphasis on the hybrid nature of ourselves and the entities that surround us; but—as we claim—it can lead to giving priority to being over Being and to stressing the historicity of being. The precedence of being means that no particular reality—for example, nature—is absolutely prior to other entities or independent of them. Nothing exists in-itself and by-itself; everything exists relationally and through chains of mediation. Consequently, any form of extreme realism is untenable from our perspective. The precedence of being means also that human constructs—for example, scientific knowledge, must be considered contextually within the subpractice that produces them and related to other forms of practice and their correlatives. Consequently, all versions of realism and antirealism that decontextualize scientific knowledge are untenable. The concept of free-floating meaning and truth (ascribed to certain sentences independently of our will and action) must be replaced with the idea that meaning is relative to cognitive activity and that truth is a value located within the context of commitment, responsibility, and trust. Only then can truth be understood—to use Fleck's expression—as "an event in the history of thought,"[197] as a feature of statements that are made by those connected with a thought-style and, thus, understood identically, saturated with their commitment, and based on their mutual trust.

Furthermore, any realism that describes scientific knowledge while merely stressing its relation to reality and concealing its relation to scientists, is untenable. For example, talking about a scientific discovery in which "uncovering and revealing something which had been there all along" is crucial, obscures the fact that the novelty and significance of discovery is related to discoverers and their audience.[198] The objectivity of scientific knowledge must be viewed as the objectivity of research—in particular, of experiential and manipulative activity. Moreover, it must be viewed in the context of subpractices external to science.

There are two important lessons, therefore, that one should learn from contextualism. First, a thing cannot be extracted from one context—for instance, the everyday life-world—and considered objectively as if it were apart from all contexts, since its local decontextualization means, unavoidably, its local contextualization within another life-world, for instance that of science. This view conveys the danger of treating contexts as pre-given, prior, and fixed. They are, however, always local and historically changeable. The real task, therefore, is to grasp the *dynamics of entities* as they move from one context to another, together with the *dynamics of contexts*. The second lesson is to see that the problem of objectivity cannot be posed "analytically" in reference to a piece of (scientific) knowledge, say, to a theory and its referents taken to exist by themselves. It must be formulated in a holistic way, it must consider theories and their referents as belonging to their own contexts, to local research and its world. In other words, it requires a consideration of contextualized theories and innerworldly referents.

OBJECTIVITY AND CREATIVITY OF SCIENCE

From our perspective an entity or an event is always inner-worldly in the sense clarified by Heelan: "It is one capable of showing its presence in the world, i.e., in the historical cultural arena in which human life is lived and toward which it always turns."[199] Unlike Heelan, who remains faithful to Husserl and Heidegger's perspective, we will consider innerworldliness, or—using Heidegger's term—worldliness, not only in a pluralistic but also in a relational fashion.

Worldliness is *relational*. To be an inner-worldly entity or event does not simply mean that it shows itself *by-itself,* by virtue of its own features and activity. As the being of an entity is always being in mediation (in interrelation to other becomings within practice), so showing its presence is not a matter of pure self-presentation or simply an

interaction with human perception; it arises from interactions between itself and other entities (people, in particular) within practice.
Worldliness is *pluralistic* because it is not constituted simply by entities
and events having being as such within the world of everyday life. As
we shape ourselves in different forms of practice, so there are various
worlds (and each is a life-world), that is, worlds in which humans dwell
in virtue of various forms of practice. Moreover, contrary to Heidegger, we think the question whether all such worlds are derivative with
respect to the world of everyday life, the human life-world itself, is not
an ontological issue; it is an ontic problem to which ontological considerations do not apply. If such worlds derive from the world of everyday life this is a historical fact but not an ontological necessity.[200]
Moreover, it is not true, historically, that the life-world of scientific research grew exclusively out of the everyday life-world; it grew out of
the intersection of the life-worlds of everyday practice, philosophy,
and theology. The everyday life-world was *then,* historically speaking,
prior and external to the world of scientific research.

As those worlds are worlds of life and practice and not exclusively
of thought, as they are more than what we can know of them, their relations are a matter of practice and not only of thinking. All relations
and interactions between them, all cases of one growing out of another
are worked out in practice; they occur in its development. Does this
mean that our thinking is only *about* worldly entities, life-worlds and the
relations between entities or worlds as they are constituted in sociohistorical practice, and that, at least in this sense, they are objective?
Unfortunately not; thinking is always in acting. As our thinking and experiencing are part of practice, so they participate in every act that establishes, changes, or breaks relations between entities or worlds.

This remark permits us to recognize one of the tacit presumptions of both semantic realism and antirealism. The extreme (naive)
version of semantic realism presupposes that (scientific) language
works representationally, that it is descriptive *by-itself* and undertakes
the reconstruction of its use in science as an instrument for describing
the world.[201] The critical (weak) version undertakes the reconstruction of the use of language for representation, admitting that language is not descriptive by-itself but in virtue of the conceptual effort
of scientists. On the other hand, antirealists (instrumentalists) presuppose simply that language is used in science, and they try to show how
it is used and for what purposes. Thus, for them, either language does

not represent (as for linguistic instrumentalists), or the descriptive use of language is worked out in scientific activity. Only this latter approach can engage the problem of the constitution of objectivity understood as the relation between language and the life-world(s).

We can, of course, admit that entities, events, and phenomena migrate between life-worlds since we experience their migration. In changing worlds they change their way of being: they switch, for instance, from furnishing our landscape, serving us as tools, being worshipped by us, from threatening or puzzling us, and so forth, to being thematized and studied in ways proper to local scientific research at a given time. Having been thematized by science and placed within a certain theoretical context, they give birth to other becomings, the existence of which is not suspected by nonscientific subpractices. Next, they change their way of being again when they are projected outside science. This projection usually involves—as Latour and Rouse argue—"transferring the conditions of the laboratory itself out into the world";[202] in this sense scientific objects undergo yet another recontextualization. They enter technological contexts and begin to generate phenomena not expected by science. This process, however, cannot be portrayed in an objectivist, realistic way as a march of self-sustaining entities through different contexts, correlated with various forms of representing and manipulating. As Latour and Shapin and Schaffer show the constitution of the objective is a social process: it takes place in social networks that link science, technology, political power, media, and so on. Furthermore, the decontextualization of entities that belong previously to other life-worlds, their contextualization within the local settings of science, and their recontextualization outside science's life-world should not be understood purely linguistically. Although, as hermeneutics insists, "what is the object of knowledge and statement is always already within the world horizon of language,"[203] we should not forget that language is not a self-sustaining ground of being and that cognition itself is but one aspect of practice.

If we take into account migration between semiautonomous spheres of practice and differences in the ways of being of entities within particular forms of practice, we may understand such phenomena as scientific discoveries or the fact that certain entities, processes, or events exist only within the life-world of science. Something (an object as well as a problem) that moves from a nonscientific sphere of practice into a scientific context may necessitate changes in an existing theoretical context enabling it to absorb the newcomer. These changes may, in turn, lead to the *discovery* of other entities unknown not only to other coexisting forms of practice but also to scientific research. Such changes may also necessitate the postulation of objects that must exist in the life-world of science if it is to be an intelligible whole containing newly contextualized objects, events, phenomena, and so forth. Objects discovered exist at the moment of their discovery only for the experiential part of scientific research, only within the empirical part of the life-world of science. This is why they require further conceptual elaboration to become present for theorizing. Moreover, they must be included into the system of theorized objects, into the theoretical part of the life-world of science, in order to become available for manipulation. Objects postulated by a theory, that initially exist only within its world, must become, as Heelan says, embodied—that is, included in the empirical and manipulable part of the scientific life-world in order to obtain scientific being. Such transitions have a number of predictable and unpredictable or, at least, uncontrollable effects.

A further complication of this picture stems from the fact that perception, theorizing, and manipulation, as well as relations linking them together into scientific research, are historically changeable. Any one may dominate for a time and organize an entire research activity. The question whether this changeability may be subsumed under a general regularity is an empirical issue that cannot be settled without reference to the history of science. It is equally possible that the known part of the history of science is a unique process of arriving at a contemporary form of the interrelation of perception, theorizing, and manipulation and, thus, is not arranged according to a repeatable general pattern.

The holistic approach clearly shows that none of the essentialist views of science, neither the experience-centered view of empiricists nor the theory-centered view of rationalists, can be accepted as a completely insightful picture of science as a cognitive subpractice. Not only do they ignore the fact that experiencing, theorizing, and experimentation are linked into a whole, but they also relegate to oblivion the phenomenon of scientific *creativity*. If scientists are determined and limited by empirical data, they do not create anything, they only synthesize the data available to them. If scientists are conditioned by sensual forms and intellectual categories, or by conceptual (linguistic or theoretical) frameworks, they see and understand only what those structures allow them to perceive and know. If scientists are determined by the means and rules of scientific experimentation, they do not create anything, they simply manufacture scientific objects out of the raw material of reality. Finally, if scientists are not entirely determined and limited by any of these factors—as realists presuppose—but are determined by the world, they may create theories and evidence, models and experimental worlds, but they do not create the objective, they only discover it. This last view is acceptable to scientism and naturalism. First, it allows adherents of these positions to understand and evaluate scientific knowledge in terms of truth (or empirical adequacy), so it always and necessarily (by definition) looks better than other forms of cognition. Second, it is coherent with the scientific view of ourselves as biological creatures who adapt to environments.

This view has, however, philosophical consequences far from obvious and acceptable. They are connected with science's practical function and with its technological involvement. Even if the "interior" of science may be studied objectively, in a purely methodological or rationalist way without (say) ethical categories, this approach is not germane to its technological involvement; it is no longer possible, given the death of the myth of science, as a panacea for all human misfortunes. The realist portrait of the scientist as a discoverer of the order of the World is part and parcel of this myth. Contrary to what Popper declares in his comments on rational criticism, his picture of scientific research as a discovery of objective knowledge governed by the Method of science does not permit an ethical evaluation of *how* scientists proceed and *what* they discover. It replaces the ethical evaluation of *how* scientists proceed with the ideal of scientific method. It authorizes nonresponsibility on the part of scientists for *what* they study and for *how* they study it. It proclaims absolute autonomy from society and absolute freedom from any ethical responsibility, at the price of absolute methodological slavery.

The view that there is a fixed method of scientific pursuit not only prohibits an ethical appraisal of research, veils the historicity of norms that regulate it, and fails to see the historicity of this conditioning itself, it also obscures scientific *creativity*. Creativity, like self-making, cannot be understood without presupposing situatedness and the dialectical tension between thrownness and futurity. If creative activity is supposed to obey ahistorically strict rules, creativity disappears. If scientific research is directed by the rule-bound method of science or the principles of universal epistemic rationality, then it does not create anything. It is simply—as in Popper's conception of the third world— the routine of discovering. Furthermore, the identification of scientific research with discovery requires the presupposition that discovering different things—for example, a law of free fall, the laws of mechanics, the principles of thermodynamics, or the theory of evolution—always proceeds in the same way.

If we want to elaborate, however, the idea that scientific research is creative, that, considering only one aspect of creativity, it conveys us from a state of not knowing to a state of knowing, or, more concretely, from an existing body of knowledge (beliefs) to a new one, then we cannot presuppose that it is done according to certain invariant rules. To put the point more precisely: it is possible to claim that creative research is done according to stable rules, but this would mean that every case of surmounting existing beliefs proceeds in the same way, regardless of the differences in the content of those beliefs, in experimental equipment, in mathematical models, and so on. Against this view, as well as against the dogmatic methodological or rationalist view, the following objection can be raised: "since we have no key to the treasury of transcendental rationality, all restrictions imposed on the implicit everyday criteria of meaning are royal commands issued *ex nihilo* by philosophers and carry no other legitimacy."[204] Kolakowski's point also applies to epistemic norms. More moderate views of scientific rationality try to avoid imposing "royal commands" on scientists. However, an objection can also be raised against Kuhn and McMullin's view on standards of scientific assessment that are somehow universal, though their content can change in the course of history. This view sustains the separation between methodological standards (of theory-choice) and the content of scientific theories. It seems, however, that they cannot be entirely separated. There are good reasons for agreeing with Shapin that: "Experimental culture shared norms insofar as its members shared a view of reality."[205] Moreover, it is not only a decision whether a report is a reliable testimony or whether someone is a "genuine investigator" that requires reference to reality and knowledge of reality. Evaluations concerning the accuracy of a theory, the broadness of its scope, or its fruitfulness, also require reference to reality. And—Shapin is right—differences in the way reality is viewed may well generate variations in evaluations. This much Kuhn and McMullin would probably admit. However, in admitting this and recognizing the historical variability in the content of values, Kuhn is on weak grounds to the extent he believes in their historical universality. He should rather acknowledge, as McMullin does, that scientific values "have gradually been shaped over the experience of many centuries."[206] Thus, as constructivists observe: "Criteria of validity used in scientific discourse are cultural resources whose meaning has to be re-interpreted and re-created constantly in the course of social life."[207] When scientists use certain rules, their application always depends on their using "supplementary interpretative procedures which can never themselves be formally specified in full. It is this open-endedness of rules which enables scientists, like other actors, to bring various apparently divergent actions within the scope of a given rule and to derive a given action from quite different rules."[208]

Moreover, as Feyerabend emphasizes, methodological standards and rules "make sense only in a world that has a certain structure. They become inapplicable, or start running idle in a domain that does not exhibit this structure." For instance, the Popperian demand for infinite content increase "has no point in a finite world that is composed of a finite number of basic qualities." This is why it appeared in science only after the scientific world had been made infinite and populated with new continents as a result of geographical discoveries.[209] Realizing the close connection between scientific methodologies and cosmologies leads to the conclusion that methodological rules are parts of wholes containing a particular form of scientific research and its world. To accept this is to look for the process of constituting the creativity and objectivity of scientific cognition within the historical span of research and to recognize three hermeneutic circles within which this process proceeds: the normative-hermeneutic, the historical-hermeneutic, and the hermeneutic-ontological circles.

As we argued earlier, epistemic and nonepistemic (moral) values exist and function within a complicated interplay between scientists, the instrumentation of research, and (scientific) institutions; between, in fact, personal acts, acts of undertaking responsibilities, acts of trust invested in fellow scientists and instruments, and social institutions that support values and standards in the sphere between persons. Values exist as elements of the normative-hermeneutic circle that links them, the institutional locus of authority, and scientists who enact personal acts of responsibility and trust within the nexus of power-knowledge relations. Within this circle values are dynamic: they switch their way of being back and forth from correlatives of personal acts to external standards and norms. That methodological norms and epistemic values appear to be objective, absolute, and coercive does not mean that their existence and legitimacy is independent of the commitment of individual members of the scientific community. Only obeying rules and honoring values constitute them as objective and give them legitimacy: they "derive their authority from their use by those persons who already count as doing science and as belonging to a particular field of work."[210] At the same time, within those acts, the purely factual actions of scientists gain meaning and can be evaluated.

Within the normative-hermeneutic circle scientists are (morally) responsible for *how* they proceed. The circle shows that as there is no freedom without responsibility, so there is no value without commitment. The rejection of methodological slavery or a depersonalized subordination to rules of scientific rationality allows us to find the conditions of responsibility for scientific research and, consequently, of its freedom within the normative-hermeneutic circle. In the light of this circle we can also consider scientific *creativity* and *objectivity*. They cannot be attributed either to theorizing, or experience, or experimental manipulation alone; they are characteristics of scientific subpractice as a whole. Moreover, they are features of the historicity of science. Historicity is marked by creativity in the sense that historically developing scientific research produces *novelties;* and it is marked by objectivity in the sense that the subject matter of research becomes *the objective.* The responsibility of scientists for *what* they discover can also be established within the normative-hermeneutic circle by showing that there is no (cognitive) creation independent of a limiting objectivity. However, considering creativity and objectivity requires uncovering two other circles: the historical-hermeneutic and the hermeneutic-ontological. Let us concentrate first on creativity and its relation to the historical-hermeneutic circle.

An ontological condition for producing novelties is *interruption of continuance.* Breaks in continuity cannot be recognized, however, from inside the very process of creation; from inside there is only continuity through modification of what already exists. So novelties exist only for the externality of the creative process; they can be recognized from outside, and external activity is necessary for their affirmation — that is, for the constitution of their being. The creative process, therefore, must be *relatively autonomous*. It can be connected with what is outside it (in fact, with other productive processes) only at certain moments—that is, at moments of interaction between them. The recognition of novelties requires the existence of the outer perspective from which the process appears to contain novelties, products of its inner creativity. Yet, the affirmation of novelties as creations from the perspective of this external productive process requires something more. It requires the opposition of novelties to something not created, to something already existing, and the process that embodies them into what already exists and, thus, makes them *real*. So novelties exist within a historical-hermeneutic circle that links the creative process and its externality. This circle is historical, since novelties are first and foremost changes, and it is hermeneutic, since it contains the understanding of changes as both novel and real. Let us detail this analysis by considering scientific theorizing and experimentation.

The creativity of scientific theorizing cannot be recognized from within theorizing itself, since from this perspective every new theory may be understood as a transformation of earlier theories. It exists only within the historical-hermeneutic circle that links theorizing with, for instance, scientific experience; for example, theoretical novelties, new entities postulated by a theory, become visible, recognized, and made real from the perspective of scientific experience. From this angle, the life-world of the theoretician is populated by entities unknown to observers and thus nonexistent for them—namely, fictitious creatures, models, symbolic generalizations, and other unreal constructs of the imagination; in general, by *artifacts* that lack flesh and blood. All these constructs of the imagination can be affirmed as novelties and made real only by turning them into new empirical facts by confirming the theoretical postulates that refer to their presence through observation and experimentation. Similarly, the creativity of experience, its ability to discover new phenomena, and the very fact of discovery can be recognized from the perspective of theorizing. This time it is experience (experimentation) that enters a historical-hermeneutic circle with theorizing external to it. From the perspective of theoreticians, the life-world of observers looks as if it were populated by scattered islands of data which resist theories and appear anomalous. In general, they appear as lacking intelligibility, coordination, and order. All these raw *facts* can be affirmed as novelties only by transforming them into new ideas. Analogous circular relations hold between observation and experimental manipulation, or between manipulation and theorizing: in each case their internally constructed novelties must become mutually recognized and affirmed.

Just as creativity is a historically viewed feature of scientific theorizing, experiencing, and, in general, of research, so too is objectivity. It is constituted within the same historical-hermeneutic circle which joins theorizing, experiencing, and manipulating. Transforming theoretical artifacts into empirical, observable, and manipulable scientific facts constitutes their objectification: they cease to be pure theoretical creations that exist exclusively as correlatives of theorizing or as the intentional referents of human subjectivity. They begin their empirical existence, the existence for human experience, even if initially only within the life-world of science. Observability or manipulability are not, however, features that are only added to the way of being ascribed to artifacts by theorizing. The change of placement from the theoretical to the empirical (experimental) context means that to the extent their way of being transforms, entities also change. Consequently, they become objective for theorizing. Exactly the same phenomenon is realized when empirical facts, discovered incidentally by experimenters or appropriated by scientific observers from other life-worlds, undergo theoretical conceptualization. Placing them within a theoretical context does not mean an addition of theoretical clothing. Their way of being is now theoretical; they exist within theorizing, and this transforms them into ideas. After such theoretical metamorphosis they become objective for observation and experimentation. The only reason for rejecting an understanding of this transformation as objectification stems from an "empiricist prejudice," from the belief that ways of being are ordered in a stable, unhistorical hierarchy in which empirical being is more fundamental and always prior to conceptual being. However, if we recognize and affirm the circular character of the relationship between experiencing and conceptualizing, we must reject the empiricist prejudice.

Analyzing the phenomenon of the constitution of different forms of objectivity within research does not exhaust the analysis of objectivity. All these forms are internal objectivities: the objectivity of facts for theorizing, the objectivity of theoretical entities for experiencing, and the objectivity of entities or processes put into use within scientific experimental production for theorizing and observing. The recognition and affirmation of the creativity of science requires a perspective external to science as a whole, the perspective of other spheres of practice. It must be shown to be an element of the hermeneutic-ontological circle in which the internal interrelation of creativity and objectivity becomes linked to the *external objectivity* of scientific research. The consideration of science that refers to reality outside the life-world of science requires making reference to relations between different life-worlds.

As Husserl and Heidegger emphasize, the world of everyday practice and the objective world of science are related to each other. For them, the former is always prior to the latter and constitutes its foundation. Moreover, the everyday life-world is the only world that "remains unchanged as what it is, in its own essential structure and its own concrete causal style, whatever we may do with or without techniques."[211] Even science cannot change it; it is everyday life, with its world, that provides the ultimate "context of discovery" for science, and this is why it "invests scientific claims with their significance."[212] For us, however, neither experience is the most fundamental form of our encountering the world, nor is the everyday life-world the most fundamental reality, the world in itself, that has no alternative. It is true that we cannot encounter the life-world as an inner-worldly object; nor is it, contrary to Husserl, "the indefinitely open horizon within which we experience objects, including the ideal objects constituted by scientific theorizing."[213] In other words, as Heidegger argues, we neither encounter the everyday life-world cognitively nor is it simply a horizon within which we experience objects.

We are in the world, or—more adequately—we are together with the everyday life-world, and also—if we do not presuppose its exclusiveness—the life-world of any domain of practice. That we are-together-with-these-worlds means that they are for us in the dialectic of different forms of presence and absence. The everyday life-world manifests itself to us through the web of practical meanings in the dialectic of the opposition between handiness and unserviceability. The world of science manifests itself to us through the web of cognitive meanings in the dialectic of intelligibility, evidentiality, and manipulability, on the one hand, and incomprehensibility, empirical and manipulative inaccessibility, on the other. The existence of these worlds for us means something else: in existing for us they exist for each other. The possibility of transferring entities from one life-world into another and appropriating them by practices associated with this other world means that life-worlds exist for each other; they must be capable of showing their presence to practices connected with (some) other worlds, and practices connected with them must be capable of recognizing the presence of (some) other worlds. However, since life-worlds are not worldly entities, their mutual existence cannot be described in the same way as the existence of inner-worldly entities. Analyzing relations between life-worlds demands an assumption opposite to that required by considering the migration of entities between life-worlds. To consider their migration and the emergence of new inner-worldly entities we must presuppose the (temporal) stabilization of spheres of practice connected with those life-worlds and treat them as if they were filters which entities can enter, go through, and leave modified. To analyze relations between life-worlds we must regard domains of practice as internally dynamic, historical, and mutually interrelated. In other words, various forms of practice and their life-worlds must be seen, first, as emerging from other spheres of practice, which are left transformed; next, as establishing autonomy from their source as well as from other forms of practice and their life-worlds; and, finally, as interacting with those other forms and worlds now clearly separated. Let us again restrict analysis of this line of reasoning to scientific research.

When the life-world of science emerged from the complex of everyday life, philosophical research, and the theological understanding of the world, it took the intersection of their life-worlds as given, as the World, which thereafter scientific research attempted to understand in its own way. Establishing itself as an independent domain of practice, it mobilized epistemic ideals of knowing and learning present in everyday life and in other forms of cognition, though not central to them, such as mathematics, argumentation, thought-experiments, systematic observation connected with manipulation, and the ideal of explanation. Carving out its independence and autonomy, scientific research modified these resources and introduced others, such as strictness of procedure or the idea that its forms of research do not modify reality. As we have argued, the final result was the rejection of the participant's perspective and the emergence of the *disengaged observer*. There are no good reasons, it seems, to claim that this is anything more than a matter of historical contingency; searching for the ontological conditions of scientific research may reveal the conditions of its possibility but not of its necessity; indeed, even this possibility exists only as a correlative of practice, as possibility-already-actualized.

The world of everyday life, modified by philosophy and theology, was the given reality for scientific research; without liberating itself from this background, scientific cognition could not have constituted its claim to epistcmic exclusiveness, to truth. The establishment of the *objective world,* of a reality stripped of relativization to scientific (or even, generally, to human) cognition and noncognitive activity, completed the process of the self-constitution of scientific research. Imposing this world on the everyday life-world was the capping-stone that gave science the status of a totalizing form of cognition. Realism is the appropriate epistemic attitude for this situation, and philosophical realism in reference to science attempts to justify this fact.

EPILOGUE

INSTEAD OF SUMMARIZING our argument, we conclude with a brief discussion of the concept of the hermeneutic circle that has appeared under various forms. This concept is a key discursive device for any attempt at overcoming the subject/object opposition basic to cognitive activity in many fields, in everyday cognition, in science and metaphysics, and typically in epistemology. It manifests itself most visibly in the realist attitude toward the world, and this attitude is a natural, self-evident disposition directing most of our cognitive enterprises. However, when understanding cognition becomes an issue, this attitude is no longer self-evident; it itself becomes a problem requiring understanding.

Understanding how it is possible to be within the subject/object opposition and conceive the world as a system of objects which exist out-there cannot be achieved from inside that opposition. Cognizing based on the realist attitude is always already objectifying, even when it makes itself its own subject matter. That we must somehow suspend our natural realistic attitude does not mean that we can step outside ourselves and see the entire opposition from a nonhuman perspective. It requires thematizing our being within the subject/object opposition from the perspective of this opposition itself by elaborating its own (that is our own) self-understanding. To elaborate such self-understanding is not to reject one or other of the poles of the subject/object opposition. Nor is it to carve a middle way between its two extremes. As we have indicated earlier, it requires sublation—namely, the move of simultaneous negation and affirmation of both poles during which both become reconstituted. In other words, the idea of sublation allows us to avoid two reductive moves: the reduction of the subjective to the objective, as is the case in naturalizing approaches to cognition, and the reduction of the objective to the subjective, as is the case with idealistic accounts. Sublation, in its turn, is not for us the Hegelian move of revealing absolute Spirit behind human cognition as performed within the subject/object opposition. In place of Hegel's metaphysical solution, we appeal to the ontology of being and to hermeneutics. Their way of dealing from within with the problem of the subject/object opposition has one feature in common with that of Hegel: the central role that circularity plays. As we stated in chapter 1, the hermeneutic circle is neither a logically vicious circle nor simply a methodological circle composed of anticipatory prejudgments and judgments that actualize them. It is—for Heidegger and Gadamer— the internal and central structure linking human being and understanding. Before we understand the world, in any conceptual and objectified way, we are already in the world, and our being, in all its forms, itself constitutes the understanding of being. Before we understand history we already belong to history, our consciousness is conditioned and limited by history; but history is not a self-sufficient, naturelike process, it proceeds in virtue of our (interpretative) activity. In the course of our analysis, which goes beyond the position of Heidegger and Gadamer, we have argued that the hermeneutic circle is dynamically composed of several circles.

Understanding our involvement in the subject/object opposition from the hermeneutic-ontological perspective allows us to avoid a difficulty stated by Kolakowski: if we want to understand ourselves as belonging to "an already given situation," we must "understand the presence of the world as precisely an existent"—that is, as existing in an unconditioned way. "The category of existence in the unconditioned sense cannot, of course, be the work of successive abstractions, as was imagined by empiricists confident in the natural validity of the Aristotelian hierarchy of predicates; for it is clear that stages of abstraction are stages in the scope of sets, but in no way do we reach a position where the act of existence reveals itself."[1] In other words, the unconditioned existence of the world cannot be given to us as the result of a search for conditions of conditions because this search is either endless or must be broken by the postulate of the existence of a nonconditional condition. The possibility of an experience in which "the act of existence reveals itself" requires, therefore, the suspension of the objectivist, realist way of using concepts. Yet, it is not clear whether, even after this suspension, an unconditional existence can give itself to us in a discursive (conceptual) cognitive act. The solution of Heidegger and Gadamer is different. Neither the world nor ourselves can be given to us as an ultimate unconditioned existence; both are given simultaneously as mutually conditioned; our existence is our-being-in-the-world and the existence of the world is the being-of-our-world. In other words, neither is the existing-by-itself but is always the existing in relativization. This is why the presence of the world for us is never absolute and direct but always partial and relative. Moreover, presence is dialectically connected with absence. The world presents itself as this or that system of beings, as their reference frame, as a whole of being, and so forth, and it presents itself to us according to how we perceive it in our everyday life, or how we experience it in a mythical or religious experience, or how we conceptualize and manipulate it in scientific investigation, or speculate about it in philosophy. There can be no absolute, nonarbitrary way of separating the presence of the world from our modes of conceptualizing it.

The first ontological circle that reveals itself when the analysis of being goes beyond the individualism of Heidegger's ontology is the circle of participation and embracement. When seen from an individual perspective, it links each of us with communities that embrace us, in which we are shaped, influenced by others, and within which we interact with them. When seen from the perspective of communities, it links them with individuals who participate in them, constitute, and maintain them. A particular form of the actualization of this ontological circle is a socio-hermeneutic circle that connects individual understanding and social knowledge behind which there are other subjects. As we cannot understand ourselves other than in a socially conditioned way, so communities cannot build their self-understanding other than through the individual activity of their members. This is because, in general, "the self-knowledge of a human individual exists only in the process of communication and mutual understanding."[2]

The socio-hermeneutic circle has its source in an ontological circle of the sociability and intentionality of individual consciousness. Our consciousness is intentional; it is the consciousness-of-something, neither by itself nor by virtue of the world that "gives itself to us," but in virtue of what we initially learn from others about the existence and internal structure of the world. On the other hand, without intentionality the sociability of consciousness could not be constituted: to learn from other people we must recognize their consciousness as different from ours and such that we can direct our own consciousness toward theirs.

The ways of being of each of us, seen as participation in various communities, also have a circular structure; they are constituted within an ontic circle of individuality and conformity. This circle is founded on human natality and manifests itself as an interplay between human personality, with all its idiosyncrasies underlying creativity, and the relative uniformity of members of a given community, who have been shaped in a certain way. The fact that our being is always being-together is also conditioned by a circle—namely, by an ontological circle of plurality and being-together. As plurality is the necessary ontological condition of our being-together, so human plurality requires our reciprocal recognition of individual independence, without which we would not be individualized and could not interact. The circle of individuality and conformity as well as the circle of plurality and being-together are embedded in practice; they are realized in practice and the internal ontological structure of human practice conditions them: human practice is itself circular.

There are three ontic-ontological circles that link: first, practice and its ontic structural web, the network of factual social relations, sociocultural institutions, systems, and so on; second, practice and meaningfulness that are mutually interrelated in an analogous way; and, third, practice and the world, which it constantly converts into reality and objectifies and which reciprocally influences it.

History is also ontologically circular, since it simultaneously situates human practice as our self-making and is a product of that practice. The most fundamental circular structure of history is the ontological circle of human individuality and plurality. Plurality, which manifests itself ontically as the succession of human generations, situates individual human beings within history; individuality is, on the other hand, a condition of plurality. Individuality and plurality are the basis of two further ontological circles constitutive for historicity. They connect temporalizing with temporality and historizing with historicity. As the historizing of practice and its world is the source of the historicity of our individual being, so our individual temporalizing is the source of the temporality of practice and its world. Through these circles the concepts of historicity and temporality join the ontological and ontic levels of our self-understanding within an ontic-ontological circle. Within this circle, any search for ontological structures and conditions must be supplemented by the reconstitution of the ontic dimension of our being in a hermeneutical historical narrative, since these ontological structures and conditions are always historically situated.

Finally, any form of cognition should be considered as involved in three hermeneutic circles that link its normative, historical, and ontological aspects. In particular, as we argued in chapter 7, cognitive creativity and objectivity reveal themselves within these hermeneutic circles. The normative-hermeneutic circle links values, the institutional locus of authority, and scientists who undertake responsibility for their cognitive acts. The historical-hermeneutic circle links novelties, the creative process of cognition, and the external context, which allows us to recognize novelties. Also objectivity, as a function of certain forms of cognition, for instance, of scientific research, exists within the historical-hermeneutic circle. The hermeneutic-ontological circle embraces scientific research and other spheres of practice and their worlds. Within this circle, the scientific deconstruction of external human reality and the projection of scientific reality outside science takes place.

Localizing science within these circles—in particular, within the last three hermeneutic circles—enables us to consider scientific research in a way that does not presuppose the subject/object distinction and, consequently, is neither idealistic nor realistic. This is because we reject two fundamental assumptions shared, paradoxically, by idealists, such as Berkeley or Kant, and (most) realists, in particular, those who accept the picture of reality as causally impacting on our cognitive capacities. First, we do not treat cognition as an activity performed by an individual cognizer, who ultimately (from a radical epistemological viewpoint) can be considered as entirely isolated from any social milieu; and second, we do not isolate (scientific) cognition from other human activities, both individual and collective. This means that we do not presuppose any opposition, or even any distance between participation in cognition and the activity of being. Only on the basis of the first assumption can we consider learning and cooperating with other people in order to establish knowledge as an unnecessary element of cognition. Only on the basis of the second assumption can the subject, whose only activity is cognition, and an object (the world), which is prior and entirely independent of this activity, be the initial conditions of a theory of cognition. If both assumptions are rejected and the consequences of their rejection elaborated, the idealist position is overcome; its fundamental problem of guaranteeing the intersubjectivity of knowledge cannot appear because the very process of cognition is social interaction. The realist position, on the other hand, can be recognized as establishing illusory solutions to its own problems. First of all, the objective, external validity of knowledge does not require any justification; we are in the world before we begin to cognize it in an objectifying fashion. It is not the world that is absolutely prior to our cognition; it is our objectifying cognition that conceptualizes the world as existing independently of this very conceptualization. Second, objectifying cognition certainly presupposes a realist attitude, but this attitude cannot be justified by appealing to the causal impact of the world upon our senses. There is no nonarbitrary way of cashing out the causal relationship that is presupposed. Even if the world exercises such an impact, it does not supply realists with an epistemological justification; it was already argued by Fichte that the representational character of our perceptions does not follow from the fact that they are effects of external causes. The relation of representing belongs to epistemological or semantic discourse, whereas the relation of causality belongs to ontic discourse. The realist attitude is not, however, simply a direct result of the nature of our senses (our receptivity) or a fundamental epistemological structure of our subjectivity. It is a matter of engagement and normative commitment, the nature of which can be grasped with the help of hermeneutic considerations, since it exists within the normative-hermeneutic circle rather than through any objectifying or naturalizing prism.

That the realist commitment is natural in everyday life and in science, since for both the world is external and prior, does not mean that the idea of an independent and prior reality can and should be extended on purely ontological considerations. The concept of reality that is absolutely independent of the entirety of human practice and prior to it does not make sense if we agree that neither can we be nor can we understand our being in any way other than from within the world, and that the world can neither be nor can it be understood by us differently than it is as *our* world. However, the fact that we undertake a realist commitment when we cognize within the subject/object opposition reveals the hermeneutic-ontological circle. It links being and our understanding of being: as our understanding requires (our) being so (our) being requires understanding. In traditional terms, the existence of the hermeneutic-ontological circle means that epistemology and ontology condition each other: metaphysics presupposes certain epistemological claims, and epistemology presupposes ontological prejudgments. Neither is absolutely fundamental and ungrounded.

This result, which can be recognized from the perspective of our fundamental ontology of being that we identify with human practice, allows us to answer a question posed by Kolakowski:

> If it is a chimerical hope for man to shed his own skin; if the world is given only as a world endowed with meaning, and meaning is the outcome of man's practical project; if man is unable to understand himself by placing himself in a premeaningful, prehuman world . . . the metaphysical inquiry and the epistemological inquiry are annulled at a stroke. Is this therefore the end of philosophy?[3]

For us, as well as for him, the answer is in the negative. Even if foundational ontology and epistemology are annulled philosophy does not cease to exist; it remains as an effort to understand both our being-within the subject/object opposition and our attempt to think it, an effort that is our very act of being. It is precisely in this sense that our study is philosophical.

NOTES

Introduction

1. There are authors who locate their views regarding science in metascience, science of science, psychology of science, etc. To the best of our knowledge, these fields are in the minority, and we will not discuss their achievements.

2. We do not refer here to studies of the foundations of physics or biology. Whether they are philosophical or not is not an issue for us.

3. What we have in mind are overviews, such as those offered by F. Suppe, "The Search for Philosophic Understanding of Scientific Theories," in *The Structure of Scientific Theories*, ed. F. Suppe (Chicago: University of Illinois Press, 1977), pp. 3–232, or M. Lynch, *Scientific Practice and Ordinary Action: Ethnomethodology and Social Studies of Science* (Cambridge: Cambridge University Press, 1993). In fact, we have both provided views of the central issues in the philosophy of science (J. E. McGuire, "Scientific Change: Perspectives and Proposals," in *Introduction to the Philosophy of Science: A Text by Members of the Department of the History and Philosophy of Science of the University of Pittsburgh*, by M. H. Salmon, J. Earman, C. Glymour, J. G. Lennox, P. Machamer, J. E. McGuire, J. D. Norton, W. C. Salmon, K. F. Schaffner [Englewood Cliffs, N.J.: Prentice Hall, 1992], pp. 132–78; B. Tuchańska, "The Methodological Problem of the Development of Science versus the Historical Problem of How Science Performs Its Social Functions," *Polish Sociological Bulletin* 3 [1980]: 5–24.)

4. J. Weinsheimer, *Gadamer's Hermeneutics: A Reading of Truth and Method* (New Haven: Yale University Press, 1985), p. 5.

5. H.-G. Gadamer, *Truth and Method*, trans. J. Weinsheimer, D. G. Marshall (New York: Crossroad, 1989), p. 552.

6. J. A. Coffa, *The Semantic Tradition from Kant to Carnap: To the Vienna Station*, ed. L. Wessels (Cambridge: Cambridge University Press, 1991), p. 233.

7. H.-J. Rheinberger, *Toward a History of Epistemic Things* (Stanford: Stanford University Press, 1997); P. Galison, *Image and Logic: A Material Culture of Microphysics* (Chicago: University of Chicago Press, 1997).

8. P. Heelan, "Hermeneutical Phenomenology and the Philosophy of Science," in *Gadamer and Hermeneutics: Science, Culture, Literature. Plato, Heidegger, Barthes, Ricoeur, Habermas, Derrida*, ed. H. J. Silverman (New York: Routledge, 1991), p. 213.

9. R. P. Crease, "The Play of Nature: Experimentation as Performance." In *Continental and Postmodern Perspectives in the Philosophy of Science*, B. E. Babich, D. B. Bergoffen, S. V. Glynn (Avebury: Aldershot, 1995), pp. 70–83; P. Heelan, "Natural Science as Hermeneutic of Instrumentation," *Philosophy of Science* 50 (1983): 182; P. Heelan, *Space Perception and the Philosophy of Science*

(Berkeley: University of California Press, 1983), p. 5; Heelan, "Hermeneutical Phenomenology," pp. 215–18; P. Heelan, "Why a Hermeneutical Philosophy of the Natural Sciences," *Man and World* 30 (1997): 274. J. J. Kockelmans, "Beyond Realism and Idealism: A Response to Patrick A. Heelan," in *Gadamer and Hermeneutics: Science, Culture, Literature. Plato, Heidegger, Barthes, Ricoeur, Habermas, Derrida,* ed. H. J. Silverman (New York: Routledge, 1991), pp. 229–30.

10. Heelan, "Hermeneutical Phenomenology," p. 225.

11. Ibid., p. 213.

12. E. Gilson, *Being and Some Philosophers,* 2d ed. (Toronto: Pontifical Institute of Mediaeval Studies, 1961), p. 113.

13. R. J. Bernstein, *Beyond Objectivism and Relativism: Science, Hermeneutics, and Praxis* (Philadelphia: University of Pennsylvania Press 1983), p. 123.

14. A note of caution is necessary here. We distinguish between *Fundamentalontologie* and *fundamentale Ontologie:* the first is Heidegger's term of art referring to his inquiry into the *existentialist conditions of human being.* Although we mobilize his ontology, we reject its existentialist slant. Our fundamental ontology *(fundamentale Ontologie)* is the enterprise of mapping out the *ontological conditions and structures of our sociocultural-historical being.* We owe the importance of making this distinction to Lorenz B. Puntel.

15. I. Kant, *Critique of Pure Reason,* trans. N. Kemp Smith (New York: St. Martin's Press, 1965), B75.

16. R. P. Crease, "Hermeneutics and the Natural Sciences: Introduction," *Man and World* 30 (1997): 262.

17. L. Kolakowski, *The Presence of Myth* (Chicago: University of Chicago Press, 1989), p. 7.

18. R. Campbell, *Truth and Historicity* (Oxford: Oxford University Press 1992), p. 412.

19. L. Kolakowski, "The Priest and the Jester," in *Toward a Marxist Humanism: Essays on the Left Today,* trans. J. Zielonko Peel (New York: Grove Press, 1968), p. 29.

20. Ibid., pp. 29, 24; also J. Rouse, *Knowledge and Power: Toward a Political Philosophy of Science* (Ithaca: Cornell University Press, 1987), p. viii.

21. Kolakowski, "The Priest," p. 34.

22. Ibid., p. 24.

23. Ibid., p. 25.

24. J. Grondin, "Hermeneutical Truth and Its Historical Presuppositions: A Possible Bridge between Analysis and Hermeneutics," in *Anti-Foundationalism and Practical Reasoning,* ed. E. Simpson (Edmonton: Academic Printing & Publishing, 1987), pp. 54–55.

25. Crease, "Hermeneutics and the Natural Sciences," p. 259; Heelan, "Hermeneutical Phenomenology," p. 213; Rouse, *Knowledge,* p. ix.

26. B. R. Wachterhauser, "Introduction: History and Language in Understanding," in *Hermeneutics and Modern Philosophy,* ed. B. R. Wachterhauser (Albany: SUNY Press, 1986), pp. 7–8.

27. Ibid., p. 8.

28. L. Kolakowski, "Looking for the Barbarians: The Illusions of Cultural Universalism," in L. Kolakowski, *Modernity on Endless Trial,* trans. A. Kolakowska (Chicago: University of Chicago Press, 1990), pp. 22–23.

29. L. Kolakowski, "In Praise of Inconsistency," in *Toward a Marxist Humanism: Essays on the Left Today*, trans. J. Zielonko Peel (New York: Grove Press, 1968), p. 214.

30. R. Rorty, "Solidarity or Objectivity?" in *Post-Analytic Philosophy*, ed. J. Rajchman and C. West (New York: Columbia University Press, 1985), pp. 11–12; R. Rorty, "Cosmopolitanism without Emancipation: A Response to Jean-François Lyotard," in *Philosophical Papers*. Vol. 1, *Objectivity, Relativism, and Truth* (Cambridge: Cambridge University Press, 1991), p. 214; Kolakowski, "Looking for the Barbarians," pp. 22–23.

31. H.-G. Gadamer, "The Universality of Hermeneutical Problem," in *The Hermeneutical Tradition: From Ast to Ricoeur*, ed. G. L. Ormiston and A. D. Schrift (Albany: SUNY Press, 1990), p. 154.

32. Kolakowski, "The Priest," p. 33.

33. P. K. Feyerabend, "Wittgenstein's *Philosophical Investigations*," in *Philosophical Papers*, vol. 2, *Problems of Empiricism* (Cambridge: Cambridge University Press, 1981), pp. 129–30. As Funkenstein used to say, one does not need to be a triangle to do geometry. But, of course, if one is a triangle, one does not do geometry, one *is* geometry.

34. R. K. Merton, "The Perspectives of Insider and Outsider," in R. K. Merton, *The Sociology of Science: Theoretical and Empirical Investigations*, ed. N. W. Storer (Chicago: University of Chicago Press, 1973), p. 105.

35. H.-G. Gadamer, *Reason in the Age of Science*, trans. F. G. Lawrance (Cambridge, Mass.: MIT Press, 1981), p. 161. Florian Znaniecki much earlier characterized this approach of a participant in cultural life as taking phenomena with the *humanistic coefficient;* i.e., reconstructing the experience of people who are dealing actively with the phenomena under study; F. Znaniecki, *The Social Role of the Man of Knowledge* (New York: Octagonal Books, 1965), p. 7.

36. C. Taylor, "Overcoming Epistemology," in *Philosophical Arguments* (Cambridge, Mass.: Harvard University Press, 1995), p. 14.

CHAPTER 1

1. Gadamer, *Truth*, p. 276.

2. H. M. Collins and S. Yearley, "Epistemological Chicken," in *Science as Practice and Culture*, ed. A. Pickering (Chicago: University of Chicago Press, 1992), p. 309.

3. K. M. Ford, C. Glymour, and P. J. Hayes, "Introduction," in *Android Epistemology*, ed. K. M. Ford et al. (Menlo Park and Cambridge, Mass.: AAAI Press and MIT Press, 1995), pp. xi, xii.

4. This attitude is expressed by Johnson-Laird, who accepts a broadly computational picture of how the brain works. P. N. Johnson-Laird, *Mental Models: Towards a Cognitive Science of Language, Inference, and Consciousness* (Cambridge, Mass.: Harvard University Press, 1983, p. 165). Also for Pinker, the mind is a naturally selected neural computer. S. Pinker, *How the Mind Works* (New York: W. W. Norton & Co., 1997), p. 521.

5. True, some philosophers take another tack. McDowell, for example, attempts to steer a middle way between subjective idealism and what he calls "bald naturalism." He argues that thought is creatively spontaneous, that reality

does not lie beyond the boundary of the conceptual, that every experience is already conceptualized, and that no gap exists between thinking and what is thought. Thus, in thinking truly, what is thought *is* what is the case. He rightly rejects the myth of the unconceptualized given, a thesis to which naturalism is committed. J. McDowell, *Mind and the World* (Cambridge, Mass.: Harvard University Press, 1996), pp. 67–75.

6. The notion of objectification must be handled with care. It encompasses three different dimensions that can overlap. Perhaps the term occurs most frequently in the context of the subject/object split. In this context, objectification refers to the fact that something, which is taken to be what a cognizing act is about, is made into an occurrent entity, an object, that is cognizable from a particular perspective. It is this sense of objectification to which the phrase "objectifying cognition" refers. In a more general sense *objectification* refers to the process of putting the results of mental activity into the public sphere, and most particularly into the mechanisms of communication that achieve that. Here objectification trades on the idea that human activity externalizes itself and is expressed in its artifacts. The third sense is prevalent in science. Here the term refers to the process of creating within cognitive communities new forms of intersubjectivity and new ways of dealing with experience, in terms of research programs, theories, practices, and technologies. It is clear that each of the three forms of objectification function simultaneously in scientific activity.

7. R. N. Giere, *Explaining Science: A Cognitive Approach* (Chicago: University of Chicago Press, 1988), p. 1.

8. Nickles in W. Callebaut, *Taking the Naturalistic Turn, or How Real Philosophy of Science Is Done* (Chicago: University of Chicago Press, 1993), p. 33.

9. Dupré sees "supervenience as a very weak, but still questionable, form of reductionism." His claim is as follows: "If a level of organization *H* supervenes on a lower level *L*, and if God knew the complete state of things at level *L*, then she could infer the complete description of the state of things at level *H*. This is just the kind of thing I mean by reductionism in principle" (J. Dupré, *The Disorder of Things: Metaphysical Foundations of the Disunity of Science* [Cambridge, Mass.: Harvard University Press, 1993], p. 97.)

10. S. Pinker, "Down to Darwin," an interview by H. Blume, *Boston Book Review* 4, no. 9 (1997): 5.

11. Pinker, *How the Mind,* pp. 525, 21–44.

12. For a stimulating account of evolution and the anthropology of the hunter-gatherers, see the work of the archaeologist Tim Ingold. He argues persuasively that the standard understanding of the hunter-gatherer scenario presupposes a distinction (contrary to its avowed presumptions) between biological evolution and the possibility of the development of a distinctive *human* history. The claim of evolutionary biology that we and our predecessors differ in degree rather than kind is sustained only by attributing the movement of history (from the Pleistocene hunting and gathering to science and civilization) to social and cultural processes different in kind, not merely in degree, from the process of evolution. Darwinianism can explain natural selection, but it cannot explain Darwin. (T. Ingold, "The Evolution of Society," in *Evolution: Society, Science and the Universe,* ed. A. C. Fabian [Cambridge: Cambridge University Press, 1998], pp. 89–98.)

13. D. Sperber, *Explaining Culture: A Naturalistic Approach* (Cambridge: Blackwell Publ., 1966), pp. 14, 34, 79, 97, 146–47.

14. Crease, "Hermeneutics and the Natural Sciences," pp. 259, 268.

15. Heelan, "Why a Hermeneutical Philosophy," p. 288.

16. Heelan, "Natural Science as Hermeneutic of Instrumentation," p. 187.

17. Crease, "Hermeneutics and the Natural Sciences," p. 259.

18. Heelan, "Why a Hermeneutical Philosophy," p. 279.

19. Ibid., p. 276.

20. Ibid., p. 277.

21. T. Kisiel, "A Hermeneutics of the Natural Science? The Debate Updated," *Man and World* 30 (1997): 336.

22. Crease, "The Play of Nature"; Heelan, "Natural Science as Hermeneutic of Instrumentation," pp. 185–200; Heelan, "Hermeneutical Phenomenology,"pp. 225–28; Heelan, "Why a Hermeneutical Philosophy," p. 278; Kisiel, "A Hermeneutics of the Natural," pp. 334–36.

23. Heelan, "Why a Hermeneutical," p. 278.

24. Ibid., p. 274.

25. R. J. Howard, *Three Faces of Hermeneutics: An Introduction to Current Theories of Understanding* (Berkeley: University of California Press, 1982), p. 131.

26. M. Foucault, *The Order of Things: An Archeology of the Human Sciences* (New York: Random House, 1973), p. 318.

27. Gadamer, *Truth*, pp. 443–45.

28. See the perceptive article of C. Taylor, "Heidegger, Language, and Ecology," in *Philosophical Arguments* (Cambridge, Mass: Harvard University Press. 1995), pp. 115–19.

29. Cf. P. Grice, *Studies in the Way of Words* (Cambridge, Mass.: Harvard University Press, 1989), pp. 20–30, 219; R. B. Brandom, *Making It Explicit: Reasoning, Representing, and Discursive Commitment* (Cambridge, Mass.: Harvard University Press 1994), pp. 145–53; D. Sperber and D. Wilson, *Relevance, Communication and Cognition* (Oxford: Blackwell, 1986), pp. 28–38, 60–64.

30. J. Richardson, *Existential Epistemology: A Heideggerian Critique of Cartesian Project* (Oxford: Clarendon Press, 1986), pp. 84, 111.

31. Cf. R. Rorty, *Philosophy and the Mirror of Nature* (Princeton, N.J.: Princeton University Press, 1979), pp. 80–81.

32. Cf. N. Rescher, *Process Metaphysics: An Introduction to Process Philosophy* (Albany: SUNY Press, 1996), p. 129.

33. In accepting an inferential view of concepts we do not deny that certain of our basic concepts are non-inferentially acquired.

34. P. Kitcher, *The Advancement of Science: Science without Legend, Objectivity without Illusions* (New York: Oxford University Press, 1993), pp. 303–87.

35. Ibid., pp. 127–77.

36. P. K. Machamer, "Kitcher and the Achievement of Science," *Philosophy and Phenomenological Research* 55, no. 3 (1995): 629–36.

37. B. Barnes, *About Science* (Oxford: Basil Blackwell, 1985), pp. 40, 82; T. S. Kuhn, *The Structure of Scientific Revolutions,* 2d ed. (Chicago: University of Chicago Press, 1970), pp. 24, 32–42; McGuire, "Scientific," pp. 145–51.

38. H. Longino, "Multiplying Subjects and the Diffusion of Power," *Journal of Philosophy* 88, no. 11 (1991): 667–69.

39. Kitcher, *The Advancement*, p. 389.

40. S. E. Toulmin, *Human Understanding*, vol. 1, *General Introduction* and *Part I* (Princeton, N.J.: Princeton University Press, 1972), pp. 158.

41. T. S. Kuhn, "Afterwords," in *World Changes: Thomas Kuhn and the Nature of Science*, ed. P. Horwitch (Cambridge, Mass.: MIT Press, 1993), p. 328. Contrary to Kuhn, we do not believe that help may come from the biological conception of evolution, delivering "important clues to the sense in which science is intrinsically a community activity" (p. 329). None of the theories of evolution contains intrinsically social concepts referring to relations and interactions among individuals (cf. M. Grene, "Evolution, 'Typology' and 'Population Thinking,'" *American Philosophical Quarterly* 27, no. (1990): 237–44; K. D. Knorr-Cetina, *The Manufacture of Knowledge: An Essay on the Constructivist and Contextual Nature of Science* (Oxford: Pergamon Press, 1981), p. 13. Consequently, adherents of the evolutionary approach to the development of science are forced to supplement their perspectives with sociological categories. Toulmin introduces the (social) environment of science, functioning as a constraint on the selection of scientific concepts but simultaneously subject to evolutionary variation and selection (Toulmin, *Human*, pp. 216, 350–53). Hull conceives the mechanism of selection to be a social (ahistorical) mechanism working within small research groups. (D. Hull, *Science as a Process: An Evolutionary Account of the Social and Conceptual Development of Science* [Chicago: University of Chicago Press, 1988], pp. 15, 433; E. Gatens-Robinson, "Why Falsification Is the Wrong Paradigm for Evolutionary Epistemology: An Analysis of Hull's Selection Theory," *Philosophy of Science* 60 [1993]: 544–47, 553).

42. H. Longino, "The Fate of Knowledge in Social Theories of Science," in *Socializing Epistemology: The Social Dimensions of Knowledge,* ed. F. F. Schmitt (Lanham: Rowman and Littlefield, 1994), p. 142.

43. Ibid., pp. 142–43.

44. Cf. Giere in Callebaut, *Taking*, p. 182.

45. Knorr-Cetina in Callebaut, *Taking*, p. 183.

46. S. Woolgar, *Science: The Very Idea* (Chichester: Ellis Harwood Ltd. Publ.; London: Tavistock Publ. 1988), p. 89.

47. Ibid.

48. Cf. D. E. Chubin and S. Restivo, "The 'Mooting' of Science Studies: Research Programmes and Science Policy," in *Science Observed: Perspectives on the Social Study of Science* (London: SAGE Publ., 1983), pp. 54–59; H. M. Collins and S. Restivo, "Development, Diversity, and Conflict in the Sociology of Knowledge," *Sociological Quarterly* 24 (1983): 185; A. Pickering, "From Science as Knowledge to Science as Practice," in *Science as Practice and Culture,* ed. A. Pickering (Chicago: University of Chicago Press, 1992), p. 1; S. Woolgar and M. Ashmore, "The Next Step: An Introduction to the Reflexive Project," in *Knowledge and Reflexivity: New Frontiers in the Sociology of Knowledge* (London: SAGE Publ., 1988), p. 7.

49. Collins and Restivo, "Development," p. 196.

50. K. D. Knorr-Cetina, "The Ethnographic Study of Scientific Work: Towards a Constructivist Interpretation of Science," in *Science Observed: Perspectives on the Social Study of Science* (London: SAGE Publ., 1983), p. 117.

51. M. Mulkay and N. Gilbert, "Theory Choice," in M. Mulkay, *Sociology of Science: A Sociological Pilgrimage* (Philadelphia: Open University Press, 1991), p. 152.

52. Knorr-Cetina, "The Ethnographic," p. 117.

53. B. Latour and S. Woolgar, *Laboratory Life: The Construction of Scientific Facts,* 2d ed. (Princeton, N.J.: Princeton University Press, 1986), p. 151.

54. B. Tuchańska, "The Problem of Cognition as an Ontological Question," *Acta Universitatis Lodziensis, Folia Philosophica* 6 (1988), p. 38.

55. Knorr-Cetina, *The Manufacture,* p. 14. Because of the importance that constructivists attach to discourse and the generation of documents, they are accused of remaining inside a theory-dominated view of science (T. Lenoir, "Practice, Reason, Context: The Dialogue between Theory and Experiment," *Science in Context* 2, no. 1 [1988]: 7), of collapsing social relations into discourse (S. Aronowitz, *Science as Power: Discourse and Ideology in Modern Society* [Basingstoke: Macmillan Press, 1988], p. 298), or equating "natural science research with the sorts of literary and interpretive activities that social scientists are accustomed to doing" (Lynch, *Scientific Practice,* p. 95), even though— as Lynch admits—they mention equipment and skills. Such criticisms seem to miss the difference between the Anglo-American reading of the concept of discourse and the Continental (influencing such constructivists as Knorr-Cetina, Latour, or Woolgar). Continental philosophy, Foucault for instance, holds that "*praxis* cannot exist outside discourse" (S. Woolgar, "On the Alleged Distinction between Discourse and *Praxis,*" *Social Studies of Science* 16 (1986): 313). It is true that in effect discourse becomes the core of praxis and, paradoxically, objects, machines, and circumstances are treated as texts because "they too manifest a discourse" (p. 313). It must, however, be realized that from the Continental perspective, an attempt at purifying and separating discourse from praxis (typical of Anglo-American philosophy) leads to equally unacceptable views, such as the idea of (cognitive) discourse that does not have praxis as its core or the concept of objective things—i.e., things without any traces of (cognitive) discourse by which to relate them.

56. K. D. Knorr-Cetina, M. Mulkay, "Emerging Principles in Social Studies of Science," in *Science Observed: Perspectives on the Social Study of Science,* ed. K. D. Knorr-Cetina, M. Mulkay (London: SAGE Publ., 1983), p. 7.

57. Knorr-Cetina, *The Manufacture,* p. 20.

58. Lynch, *Scientific Practice,* p. 92.

59. G. Freudenthal, "The Role of Shared Knowledge in Science: The Failure of the Constructivist Programme in the Sociology of Science," *Social Studies of Science* 14 (1984); G. Gutting, "Conceptual Structures and Scientific Change," *Studies in History and Philosophy of Science* 4 (1973): 210–21; K. Lehrer and C. Wagner, *Rational Consensus in Science and Society: A Philosophical and Mathematical Study* (Dordrecht: Reidel, 1981), pp. 3–8; H. Longino, "Essential Tensions—Phase Two: Feminist, Philosophical, and Social Studies of Science," in *The Social Dimensions of Science,* ed. E. McMullin (Notre Dame: University of Notre Dame Press, 1992), pp. 207–12.

60. Knorr-Cetina, "The Ethnographic," p. 133.

61. N. Elias, *The Society of Individuals,* ed. M. Schröter, trans. E. Jephcott (Oxford: Basil Blackwell, 1991), p. 25. There are, of course, sociological stud-

ies in which consensus is explained by reference to social mechanisms of violence, coercion, education, authority, etc. (e.g., A. Zybertowicz, *Przemoc i poznanie: Studium z nieklasycznej socjologii wiedzy* [Violence and cognition: A study in nonclassical sociology of science]. [Toruń: UMK, 1995]).

62. Knorr-Cetina, *The Manufacture*, pp. 2–3.

63. Woolgar, *Science*, pp. 55–66.

64. Ibid., pp. 60, 65.

65. Latour, Woolgar, *Laboratory Life*, pp. 180–82; Knorr-Cetina in Callebaut, *Taking*, p. 184.

66. B. Latour, "One More Turn after the Social Turn," in *The Social Dimensions of Science*, ed. E. McMullin (Notre Dame: University of Notre Dame Press, 1992), p. 280.

67. Latour, Woolgar, *Laboratory Life*, pp. 235–38.

68. Knorr-Cetina, *The Manufacture*; K. Knorr-Cetina, "The Fabrication of Facts: Toward a Microsociology of Scientific Knowledge," *Society and Knowledge: Contemporary Perspectives in the Sociology of Knowledge*, ed. N. Stehr and V. Meja (New Brunswick: Transaction Books, 1984), p. 240.

69. Knorr-Cetina, *The Manufacture*, pp. 131–32.

70. B. Latour, *Science in Action: How to Follow Scientists and Engineers through Society* (Cambridge, Mass.: Harvard University Press, 1987), pp. 174, 180; M. Callon, B. Latour, "Don't Throw the Baby Out with the Bath School! A Reply to Collins and Yearly," in *Science as Practice and Culture*, ed. A. Pickering (Chicago: University of Chicago Press, 1992), p. 347.

71. B. Latour, *We Have Never Been Modern* (Cambridge, Mass.: Harvard University Press, 1993), p. 121.

72. Latour, *Science*, p. 198; B. Latour, "Give Me a Laboratory and I Will Raise the World," in *Science Observed: Perspectives on the Social Study of Science*, ed. K. D. Knorr-Cetina and M. Mulkay (London: SAGE Publ., 1983), p. 146.

73. Latour, *Science*, p. 13; H. M. Collins, "An Empirical Relativist Programme in the Sociology of Scientific Knowledge," in *Science Observed: Perspectives on the Social Studies of Science*, ed. K. D. Knorr-Cetina and M. Mulkay (London: SAGE Publ., 1983), p. 91; Collins and Yearley, "Epistemological Chicken"; Latour and Woolgar, *Laboratory Life*, p. 236; R. P. Crease, "The Play of Nature," p. 83.

74. Knorr-Cetina, *The Manufacture*, p. 23.

75. Rouse, *Knowledge*, p. 122.

76. Lynch, *Scientific Practice*, p. 269.

77. Heelan, "Hermeneutical Phenomenology," p. 225.

78. Latour, *We Have Never*, p. 95.

79. Ibid., p. 6.

80. Latour, "One More Turn," p. 283.

81. Latour, *We Have Never*, p. 6.

82. Ibid., p. 129.

83. Ibid., p. 135.

84. This attitude is evident in the Strong Programme with its aim of studying the distribution of beliefs in society and—most importantly—looking for their causal social determination (D. Bloor, *Knowledge and Social Imagery* [Lon-

don: Routledge and Kegan Paul, 1976], p. 3). Advocates of the Strong Programme believe that this makes the sociology of science "absolutely identical in its procedure with that of any other science" (p. 17).

85. A. Pickering, *The Mangle of Practice: Time, Agency and Science* (Chicago: University of Chicago Press, 1995), pp. 1-27, 179-86, 208-12.

86. A. C. Crombie, *Styles of Thinking in the European Tradition* (London: Duckworth, 1994), 1:7.

87. I. Hacking, "Statistical Language, Statistical Truth, and Statistical Reason: The Self-Authentification of a Style of Scientific Reasoning," in *The Social Dimensions of Science,* ed. E. McMullin (Notre Dame: University of Notre Dame Press, 1992), p. 143.

88. Kuhn, T. S., "The Road Since Structure," in *PSA 1990: Proceedings of the 1990 Biennial Meeting of the Philosophy of Science Association,* ed. A. Fine, M. Forbes, L. Wessels (East Lansing: PSA, 1991), 2:9. Kuhn's position has deeper affinities with Kantianism. Indeed, Kuhn tells us here that the view he is developing is "a sort of post-Darwinian Kantianism. Like the Kantian categories, the lexicon supplies preconditions of possible experience. But lexical categories, unlike their Kantian forbears, can and do change, both with time and the passage from one community to another." He goes on to add that "[u]nderlying all these processes of differentiation and change, there must, of course, be something permanent, fixed, and stable. But unlike Kant's *Ding an sich*, it is ineffable, undescribable, undiscussible" ("The Road," p. 12). For an insightful discussion of Kuhn's late philosophy see G. Irzik and T. Grunberg, "Whorfian Variations on Kantian Themes: Kuhn's Linguistic Turn and Its Relation to Carnap's Philosophy," *Studies in History and Philosophy of Science* 29A (1998): 207-21.

89. M. Foucault, *The Archeology of Knowledge and the Discourse on Language,* ed. A. M. S. Smith (New York: Pantheon Books, 1972), pp. 73-75.

90. Ibid., p. 128.

91. Foucault, *The Order,* pp. xiii-xiv.

92. M. Foucault, "Questions of Method: An Interview with Michel Foucault," in *After Philosophy: End or Transformation?* ed. K. Baynes, J. Bohman, and T. McCarthy (Cambridge, Mass.: MIT Press, 1987), pp. 107-8.

93. Foucault, *The Archeology,* pp. 7, 125-28, 138-40.

94. Ibid., pp. 31-76. In this context Foucault makes use of the French distinction between *savoir* and *connaissance*. Historically induced preconditions, necessary for the development of a discursive formation, constitute *savoir,* whereas an established discursive formation that embodies organized, positive knowledge is *connaissance* (pp. 181-84).

95. M. Foucault, "Nietzsche, Genealogy, History," in *The Foucault Reader,* ed. P. Rabinow (New York: Pantheon, 1984), pp. 78-79; L. Shiner, "Reading Foucault: Anti-Method and the Genealogy of Power-Knowledge," *History and Theory* 21 (1982): 387.

96. M. Biagioli, *Galileo Courtier: The Practice of Science in the Culture of Absolutism* (Chicago: University of Chicago Press 1993), p. 18.

97. R. K. Merton, *Science, Technology, and Society in Seventeenth Century England* (New York: Harper and Row, 1970), p. 136.

98. B. Shapiro, *Probability and Certainty in Seventeenth-Century England: The Relationship between Natural Science, Religion, History, Law and Literature* (Princeton: Princeton University Press, 1983); C. Webster, *The Great Instauration: Science, Medicine, and Reform 1626–1660* (New York: Holmes and Meir, 1975).

99. S. Shapin, S. Schaffer, *Leviathan and the Air-Pump: Hobbes, Boyle, and the Experimental Life* (Princeton: Princeton University Press, 1985), p. 15.

100. Ibid., p. 343.

101. F. Ringer, "The Intellectual Field, Intellectual History, and the Sociology of Knowledge," *Theory and Society* 19 (1990): 270; P. Bourdieu, "Intellectual Field and Creative Project," *Social Science Information* 8 (1969): 89–119; P. Bourdieu, "The Genesis of the Concepts of *Habitus* and of *Field*," *Sociocriticism* 2 (1985): 11–24.

102. Ringer, "The Intellectual," p. 270.

103. Ibid., pp. 271, 274, 277.

104. P. Bourdieu, *The Logic of Practice,* trans. R. Nice (Stanford: Stanford University Press 1990), p. 56.

105. Ibid.

106. P. K. Machamer, "Selection, System and Historiography," in *Trends in the Historiography of Science,* ed. K. Gavroglu et. al. (Dordrecht: Reidel Publ. Co., 1994), pp. 149–60.

107. Aristotle, "Metaphysica," in *The Works of Aristotle,* ed. W. D. Ross, vol. 8 (Oxford: Clarendon Press 1960), Z 1028a 10ff, Z 4 1037a17ff.

108. M. Frede, "Categories in Aristotle," in M. Frede, *Essays in Ancient Philosophy* (Minneapolis: University of Minnesota Press, 1987), pp. 43–48.

109. Ibid., pp. 45–46.

110. Substances, events, and processes relate to time and space in two basic ways: as continuants or occurrents (C. D. Broad, *Examination of McTaggart's Philosophy* [Cambridge: Cambridge University Press 1933], 1:138ff; W. E. Johnson, *Logic: In Three Parts,* pt. 1, *Propositions and Relations* [New York: Dover Publications, 1964], pp. 199–201). *Occurrents,* which are typically processes, events, and states, have spatial and temporal parts; *continuants,* such as things, substances, or material objects, are extended in three dimensions (and consequently take up spatial parts) and persist through time as a whole without temporal extension (cf. Broad, *Examination,* pp. 142–43; P. Simons, *Parts: A Study in Ontology* [Oxford: Clarendon Press; New York: Oxford University Press, 1987], pp. 117–19, 129–37).

111. Foucault, *The Archeology,* p. 231.

112. M. Foucault, "Theatrum Philosophicum," in *Language, Counter-Memory, Practice: Selected Essays and Interviews,* ed. D. F. Bouchard (ed.) and trans. D. F. Bouchard, S. Simon (Ithaca: Cornell University Press, 1977), p. 192.

113. P. Ricoeur, *The Reality of the Historical Past* (Milwaukee: Marquette University Press, 1984), p. 24.

114. J. Seibt, "Existence in Time: From Substance to Process," in *Perspectives on Time,* ed. J. Faye, U. Scheffler, M. Urchs (Dordrecht: Kluwer, 1996), p. 170.

115. F. Nietzsche, *The Will to Power,* ed. W. Kaufmann and trans. W. Kaufmann, R. J. Hollingdale (New York: Vintage Books, 1968), n. 484. In references to *The Will to Power,* we give the in-text numbers designating entries in the Kauf-

mann edition. One must be careful in using Nietzsche's notebook published under the title *The Will to Power* since it was posthumously subjected to selection and published. Nevertheless, the passages we quote are perfectly in keeping with the tenor of Nietzsche's thought as evident in his late writing.

116. Ibid., 485.

117. Ibid., 477, 550.

118. Ibid., 481, 484.

119. Ibid., 531, 550–51. Nietzsche explicitly adds the notion of intention to his list of invisible substrata. "I notice something and seek a reason for it: this means originally: I seek an intention in it, and above all someone who has intentions, a subject, a doer: every event a deed—formerly one saw intentions in all events, this is our oldest habit" (550).

120. Broad, *Examination*, p. 156; Rescher, *Process*, p. 57.

121. Seibt, "Existence," p. 170. Seibt's study is invaluable.

122. Rescher, *Process*, p. 56.

123. Emmet (herself an advocate of process-ontology) claims that processes are not sufficient and are in need of things to sustain them and carry them forward. She is merely repeating the Aristotelian notion that if *sustain* and *carry forward* are words suggesting activities, they must be viewed as constituents of things capable of acting and being acted upon. The point is only compelling to one who already accepts the structure of substance-ontology (D. Emmet, *The Passage of Nature* [Philadelphia: Temple University Press, 1992], p. 9). Nor are there compelling grounds for holding that the efficacy of dispositional properties presuppose a nonidentical, causal substratum inherent in things. The notion of a "bare disposition," viewed as both the source and subject of its own displays, is intuitively plausible. Processes serve this intuition well. For a strong account of dispositions all the way down, see H. Mellor, "In Defense of Dispositions," *Philosophical Review* 83 (1974): 157–81. Mellor argues that the necessary basis for dispositional ascription between displays need only be another disposition.

124. Ricoeur, *The Reality*, pp. 14–24.

125. J. Weinsheimer, *Gadamer's Hermeneutics*, p. 14.

126. C. B. Guignon, "Pragmatism or Hermeneutics? Epistemology After Foundationalism," in *The Interpretative Turn: Philosophy, Science, Culture*, ed. D. R. Hiley, J. F. Bohman, R. Shusterman (Cornell: Cornell University Press, 1991), p. 100.

127. Unlike some translators and commentators, who capitalize the Heideggerian term *being* we will use *being* for its substantive (predicative) use—i.e., when it means "existing"—and *Being* or *beings* for its substantial use. This convention will indicate the difference between Parmenidean Being and the Heideggerian concept of being that in fact carries the meaning of "ising" because "*Sein*, the verbal noun for 'being,' is with respect to its syntactic base a *process*, an activity, a 'being-there'" (G. Steiner, *Martin Heidegger* [Chicago: University of Chicago Press, 1991], p. 46). We welcome Stambaugh's translation of *Sein und Zeit*, because she concurs with our practice and also rejects the convention of capitalizing *being* (Cf. J. Stambaugh, "Translator's Preface," in M. Heidegger, *Being and Time*, trans. J. Stambaugh [Albany: SUNY Press, 1996], p. xiv).

128. For discussions of why ontic categories of logic and standard conceptions of entities do not fit Heidegger's thought, see S. Mulhall, *Heidegger and Being and Time* (New York: Routledge, 1996), pp. 8–12; Steiner, *Martin Heidegger*, pp. xviii–ix, 34–59; M. Friedman, "Overcoming Metaphysics: Carnap and Heidegger," in *Origins of Logical Empiricism*, ed. R. Giere and A. Richardson, *Minnesota Studies*, vol. 17 (Minneapolis: University of Minnesota Press, 1996).

129. M. Heidegger, *Being and Time*, trans. J. Stambaugh (Albany: SUNY Press, 1996), 2–4. In the case of this work we do not give the page number but rather—following the usual convention—the pagination of the seventh German edition of 1953, and we refer mainly to Stambaugh's translation of *Sein und Zeit*, though in several places we prefer MacQuarrie and Robinson's translation.

130. Husserlian phenomenology attributes absolute priority to transcendental subjectivity—i.e., to "universal intersubjectivity, into which all objectivity, everything that exists at all, is resolved" (E. Husserl, *The Crisis of European Sciences and Transcendental Phenomenology: An Introduction to Phenomenological Philosophy*, trans. D. Carr [Evanston: Northwestern University Press, 1970], pp. 179, 153). Husserl's attempt at grounding the world in universal intersubjectivity meets with a special difficulty: the paradox of "humanity as world-constituting subjectivity and yet as incorporated in the world itself," which, in turn, leads to the problem of the constitution of intersubjectivity (p. 182).

131. K. Rosner, *Hermeneutyka jako krytyka kultury: Heidegger, Gadamer, Ricoeur* [Hermeneutics as a criticism of culture: Heidegger, Gadamer, Ricoeur] (Warszawa: PIW, 1991), p. 25.

132. Heidegger, *Being*, 35–36; H. L. Dreyfus, J. Haugeland, "Husserl and Heidegger: Philosophy's Last Stand," in *Heidegger and Modern Philosophy: Critical Essays*, ed. M. Murray (New Haven: Yale University Press, 1978), pp. 228–31; Gilson, *Being*, p. 122.

133. Heidegger, *Being*, 16–17.

134. Ibid., 11–4, 38, 42.

135. Mulhall, *Heidegger*, p. 15.

136. Heidegger, *Being*, 50.

137. J. D. Caputo, "Hermeneutics as the Recovery of Man," in *Hermeneutics and Modern Philosophy*, ed. B. R. Wachterhauser (Albany: SUNY Press, 1986), p. 434.

138. H. L. Dreyfus, *Being-in-the-World: A Commentary on Heidegger's Being and Time, Division I* (Cambridge, Mass.: MIT Press, 1991), p. 35.

139. Heidegger, *Being*, 11–15.

140. Mulhall, *Heidegger*, p. 15; Heidegger, *Being*, 119. The Da of Dasein means both "here" and "there" or even "where," but in no case should it be confused with spatial localization, in reference to which entities are objectively present (56). Also applying the concept of presence (being-present-here or being-visible) to Dasein is inappropriate.

141. Heidegger, *Being*, 12.

142. Gadamer, *Truth*, p. 259.

143. Heidegger, *Being*, 143.

144. Ibid., 66–67.

145. T. Kisiel, "A Hermeneutics of the Natural Science?" p. 330.

146. Ibid., pp. 11, 60-62, 363; M. Heidegger, *History of the Concept of Time: Prolegomena,* trans. T. Kisiel (Bloomington: Indiana University Press, 1985), pp. 160, 162, 167; M. Heidegger, *The End of Philosophy,* trans. J. Stambaugh (New York: Harper and Row, 1973), p. 29.

147. Dreyfus, *Being-in-the-World,* pp. 49-51, 69.

148. Heidegger, *Being,* 61, 357.

149. Ibid., 68-72. For Heidegger, pragmata cannot have an essence, since what they are wholly depends on their function as determined by the totality of equipment and on their actual existence (Dreyfus and Haugeland, "Husserl," p. 225).

150. Heidegger, *Being,* 64-65, 71.

151. Ibid., 65.

152. J. DiCenso, *Hermeneutics and the Disclosure of Truth: A Study in the Work of Heidegger, Gadamer, and Ricoeur* (Charlottesville: University Press of Virginia, 1990), p. 49.

153. Heidegger, *Being,* 374, 382; W. Brock, "An Account of 'Being and Time,'" in M. Heidegger, *Existence and Being* (Washington, D.C.: Gateway Editions, 1988), p. 94. Heidegger does not understand death naturalistically, but in a phenomenological way—i.e., as a situation in which "Da-sein stands before itself in its ownmost potentiality-of-being" (Heidegger, *Being,* 250). Death is not objectively present; it is an experience that any person faces, which establishes its life as a totality belonging uniquely to this person (cf. P. Hoffman, "Death, Time, History: Division II of *Being and Time,*" in *The Cambridge Companion to Heidegger,* ed. C. B. Guignon [Cambridge: Cambridge University Press, 1993], p. 199).

154. Dreyfus, *Being-in-the-World,* p. 244; Heidegger, *Being,* 337-49.

155. Heidegger, *Being,* 37-38.

156. Caputo, "Hermeneutics," p. 423.

157. Weinsheimer, *Gadamer's Hermeneutics,* pp. 10-11.

158. Gadamer, *Truth,* p. xxiii.

159. R. Wiehl, "Heidegger, Hermeneutics, and Ontology," in *Hermeneutics and Modern Philosophy,* ed. B. R. Wachterhauser (Albany: SUNY Press, 1986), pp. 468, 470.

160. Gadamer, *Truth,* pp. 265, 295, 299; H.-G. Gadamer, "Rhetoric, Hermeneutics, and the Critique of Ideology: Metacritical Comments on *Truth and Method,*" in *The Hermeneutics Reader: Texts of the German Tradition from the Enlightenment to the Present,* ed. K. Mueller-Vollmer (New York: Continuum, 1985), p. 274; Weinsheimer, *Gadamer's Hermeneutics,* p. 5.

161. S. J. Hekman, *Hermeneutics and the Sociology of Knowledge* (Cambridge: Polity Press, 1986), p. 95. Also Gadamer, "Rhetoric," p. 286; T. Kisiel, "The Happening of Tradition: The Hermeneutics of Gadamer and Heidegger," in *Hermeneutics and Praxis,* ed. R. Hollinger (Notre Dame: University of Notre Dame Press. 1985), p. 24.

162. H.-G. Gadamer, "Text and Interpretation," in *Hermeneutics and Modern Philosophy,* ed. B. R. Wachterhauser (Albany: SUNY Press, 1986), p. 388.

163. Gadamer, *Truth,* pp. xxi, 261.

164. Ibid., p. xxiv.

165. Ibid., pp. 290, 295.

166. Ibid., p. xxxvii.

167. Ibid., pp. 378, 443; Gadamer, "Rhetoric," p. 284.

168. Weinsheimer, *Gadamer's Hermeneutics,* p. 249; Gadamer, *Truth,* p. 443.

169. Gadamer, *Truth,* pp. 443, 447.

170. Weinsheimer, *Gadamer's Hermeneutics,* pp. 11–12; J. D. Caputo, "The Thought of Being and the Conversation of Mankind: The Case of Heidegger and Rorty," in *Hermeneutics and Praxis,* ed. R. Hollinger (Notre Dame: University of Notre Dame Press, 1985), p. 253.

171. Kisiel, "The Happening," p. 16.

172. Gadamer, *Truth,* p. 302; Bernstein, *Beyond,* p. 143; Weinsheimer, *Gadamer's Hermeneutics,* p. 39; Westphal, M., "Hegel and Gadamer," in *Hermeneutics and Modern Philosophy,* ed. B. R. Wachterhauser (Albany: SUNY Press, 1986), p. 80.

173. Weinsheimer, *Gadamer's Hermeneutics,* p. 39.

174. Gadamer, *Truth,* p. 298.

175. Ibid., pp. 304, 302.

176. Ibid., pp. 304, 306.

177. Gadamer, "Rhetoric," p. 282.

178. Ibid., p. 282.

179. Gadamer, *Truth,* pp. 276–77.

180. Ibid., p. 300; DiCenso, *Hermeneutics,* p. 80.

181. Gadamer, *Truth,* p. 301.

182. Ibid., p. xxxiv.

183. Ibid., p. 299.

184. Ibid., pp. 293, 282, 306.

185. Ibid., p. 293; Bernstein, *Beyond,* p. 137.

186. Gadamer, *Truth,* p. 293.

187. Ibid., p. 295.

188. Bernstein, *Beyond,* p. 140.

189. Ibid., p. 142.

190. Gadamer, *Truth,* pp. 297–98.

191. R. Bernasconi, *Heidegger in Question: The Art of Existing* (New Jersey, N.J.: Humanities Press, 1993), p. 175.

192. Ibid., p. 293; Bernstein, *Beyond,* pp. 140–42; D. Misgeld, "On Gadamer's Hermeneutics," in *Hermeneutics and Praxis,"* ed. R. Hollinger (Notre Dame: University of Notre Dame Press, 1985), p. 149.

193. Gadamer, *Truth,* p. 281.

194. Bernstein, *Beyond,* p. 139.

195. Gadamer, *Truth,* p. 295.

196. Bernstein, *Beyond,* p. 137; G. Warnke, *Gadamer: Hermeneutics, Tradition and Reason* (Cambridge: Polity Press, 1987), p. 87.

197. Gadamer, *Truth,* pp. 4–5.

198. Misgeld, "On Gadamer's Hermeneutics," p. 156.

199. Gadamer, *Truth,* pp. 284–85, 261, 528.

200. Ibid., pp. xxxviii, 327.

201. Ibid., p. 397.

202. J. Grondin, "Hermeneutics and Relativism," in *Festivals of Interpretation: Essays on Hans-Georg Gadamer's Work*, ed. K. Wright (Albany: SUNY Press, 1990), p. 47.

203. H.-G. Gadamer, "Reply to My Critics," in *The Hermeneutical Tradition: From Ast to Ricoeur*, ed. G. L. Ormiston and A. D. Schrift (Albany: SUNY Press, 1990), p. 283.

204. D. C. Hoy, "Is Hermeneutics Ethnocentric?" In *The Interpretative Turn: Philosophy, Science, Culture*, ed. D. R. Hiley, J. F. Bohman, R. Shusterman (Ithaca: Cornell University Press, 1991), p. 163; K. Jaspers, *Philosophy of Existence*, trans. R. F. Grabau (Philadelphia: University of Pennsylvania Press, 1971), p. 17.

205. Gadamer, *Truth*, pp. 528, 296–97.

206. Hoy, "Is Hermeneutics," p. 159; D. Ingram, "Hermeneutics and Truth," in *Hermeneutics and Praxis*, ed. R. Hollinger (Notre Dame: University of Notre Dame Press, 1985), pp. 36–37; T. Rockmore, "Epistemology as Hermeneutics," *Monist* 73 (1990): 115–33; Warnke, *Gadamer*, p. 77.

207. Grondin, "Hermeneutics," p. 46; Ingram, "Hermeneutics," p. 42; Warnke, *Gadamer*, pp. 82–87.

208. Gadamer, *Truth*, pp. 397, 529.

209. Ibid., p. 397; Grondin, "Hermeneutics," p. 59.

210. Ingram, "Hermeneutics," pp. 40, 44.

211. Kisiel, "The Happening," p. 8.

212. Gadamer, *Truth*, p. 265.

CHAPTER 2

1. This assumption may be the reason why he, in fact, did not conceive his ontology as a regional ontology, but as ultimate and fundamental.

2. Cahoone, L. E., *The Dilemma of Modernity: Philosophy, Culture, and Anti-Culture* (Albany: SUNY Press 1988), pp. 156, 173.

3. As Vogel emphasizes, the subject/object dichotomy is presupposed not only in the view of the human being as an observer standing against the objective world, but also in the ethical view of a worldless ego standing against the axiologically (in particular, morally) neutral world and freely projecting values on value-free facts (L. Vogel, *The Fragile "We": Ethical Implications of Heidegger's "Being and Time"* [Evanston, Ill.: Northwestern University Press, 1994], p. 43). Heidegger rejects both views of human beings, and in this sense he is neither a metaphysical rationalist nor a metaphysical existentialist of the Sartrean type. We agree with Vogel that Heidegger's monadic view of Dasein leads regrettably to moral nihilism. But we also agree that his fundamental ontology can be reinterpreted to embrace a genuine morality articulating our being-with-others. The primordiality of our being-together-in-the-world is also key for understanding human cognition, a position we argue in detail.

4. Cf. Z. Bauman, *Hermeneutics and Social Science: Approaches to Understanding* (Aldershot: Gregg Revivals 1992), p. 169; J. Habermas, *The Philosophi-*

cal Discourse of Modernity: Twelve Lectures, trans. F. G. Lawrence (Cambridge, Mass.: MIT Press, 1993), pp. 149–50; E. Levinas, *Totality and Infinity: An Essay on Exteriority,* trans. A. Lingis (Pittsburgh: Duqucsne University Press, 1969), pp. 45, 77; J.-P. Sartre, *Being and Nothingness* (New York: Gramercy Books, 1994), pp. 244–47; Vogel, *The Fragile,* p. 7.

5. B. W. Ballard, *The Role of Mood in Heidegger's Ontology* (Lanham: University of America Press 1991), pp. 88–91; R. B. Brandom, "Heidegger's Categories in *Being and Time,*" *Monist* 66 (1983): 387–409; Dreyfus, *Being-in-the-World,* pp. 158–62; C. B. Guignon, *Heidegger and the Problem of Knowledge* (Indianapolis: Hackett Publ. Co., 1983), pp. 86–87, 103–15; J. Haugeland, "Heidegger on Being a Person," *Noûs* 16 (1982).

6. Levinas, *Totality,* pp. 45–48. For Levinas, Heidegger has relativized the Other to the subjectivity of Dasein because he grants priority to being over beings and declares that the understanding of being is constitutive of being itself (p. 45).

7. R. J. Bernstein, "Heidegger's Silence: *Ethos* and Technology," in *The New Constellation: The Ethical-Political Horizons of Modernity/Postmodernity* (Cambridge, Mass.: MIT Press, 1993), pp. 124–25; E. L. Fackenheim, *To Mend the World: Foundations of Future Jewish Thought* (New York: Schocken Books, 1982), p. 153; C. Fynsk, *Heidegger: Thought and Historicity* (Ithaca: Cornell University Press, 1986), pp. 28–31; M. Grene, *Dreadful Freedom: A Critique of Existentialism* (Chicago: University of Chicago Press, 1948), p. 69; Habermas, *The Philosophical,* p. 150; Hoffman, "Death," p. 198; G. Lukács, "Existentialism or Marxism?" in *Existentialism versus Marxism: Conflicting Views on Humanism,* ed. G. Novack (New York: A Delta Book, 1966), pp. 138–39; J. Taminiaux, *Heidegger and the Project of Fundamental Ontology* (Albany: SUNY Press, 1991), p. 131; J.-P. Sartre, *Being,* p. 247; K. Wright, "Gadamer: The Speculative Structure of Language," in *Hermeneutics and Modern Philosophy,* ed. B. R. Wachterhauser (Albany: SUNY Press, 1986), p. 198.

8. Heidegger, *Being,* 117–25; Heidegger, *History,* pp. 237–43.

9. Bauman, *Hermeneutics,* p. 169.

10. Habermas, *The Philosophical,* p. 149.

11. Guignon, *Heidegger,* p. 110. Heidegger's position, characterized in this way, seems almost undistinguishable from Marx's position, according to which the individual is a residue of social relations.

12. The concept of *das Man,* of anyone, of an anonymous average human being (misleadingly translated as *the they*), does not change anything. Heidegger introduces it to talk about the "content" of Dasein as thrown into everyday life. *Das Man* does not have—from the ontological point of view—its own way of existence. It is merely an ontological notion that refers to the existential structure of Dasein and allows Heidegger to make a distinction between they-self, an inauthentic self of everyday Dasein absorbed in its world, and authentic self, "the self which has explicitly grasped itself" (Heidegger, *Being,* 129).

13. Brandom, "Heidegger's Categories," p. 397.

14. Guignon, *Heidegger,* p. 93.

15. Sartre, *Being,* p. 247; Heidegger, *Being,* 118.

16. M. Heidegger, *Being and Time*, ed. J. MacQuarrie and trans. E. Robinson (New York: Harper and Row, 1962), 384.

17. M. Okrent, *Heidegger's Pragmatism: Understanding, Being, and the Critique of Metaphysics* (Ithaca: Cornell University Press, 1988), p. 49.

18. Heidegger, *Being* (1962), 437.

19. Heidegger, *History*, p. 239.

20. Ibid., pp. 242, 240; T. Kisiel, *The Genesis of Heidegger's* Being and Time (Berkeley: University of California Press, 1993), pp. 381-38.

21. Heidegger, *History*, p. 242.

22. Heidegger, *Being*, 118.

23. Ibid., 121, 124.

24. Ibid., 122.

25. Taminiaux, *Heidegger*, p. 132.

26. Heidegger, *Being*, 122; Vogel, *The Fragile*, p. 79.

27. Vogel, *The Fragile*, pp. 73-76, 82.

28. We use here the term *community* in the broadest possible sense, as embracing all types of human groups or gatherings, from a friendship, family, local community, or research team, through historical societies, to the entirety of humankind.

29. Even when, as in later writings—e.g., in his *Rectoral Address*—Heidegger speaks about Dasein of a people and of the State, he understands it in a monadic, solipsistic way. His positing of the people-State as Dasein "amounts to the negation of plurality, the cancellation of the pluralistic sharing of deeds and words, and to the replacement of the pluralistic debate regarding what appears to each and every one *(dokei moi)* with a unanimous passion for the Being of all beings" (Taminiaux, *Heidegger*, p. 133).

30. Cf. E. A. Tiryakian, *Sociologism and Existentialism: Two Perspectives on the Individual and Society* (Englewood Cliffs, N.J.: Prentice-Hall, 1962), p. 152.

31. Lukács, "Existentialism," pp. 138-39; also Cahoone, *The Dilemma*, pp. 168-71; Taminiaux, *Heidegger*, p. 131.

32. Heidegger, *History*, p. 246.

33. H. Dreyfus, *Being-in-the-World*, p. 145; Ballard, *The Role*, pp. 88-90; Guignon, *Heidegger*, pp. 105-9.

34. Mulhall, *Heidegger*, p. 66.

35. Heidegger, *Being*, 51-52.

36. DiCenso, *Hermeneutics*, p. 46; also D. Carr, *Time, Narrative, and History* (Bloomington and Minneapolis: Indiana University Press 1991), pp. 86-94, 107; Taminiaux, *Heidegger*, pp. 130-32; Tiryakian, *Sociologism*, pp. 127-30, 152; Vogel, *The Fragile*, pp. 38-40.

37. H. Ruin, *Enigmatic Origins: Tracing the Theme of Historicity through Heidegger's Works*, Acta Universitatis Stockholmiensis, Stockholm Studies in Philosophy 15 (Stockholm: Alquist and Wiksell International, 1994), p. 95.

38. Dreyfus, *Being-in-the-World*, pp. 151-60.

39. Heidegger, *Being*, 364.

40. Ibid., 53, 64.

41. Ibid., 83.

42. *Significance*, the system of relations interlocked among themselves that contains for-the-sake-of-whichs, in-order-tos, what-fors, what-ins, and what-withs is—as Heidegger emphasizes a few passages later—constitutive for worldliness (Heidegger, *Being*, 88).

43. Ibid., 87.

44. Ibid., 88.

45. Ibid., 118; Cahoone, *The Dilemma*, pp. 163–72.

46. Dreyfus, *Being-in-the-World*, p. 158.

47. In his later philosophy Heidegger replaces the ontological conditions of Dasein with the ontological conditions of being itself, which is "hypostatized and somehow located *outside* history and language" (DiCenso, *Hermeneutics*, p. 77). Consequently, the concept of Dasein loses its ontological centrality, and Heidegger merely articulates a new version of traditional, essentializing ontology, the novelty of which amounts to rejecting the search for grounding in favor of experiencing and narratively presenting Being (Habermas, *The Philosophical*, pp. 152–53).

48. Taminiaux, *Heidegger*, pp. 124–25.

49. Ibid., p. 113.

50. W. Pannenberg, "Hermeneutics and Universal History," in *Hermeneutics and Modern Philosophy*, ed. B. R. Wachterhauser (Albany: SUNY Press, 1986), p. 121.

51. M. Heidegger, "Hölderlin and the Essence of Poetry," in M. Heidegger, *Existence and Being*, trans. D. Scott (Washington, D.C.: Gateway Editions, 1988), pp. 280–88.

52. P. Connerton, "Gadamer's Hermeneutics," *Comparative Criticism* 5 (1983): 117.

53. S. Krolick, *Recollective Resolve: A Phenomenological Understanding of Time and Myth* (Macon, Ga.: Mercer University Press, 1987), p. 54.

54. Rorty, *Philosophy*, pp. 167–68; also Hekman, *Hermeneutics*, pp. 100–101.

55. Gadamer, *Truth*, p. 290.

56. Gadamer, "Reply," p. 277.

57. Ibid., p. 277.

58. Heidegger, *Being*, 378–79; C. O. Schrag, "Heidegger on Repetition and Historical Understanding," *Philosophy East and West* 20 (1970): 293–94.

59. B. P. Dauenhauer, "History's Point and Subject Matter: A Proposal," in *At the Nexus of Philosophy and History*, ed. B. P. Dauenhauer (Athens: University of Georgia Press, 1987), p. 159.

60. Heidegger, *Being*, 381.

61. The meaning of the term *to historize [geschehen]* is discussed in detail by Macquarrie and Robinson in their translation of *Sein und Zeit* (p. 41 n1). The term is their coinage to show the etymological kinship between *geschehen* (to happen) and *geschichte* (history). Hence "historize" refers to events happening in a historical way and connects with "historicity." In what follows we use this coinage rather than "historicizing," since the latter relates to the doctrine of "historicism" which we reject.

62. Heidegger, *Being* (1962), 389.

63. Ibid., 384.

64. P. Ricoeur, *Time and Narrative*, ed. K. Blamey and trans. D. Pellauer (Chicago: University of Chicago Press, 1988), 3:75.

65. Heidegger, *Being*, 386.

66. Ibid., 385.

67. C. O. Schrag, "Heidegger," p. 289; Hcidegger, *Being*, 386.

68. P. Ricoeur, "Narrative Time," in *On Narrative*, ed. W. J. T. Mitchell (Chicago: University of Chicago Press, 1981), p. 178.

69. Heidegger, *Being*, 325.

70. DiCenso, *Hermeneutics*, pp. 49–50; Heidegger, *Being*, 383.

71. Dauenhauer, "History's Point," p. 162.

72. Ibid., p. 162.

73. DiCenso, *Hermeneutics*, p. 49; Ricoeur, *Time*, p. 75.

74. Heidegger, *Being*, 247.

75. Heidegger, *Being* (1962), 255.

76. P. Berger and T. Luckmann, *The Social Construction of Reality: A Treatise in the Sociology of Knowledge* (Garden City, N.Y.: Doubleday 1966), pp. 93–94.

77. M. Eliade, *The Myth of the Eternal Return* (New York: Pantheon Books, 1954), pp. 158–59.

78. Ibid., pp. 157–58.

79. E. L. Fackenheim, *Metaphysics and Historicity* (Milwaukee: Marquette University Press, 1961), p. 80.

80. Heidegger, *Being*, 247.

81. Fackenheim, *To Mend*, p. 165.

82. Wachterhauser, "Introduction," p. 35.

83. Caputo, "The Thought of Being," p. 256.

84. Gadamer, "The Universality," p. 147.

85. Caputo, "The Thought," p. 256; Grondin, "Hermeneutical Truth," p. 46; Hekman, *Hermeneutics*, p. 94.

86. Howard, *Three Faces*, p. 154.

87. Gadamer, "Rhetoric," p. 282.

88. Ibid., p. 284.

89. H.-G. Gadamer, "On the Scope and Function of Hermeneutical Reflection," trans. G. B. Hess and R. E. Palmer, in *Hermeneutics and Modern Philosophy*, ed. B. R. Wachterhauser (Albany: SUNY Press, 1986), p. 289.

90. Weinsheimer, *Gadamer's Hermeneutics*, p. 248.

91. Gadamer, *Truth*, p. 368.

92. Weinsheimer, *Gadamer's Hermeneutics*, p. 205.

93. Ibid., p. 205.

94. Kisiel, "The Happening," p. 10.

95. Gadamer, *Truth*, p. 363.

96. Ibid., pp. 369–70.

97. Ibid., p. 377.

98. Grondin, "Hermeneutics," p. 57.

99. Ibid., pp. 57–58.

100. Gadamer, *Truth*, p. 307.

101. Connerton, "Gadamer's Hermeneutics," pp. 122–23.

102. Misgeld, "On Gadamer's Hermeneutics," p. 161.

103. H. L. Dreyfus, "Holism and Hermeneutics," in *Hermeneutics and Praxis,* ed. R. Hollinger (Notre Dame: University of Notre Dame Press, 1985), p. 234.

104. Kisiel, "The Happening," p. 11.

105. When Heidegger emphasizes that cognition is a mode of human being-in-the-world and not a relation between a worldless subject and an object (Heidegger, *History,* pp. 162, 165), his view is seemingly close to naturalist epistemologies or to sociobiology. However, Heidegger does not understand "being-in" in natural (biological) terms, but in ontological-existentialist terms, and he criticizes the first understanding as realistic (pp. 166–67).

106. Heidegger, *Being,* 160–61.

107. Knorr-Cetina, *The Manufacture,* p. 142; Rouse, *Knowledge,* pp. 76, 95.

108. Heidegger, *Being,* 65.

109. Rouse, *Knowledge,* p. 75.

110. Knorr-Cetina, *The Manufacture,* p. 142.

111. Heidegger, *Being,* 358.

112. Rouse, *Knowledge,* pp. 50–58, 79–81, 108.

113. Ibid., p. 79.

114. Knorr-Cetina, *The Manufacture,* pp. 142–43.

115. J. D. Caputo, "Heidegger's Philosophy of Science: The Two Essences of Science," in *Rationality, Relativism and the Human Sciences,* ed. J. Margolis, M. Krausz, R. M. Burian (Dordrecht: Martinus Nijhoff Publ., 1986), p. 43. One should be careful, however, in proclaiming affinities between either Heidegger's or Gadamer's picture of science and conceptions current in post-empiricist philosophy of science. For instance, Gadamer's notion of tradition cannot be identified with the Kuhnian concept of a paradigm or with the Lakatosian category of the scientific research program. As Misgeld points out, "Gadamer has not examined the peculiar status of traditions of inquiry present in the development of natural science." Hermeneutics must characterize tradition as something we depend upon, whereas, the core of Lakatosian methodology is a concept of the scientist as a free-agent—that is, free either to choose a research program or reject it "in terms of criteria that place him outside any historical context" (Misgeld, "On Gadamer's Hermeneutics," p. 149).

116. R. Rorty, "Inquiry as Recontextualization: An anti-Dualist Account of Interpretation," in *Philosophical Papers,* vol. 1, *Objectivity, Relativism, and Truth* (Cambridge: Cambridge University Press, 1991), pp. 97–98, 101; Dreyfus, *Being-in-the-World,* p. 207.

117. Dreyfus, *Being-in-the-World,* p. 207.

118. Heidegger, *Being,* 10.

119. Ibid., 11.

120. Dreyfus, *Being-in-the-World,* p. 207.

121. Cf. Hekman, *Hermeneutics,* p. 107; Misgeld, "On Gadamer's Hermeneutics," p. 161; Weinsheimer, *Gadamer's Hermeneutics,* pp. 3, 20–23. Gadamer himself finds it unsatisfactory (cf. *Truth,* pp. 283–85).

122. Weinsheimer, *Gadamer's Hermeneutics,* p. 15.

123. Gadamer, *Truth,* pp. 552–53; M. Heidegger, "Science and Reflection," in *The Question Concerning Technology and Other Essays,* trans. W. Lovitt (New

York: Harper and Row, 1977), pp. 176–77; Weinsheimer, *Gadamer's Hermeneutics*, pp. 32, 57–58.

124. Dreyfus, "Holism," p. 239; Weinsheimer, *Gadamer's Hermeneutics*, p. 248.

125. Weinsheimer, *Gadamer's Hermeneutics*, p. 248.

126. Gadamer, *Truth*, p. 447.

CHAPTER 3

1. Collins and Yearley, "Epistemological Chicken," p. 308.

2. Knorr-Cetina in Callebaut, *Taking*, pp. 184–85.

3. Latour, *We Have Never*, p. 85.

4. Ibid., pp. 88, 67.

5. M. Foucault, "What Is Enlightenment?" in *The Foucault Reader*, ed. P. Rabinow (New York: Pantheon, 1984), p. 48.

6. Contrary to the tradition established in English translations of Foucault, we prefer to link the terms *thought, action, power,* and *knowledge* by hyphens and not by slashes. The reason is simple and indeed crucial: in both cases Foucault is emphasizing the *unity* of thought and action or power and knowledge and not their opposition.

7. M. Foucault, "Afterward: The Subject and Power," in *Michel Foucault: Beyond Structuralism and Hermeneutics: With an Afterword by Michel Foucault*, ed. H. L. Dreyfus and P. Rabinow (Chicago: University of Chicago Press, 1982), pp. 208–26.

8. M. Foucault, "Preface to the Second Volume of *The History of Sexuality,*" in *The Foucault Reader*, ed. P. Rabinow (New York: Pantheon, 1984), p. 335.

9. Ibid.

10. M. Foucault, "Clarifications on the Question of Power," in *Foucault Alive (Interviews, 1966–84)*, ed. S. Lotringer and trans. J. Cascaito (New York: Semiotext(e), 1989), p. 188.

11. Foucault, "Afterward," p. 219.

12. M. Foucault, "The Eye of Power," in M. Foucault, *Power/Knowledge: Selected Interviews and Other Writings, 1972–1977,* ed. C. Gordon and trans. C. Gordon, L. Marshall, J. Mepham, K. Soper (New York: Pantheon Books, 1980), pp. 156–61.

13. Foucault, "Afterward," p. 219.

14. Ibid., p. 221.

15. M. Foucault, "Two Lectures," in M. Foucault, *Power/Knowledge: Selected Interviews and Other Writings, 1972–1977,* ed. C. Gordon and trans. C. Gordon, L. Marshall, J. Mepham, K. Soper (New York: Pantheon Books, 1980), p. 98.

16. M. Foucault, *Discipline and Punish: The Birth of the Prison,* trans. A. Sheridan (New York: Vintage Books, 1995), p. 27.

17. G. Deleuze, *Foucault,* trans. S. Hand (Minneapolis: University of Minnesota Press 1988), p. 81; M. Foucault, "Clarifications," pp. 183–86.

18. M. Foucault, "Truth and Power," in M. Foucault, *Power/Knowledge: Selected Interviews and Other Writings, 1972–1977,* ed. C. Gordon and trans. C.

Gordon, L. Marshall, J. Mepham, K. Soper (New York: Pantheon Books, 1980), p. 112.

19. Foucault, *Discipline,* pp. 27–28.

20. Ibid., p. 194.

21. Foucault, "Truth and Power," p. 113.

22. R. Visker, *Michel Foucault: Genealogy as Critique,* trans. C. Turner (London: Verso, 1995), p. 69; Deleuze, *Foucault,* p. 88.

23. Foucault, "Afterward," p. 217.

24. For the interpretation of the power-knowledge relation as a relation between form and content, see Deleuze, *Foucault,* pp. 60–61; for the rejection of the view of concepts as forms in favor of viewing them as the patterns of inference that stretch diachronically, see Brandom, *Making It Explicit,* pp. 618–19.

25. Foucault, *Discipline,* p. 27.

26. Heidegger, *Being,* 5.

27. Ibid., 9.

28. Heidegger, *Being* (1962), 11.

29. Heidegger, *Being,* 13.

30. Ibid., 6, 9, 38.

31. J. Barnes, "Metaphysics," in *Cambridge Companion to Aristotle,* ed. J. Barnes (Cambridge: Cambridge University Press, 1995), pp. 69–70.

32. Cf. Aristotle, "Metaphysica," 1041b1-5, 1053b15-25, 1060b30-1061a10.

33. Ibid., 1042b15-1043a; Barnes, "Metaphysics," pp. 74–77. Aristotle's view on "being *qua* being" contains a vexed question addressed by Michael Frede and Owen. Frede notes two radically different conceptions of metaphysics in Aristotle: a general science of being, a "metaphysica generalis," and the study of special kinds of beings, namely, supra-sensible beings that come first in the order of being (Aristotle 1960, iv 1003a21ff and vi, 1026a26ff). Since Aristotle sees no conflict, Frede offers a reconciliation: supra-sensible beings have the highest way of being with respect to which "all other ways of being have to be explained" (M. Frede, "The Unity of General and Special Metaphysics: Aristotle's Conception of Metaphysics," in *Essays in Ancient Philosophy* [Minneapolis: University of Minnesota Press, 1987], p. 84). Owen, armed with the notion of focal meaning, charts the course by which Aristotle converts "a special science of substance into the universal science of being, universal just inasmuch as it is primary" (G. E. L. Owen, "Logic and Metaphysics in Some Earlier Works of Aristotle," in *Logic, Science and Dialectic,* ed. M. Nussbaum [Ithaca: Cornell University Press, 1986], p. 184).

34. Aristotle, "Metaphysica," 1028b1-5; Barnes, "Metaphysics," pp. 78–81.

35. Gilson, *Being,* p. 46.

36. Ibid., pp. 91, 110; J.-L. Nancy, "Introduction," in *Who Comes After the Subject?* ed. E. Cadava, P. Connor, J.-L. Nancy (New York: Routledge, 1991), p. 6.

37. Gilson, *Being,* p. 154.

38. Heidegger, *Being,* 2; Aristotle, "Metaphysica," 1053b20-25.

39. W. V. O. Quine, "On What There Is," in *From the Logical Point of View* (Cambridge, Mass.: Harvard University Press, 1953), pp. 13–14; W. V. O. Quine, *World and Object* (Cambridge, Mass.: MIT Press; New York: Jon Wiley & Sons, 1960), pp. 131, 176.

40. Heidegger, *Being*, 9.

41. The problem of being and the idea that being cannot be separated from activity has been discussed—in fact—by philosophers and mystics, such as Plotinus, Eriugena, Eckhart, Böhme, Spinoza, and Hegel, who elaborate holistic, pantheistic ontologies (Kolakowski, "The Priest," pp. 26–27; L. Kolakowski, *Main Current of Marxism*, vol. 1, *The Founders* [Oxford: Oxford University Press, 1982], pp. 12–39, 56–80; J. Taminiaux, "Finitude and the Absolute: Remarks on Hegel and Heidegger as Interpreters of Kant," in *Dialectic and Difference: Finitude in Modern Thought*, ed. J. Decker and R. Crease, trans. T. Sheehan [Atlantic Highlands, N.J.: Humanities Press, 1985], p. 67 in reference to Hegel). To analyze their conceptions as revealing different ways of being would be a fascinating task, but evidently surpassing the scope of our book.

42. Richardson, *Existential Epistemology*, p. 48.

43. Nietzsche, *The Will*, p. 551; H. Lawson, *Reflexivity: The Post-Modern Predicament* (London: Hutchinson, 1985) p. 36.

44. A. Nehamas, *Nietzsche: Life as Literature* (Cambridge, Mass.: Harvard University Press 1985), p. 82.

45. Nietzsche, *The Will*, p. 558.

46. Frede, D., "The Question of Being: Heidegger's Project," in *The Cambridge Companion to Heidegger*, ed. C. B. Guignon (Cambridge: Cambridge University Press, 1993), p. 63.

47. Ibid., p. 63.

48. Heidegger, *Being*, 57, 193–94.

49. Rouse, *Knowledge*, pp. 108, 95.

50. Heidegger, *Being*, 50.

51. Heelan, *Space-Perception*, p. 13.

52. Okrent, *Heidegger's Pragmatism*, p. 32.

53. Nietzsche, *The Will*, p. 557; italics added.

54. Richardson, *Existential Epistemology*, p. 107.

55. M. Heidegger, *Basic Problems of Phenomenology*, trans. A. Hofstadter (Bloomington: Indiana University Press, 1982), pp. 164, 293.

56. J. Margolis, "Praxis and Meaning: Marx's Species Being and Aristotle's Political Animal," in *Marx and Aristotle: Nineteenth-Century German Social Theory and Classical Antiquity*, ed. G. E. McCarthy (Lanham: Rowman & Littlefield, 1992), p. 333.

57. B. Tuchańska, "The Marxian Conception of Man," *Dialectics and Humanism* 3 (1983): 131; Margolis, "Praxis," p. 333.

58. Elias, *The Society*, p. 11.

59. Ibid., p. 55.

60. J. Margolis, *Texts without Referents: Reconciling Science and Narrative* (Oxford: Basil Blackwell, 1989), p. 46.

61. There are, in fact, many different collectives and not only one *we* that can be identified with the entirety of humanity. However, the plurality of groups to which a person belongs, relations among those groups, variances in the strength of one's identification with different groups, possible conflicts of identification, and other similar empirical phenomena do not need to be considered on the ontological level.

62. A. Schutz, "Common-Sense and Scientific Interpretation of Human Action," in *Collected Papers*, vol. 1, *The Problem of Social Reality*, ed. M. Natanson (The Hague: Martinus Nijhoff, 1971), p. 12.

63. Carr, *Time*, pp. 144-45; R. J. Bernstein, *Praxis and Action: Contemporary Philosophies of Human Activity* (Philadelphia: University of Pennsylvania Press 1971), pp. 25-28.

64. A different way of constituting the internal structure of our being-together appears in the philosophy of dialogue. For Buber or Levinas, a personal, reciprocal, and dialogical relation between *I* and *the Other,* or *Thou,* is different from any relation between *I* and *it;* it constitutes "the ultimate event" (Levinas, *Totality*, p. 221; M. Theunissen, *The Other: Studies in the Social Ontology of Husserl, Heidegger, Sartre and Buber* (Cambridge, Mass.: MIT Press, 1984), pp. 257-344, 361-84).

65. H. Arendt, *The Human Condition* (Chicago: University of Chicago Press, 1989), pp. 7-8.

66. Ibid., p. 8.

67. That this perspective does not contradict phenomenological study is argued broadly and convincingly by Carr, *Time*, pp. 118-21, 127-33.

68. Ibid., p. 121.

69. Cahoone, *The Dilemma*, p. 25.

70. Elias, *The Society*, pp. 21, 23-24, 95-96.

71. Habermas, *The Philosophical*, p. 297; Rorty, "Cosmopolitanism," p. 214; Kolakowski, "Looking for the Barbarians," pp. 22-23.

72. Unlike German philosophy, the analytic tradition has a tendency to confuse intentionality as "aboutness" of consciousness with intentionality as a feature of action.

73. Heelan, *Space-Perception*, p. 11. These subjective intentionalities are the product of learning processes and have bodily and worldly conditions (p. 12).

74. Searle and Dennett speak of collective intentionality and consciousness as primitive phenomena. For them, however, they are evolutionary products (J. Searle, *The Construction of Social Reality* [New York: Free Press, 1995], pp. 6-27; D. C. Dennett, *Kinds of Minds: Toward an Understanding of Consciousness* (New York: Basic Books, 1989), pp. 52-53). From our standpoint, consciousness and its intentionality are neither active dispositions reducible to a base nor causally emergent natural properties.

75. E. Husserl, *Ideas: General Introduction to Pure Phenomenology,* trans. W. R. B. Gibson (London: George Allen and Unwin; New York: Macmillan, 1952), p. 106.

76. Misgeld, "On Gadamer's Hermeneutics," p. 160.

77. Gadamer, *Truth*, pp. 276-77.

78. B. R. Wachterhauser, "Must We Be What We Say? Gadamer on Truth in the Human Sciences," in *Hermeneutics and Modern Philosophy*, ed. B. R. Wachterhauser (Albany: SUNY Press, 1986), p. 225.

79. Carr, *Time*, pp. 155-56.

80. Elias, *The Society*, pp. 17-19.

81. Ibid., p. 12.

82. When Heidegger poses explicitly the problem of language's way of

being, he says only that his interpretation of language "has the sole function of pointing out the ontological 'place' for this phenomenon in the constitution of being of Da-sein" (*Being*, 66).

83. Ibid., 114.

84. J. J. Kockelmans, "Heidegger on the Essential Difference and Necessary Relationship between Philosophy and Science," in *Phenomenology and the Natural Sciences*, ed. J. J. Kockelmans and T. J. Kisiel (Evanston: Northwestern University Press, 1970), p. 153; D. C. Hoy, "History, Historicity, and Historiography in *Being and Time*," in *Heidegger and Modern Philosophy: Critical Essays*, ed. M. Murray (New Haven: Yale University Press, 1978), p. 339.

85. Haugeland, "Heidegger," p. 20. Haugeland is explicit in rejecting mineness of Dasein. For him "people are to *Dasein* as a baseball game is to baseball, as utterances are to language, as works are to literature" (p. 20).

86. Dreyfus, *Being-in-the-World*, pp. 14–15, 158. In the passage from *Being and Time*, to which Dreyfus refers, Heidegger does, in fact, say that "sciences have this being's (the human being's) kind of being," but he characterizes sciences as "ways in which human beings behave" and not as social institutions (Heidegger, *Being*, 11).

87. F. Volpi, "Dasein as *Praxis:* The Heideggerian Assimilation and Radicalization of the Practical Philosophy of Aristotle," in *Critical Heidegger*, ed. C. Macann (London: Routledge, 1996), p. 41. For further discussion, see Bernasconi, *Heidegger in Question*, pp. 2–24.

88. Taminiaux, *Heidegger*, pp. 124, 129–37; Bernstein, "Heidegger's Silence," pp. 121–28. Heidegger "ontologizes" Aristotle, so that in the end *praxis* becomes an ontological mode of the activity of Dasein (Volpi, "Dasein," p. 48). At this point the distinction between *poiesis* and *praxis* is annulled. Heidegger also omits the ethical dimension of *praxis*. The disastrous intellectual (and moral) consequences of this are discussed by Bernstein in "Heidegger's Silence."

89. Aristotle, "Ethica Nicomachea," in *The Basic Works Aristotle*, ed. R. McKeon and trans. W. D. Ross (New York: Random House, 1941), 1140b1-10.

90. Taminiaux, *Heidegger*, p. 113.

91. Ibid., p. 113.

92. Gadamer, *Reason*, p. 91.

93. Arendt, *The Human*, pp. 190–92.

94. Taminiaux, *Heidegger*, p. 113.

95. Arendt, *The Human*, p. 192.

96. Taminiaux, *Heidegger*, p. 113.

97. Arendt, *The Human*, p. 193.

98. J. E. McGuire and S. K. Strange, "An Annotated Translation of Plotinus *Ennead* iii 7: *On Eternity and Time*," *Ancient Philosophy* 8 (1988): 252; Taminiaux, *Heidegger*, p. 125.

99. McGuire, Strange, "An Annotated," pp. 252–53.

100. Arendt, *The Human*, pp. 7–9.

101. Ibid., pp. 182–83, 52–53.

102. Ibid., p. 234.

103. Ibid., pp. 220–30. Arendt evaluates this tendency negatively and wants to reestablish an ancient praxist reading of political activity.

104. A. Giddens, *Central Problems in Social Theory: Action, Structure and Contradiction in Social Analysis* (Berkeley: University of California Press, 1979), p. 216.

105. Arendt, *The Human,* p. 9.

106. Giddens, *Central,* pp. 4–5.

107. J. Rouse, "The Significance of Scientific Practices," in *Engaging Science: How to Understand Its Practices Philosophically* (Ithaca: Cornell University Press, 1996), p. 139.

108. Cf. Giddens, *Central,* pp. 4–5, 54–66, 94, 117; A. Touraine, *The Self-Production of Society,* trans. D. Coltman (Chicago: University of Chicago Press, 1977), pp. 2–9.

109. Giddens, *Central,* p. 77.

110. M. Polanyi, *Personal Knowledge: Towards a Post-Critical Philosophy* (New York: Harper & Row, 1964), p. 276.

111. Fackenheim, *Metaphysics,* p. 26.

112. Ibid., p. 27.

113. Schutz, "Common-Sense," p. 19.

114. This is all clearly visible in Searle's conception of intentionality: for him, the distinction between the mind and the world is built into the very activity of acting. He espouses this assumption while taking "a step toward Intentionalizing causation and, therefore, toward naturalizing Intentionality" (J. Searle, *Intentionality: An Essay in the Philosophy of Mind* [Cambridge: Cambridge University Press, 1983], p. 112).

115. Nietzsche, *The Will,* p. 484.

116. Heidegger, *Being,* 300

117. Brandom, *Making It,* p. 20.

118. L. Wittgenstein, *Philosophical Investigations,* trans. G. E. M. Anscombe (Oxford: Basil Blackwell, 1974), pp. 84–86, 198, 201–2, 206, 217.

119. Brandom, *Making It,* p. 20.

120. Habermas, *The Philosophical,* pp. 302–3.

121. Bernstein, *Praxis,* p. 34.

122. L. Kolakowski, "Karl Marx and the Classical Definition of Truth," in *Toward a Marxist Humanism: Essays on the Left Today,* trans. J. Zielonko Peel (New York: Grove Press, 1968), p. 43; Habermas, *The Philosophical,* p. 304; J. Taminiaux, "Marx, Art, and Truth," in *Dialectic and Difference: Finitude in Modern Thought,* ed. J. Decker and R. Crease, trans. J. Decker (Atlantic Highlands, N.J.: Humanities Press, 1985), pp. 41–44.

123. Margolis, "Praxis," p. 337.

124. N. Rotenstreich, *Theory and Practice: An Essay in Human Intentionalities* (The Hague: Martinus Nijhoff, 1977), p. 78.

125. Arendt, *The Human,* pp. 136–39.

126. T. R. Schatzki, *Social Practices: A Wittgensteinian Approach to Human Activity and the Social* (Cambridge: Cambridge University Press, 1996), pp. 91, 89.

127. Rotenstreich, *Theory,* p. 84; S. Turner, *The Social Theory of Practices: Tradition, Tacit Knowledge, Presuppositions* (Chicago: University of Chicago Press, 1994), p. 8.

128. Rouse, "The Significance," p. 135.

129. Ibid., pp. 134–37, 143, 150.

130. Ibid., p. 138.

131. In particular, a person is not—ontologically speaking—a composite of two kinds of primary entities: a subject of experiences and a subject of physical characteristics—i.e., a disembodied consciousness and a physical body (P. Strawson, *Individuals: An Essay in Descriptive Metaphysics* [London: Methuen, 1959], p. 105).

132. Husserl, *The Crisis*, p. 142.

133. Cf. Rotenstreich, *Theory*, p. 134.

134. M. Foucault, *The History of Sexuality*, vol. 2, *The Uses of Pleasure*, trans. R. Hurley (New York: Vintage Books, 1986), p. 2.

135. Heelan, Why a Hermeneutical Philosophy," pp. 289–99.

136. Lynch, *Scientific Practice*, p. 317.

137. Arendt, *The Human*, p. 246.

138. P. Bowen-Moore, *Hannah Arendt's Philosophy of Natality* (New York: St. Martin's Press 1989), p. 26.

139. Ibid., pp. 50, 77–85.

140. Arendt, *The Human*, p. 9.

141. Bowen-Moore, *Hannah Arendt's Philosophy*, pp. 102–3.

142. C. Taylor, "Interpretation and the Sciences of Man," in *Interpretive Social Science: A Reader*, ed. P. Rabinow, W. M. Sullivan (Berkeley: University of California Press, 1979), pp. 32–33.

143. Ibid., pp. 32–33.

144. B. Latour, *The Pasteurization of France*, trans. A. Sheridan, J. Law (Cambridge, Mass.: Harvard University Press, 1988), p. 161.

145. Dreyfus, "Holism," p. 235; italics in the first quotation omitted.

146. A. MacIntyre, *After Virtue: A Study in Moral Theory* (Notre Dame: University of Notre Dame Press, 1984), pp. 187–88.

147. Giddens, *Central*, p. 4.

148. Habermas, *The Philosophical*, p. 322.

149. Weinsheimer, *Gadamer's Hermeneutics*, p. 13.

CHAPTER 4

1. T. Nickles, "Good Science as Bad History: From Order of Knowing to Order of Being," in *The Social Dimensions of Science*, ed. E. McMullin (Notre Dame: University of Notre Dame Press, 1992), p. 103; Nickles in Callebaut, *Taking*, p. 213.

2. K. R. Popper, *The Open Society and Its Enemies*, vol. 2, *The High Tide of Prophecy: Hegel, Marx, and the Aftermath* (London: Routledge and Kegan Paul, 1952), p. 216; P. Kitcher, "Contrasting Conceptions of Social Epistemology," in *Socializing Epistemology: The Social Dimensions of Knowledge*, ed. F. F. Schmitt (Lanham: Rowman and Littlefield, 1994), p. 124.

3. Dreyfus, "Holism," p. 239–40.

4. Cf. Rouse, *Knowledge*, pp. 112–13.

5. Lynch, *Scientific Practice*, p. 281.

6. Heelan, "Natural Science," p. 185.

7. Ibid., p. 187.

8. Ibid., p. 188.

9. Knorr-Cetina, *The Manufacture*, p. 142.

10. The position we are outlining is in some ways reminiscent of the Sellarian distinction between the manifest image—i.e., the everyday life world —and the scientific image, insofar as Sellars stresses the complementary relationship between the two. He emphasizes that we must "directly relate the world as conceived by scientific theory to our purposes, and make it our world and no longer an alien appendage to the world in which we do our living" (W. Sellars, *Science, Perception and Reality* (London: Routledge and Kegan Paul, 1963), p. 40.

11. J. Kmita, "The Controversy About the Determinants of the Growth of Science," in *Essays on the Theory of Scientific Cognition,* trans. J. Chołowka (Dordrecht: Kluwer; Warsaw: PWN, 1991), p. 80.

12. J. Kmita, *Problems in Historical Epistemology* (Dordrecht: Reidel; Warsaw: PWN, 1988), p. 17.

13. L. Nowak, "Essence, Idealization, Praxis. An Attempt at a Certain Interpretation of the Marxist Concept of Science," *Poznań Studies in the Philosophy of the Sciences and the Humanities* 2, no. 3 (1976); L. Nowak, *The Structure of Idealization: Towards a Systematic Interpretation of the Marxian Idea of Science* (Dordrecht: Reidel, 1980), pp. 214-27.

14. Rouse, *Knowledge,* pp. 50-58, 80-81.

15. Ibid., p. 69.

16. Heelan, "Hermeneutical Phenomenology," p. 225-26.

17. Heelan, "Natural Science," p. 185.

18. Ibid., p. 184.

19. Heelan, "Hermeneutical Phenomenology," p. 226.

20. Rouse, *Knowledge,* p. 101.

21. Ibid., pp. 95-96; Rouse, "The Significance," pp. 127-33.

22. Rouse, "The Significance," p. 127.

23. Ibid., pp. 126-27. In fact, Rouse seems to waver between epistemological and non-epistemological views of science: when he considers laboratory practice as an activity whose aim is to manipulate things and regards causal efficacy as a criterion involved in this construction (Rouse, *Knowledge,* p. 102), he seems to use a non-epistemological concept of scientific experiment. That there is, however, an epistemological background is evident in his assertion that a third essential feature of laboratory practice, beside isolation and intervention, is "tracking what goes on in its constructed microworlds" (p. 102). Tracking involves "monitoring the entire process of the experiment beginning with its construction." One can expect, therefore, that the concept of monitoring has epistemological content and refers to the process of learning what happens in a given microworld; but Rouse immediately clarifies this by saying that it is not an epistemological concept. "It is a matter of seeing that things are working right rather than just seeing what the outcome is" (p. 102). He also emphasizes that laboratory "skills and procedures are aimed toward practical success (getting things to work reliably) rather than toward systematic understanding or truth" (p. 111). Finally, it turns out that what is involved in the construction of laboratory microworlds, and in acts of monitoring them, is practical "'sight' (and foresight)" and, moreover, that "microworlds

of scientific experiments are designed to permit detailed knowledge of their individual components" (p. 103).

24. R. P. Crease, "The Play of Nature," pp. 83–84.

25. N. Elias, "Scientific Establishments," in *Scientific Establishments and Hierarchies,* ed. N. Elias, H. Martins, R. Whitley (Dordrecht: Reidel, 1982), pp. 11, 25, 42.

26. S. Mennell, *Norbert Elias: Introduction* (Oxford: Basil Blackwell, 1992), p. 188.

27. Ibid., p. 192.

28. Longino, "The Fate," p. 138.

29. Ibid., pp. 138–39.

30. Sociologists conceive scientific communities as "small social systems with inherent boundaries and internal mechanisms of integration" and characterize these mechanisms in economical terms of competition (Merton), precapitalist models (Hagstrom), a capitalist market model (P. Bourdieu, "The Specificity of the Scientific Field and the Social Conditions of the Progress of Reason," trans. R. Nice, *Social Science Information* 14, no. 6 [1975], pp. 21–23; Knorr-Cetina, *The Manufacture,* pp. 70–71), or a reproductive cycle of credit and advantage (Latour and Woolgar, *Laboratory Life,* pp. 200–238; Knorr-Cetina, "The Ethnographic," p. 129). From this point of view, scientists are concerned with "making things 'work,'" so the principle of their action is "a principle of success rather than one of truth" (Knorr-Cetina, *The Manufacture,* p. 4).

31. Knorr-Cetina, "The Ethnographic," p. 119.

32. Ibid., p. 119; Knorr-Cetina in Callebaut, *Taking,* p. 182; Latour, *Science in Action,* p. 89; Rouse, *Knowledge,* pp. 100–101.

33. L. Fleck, *Genesis and Development of a Scientific Fact* (Chicago: University of Chicago Press, 1979), p. 38.

34. Ibid., p. 40.

35. Latour and Woolgar, *Laboratory Life,* pp. 160–64.

36. Bauman, *Hermeneutics,* p. 121.

37. Gadamer, *Truth,* p. 446. Gadamer makes the dialogic structure of experience explicit. Essential to having experience is being open to something's being either this or that, an awareness reflected in the questioning act itself. However, a question's openness is not boundless but has its location within a particular horizon. Acts of questioning reveal the questionability of what is questioned and bring it into the "play" of discourse. Cognition emerges as the process of considering opposites, of settling the antithesis of yes and no, of conceiving something as being this or that. Thus, what is understood through communication is embedded in a dialogic situation and expressed in terms of the dialectic of question and answer. But not only is the world orientation of speakers expressed in the event of understanding: that event also happens on the side of things, from the topics of conversation at hand (*Truth,* pp. 362–79).

38. N. Murphy, "Scientific Realism and Postmodern Philosophy," *British Journal for the Philosophy of Science* 41 (1990): 293.

39. Grondin, "Hermeneutics," pp. 57–58.

40. R. Rorty, "Pragmatism, Relativism, and Irrationalism," in *Consequences of Pragmatism: Essays, 1972–1980* (Minneapolis: University of Minnesota Press, 1982), p. 165.

41. Heidegger, *Being,* 25.

42. Ibid., 161, 162.

43. Ibid., 161.

44. Grondin, "Hermeneutics," p. 57.

45. Grondin, "Hermeneutical Truth," p. 55.

46. Bernstein, *Beyond,* p. 142.

47. Knorr-Cetina, *The Manufacture,* p. 131.

48. Ibid., pp. 13–14.

49. Ibid., p. 131.

50. Turner criticizes a view of practices as tacit, hidden regularities that underlie overt sayings, doings, and activities and poses the "problem of transmission" (Turner, *The Social,* p. 2). For Turner, practices must be transmittable so that their identity is preserved in the same way in the heads of each practitioner (pp. 58–60). Only on this condition—argues Turner—can explanations, couched either in terms of psychological or causal mechanisms, ground manifest human activities. The brunt of Turner's criticism of those who make practice the explanatory bedrock is the claim that tacit regularities underlying manifest human behavior cannot both be causal and shared (p. 123).

51. T. S. Kuhn, "Commensurability, Comparability, Communicability," in *PSA 1982: Proceedings of the 1982 Biennial Meeting of the Philosophy of Science Association,* vol. 2, *Symposia,* ed. P. D. Asquith and T. Nickles (East Lansing, Mich.: PSA, 1983), pp. 669–73, 677; Bernstein, *Beyond,* pp. 25, 31; Heelan, "Why a Hermeneutic," pp. 277–78.

52. Grondin, "Hermeneutical Truth," p. 46.

53. Fleck, *Genesis,* p. 99.

54. Gadamer, *Truth,* p. 351–52.

55. H. M. Collins, "Tacit Knowledge and Scientific Networks," in *Science in Context: Readings in the Sociology of Science,* ed. B. Barnes and D. Edge (Cambridge, Mass.: MIT Press, 1982), pp. 44–46; B. Latour, "Postmodern? No, Simply Amodern! Steps Towards an Anthropology of Science," *Studies in History and Philosophy of Science* 21 (1990): 153; J. Rouse, "The Dynamics of Power and Knowledge in Science," *Journal of Philosophy* 88, no. 11 (1991): 660–61; Rouse, "The Significance," p. 131.

56. J. Rouse, "Philosophy of Science and the Persistent Narratives of Modernity," *Studies in History and Philosophy of Science* 22 (1991): 141. To be more precise, it is open in certain places and (relatively) closed and patterned in other places. For instance, the activity of awarding scientific degrees is, in most countries, a well-patterned and historically stable procedure, whereas scientific discovery is universally perceived as unpatterned activity.

57. The ontological conditions of this circle are, of course, natality and plurality.

58. Fleck, *Genesis,* p. 49; M. Polanyi, *The Tacit Dimension* (New York: Anchor Books, 1967), pp. 78–79.

59. Polanyi, *Personal Knowledge*, pp. 173–74, 301–3; Polanyi, *The Tacit*, pp. 78–79.

60. Polanyi, *Personal Knowledge*, p. 135.

61. Polanyi, *The Tacit*, p. 80.

62. S. Shapin, *A Social History of Truth: Civility and Science in Seventeenth-Century England* (Chicago: University of Chicago Press, 1994), p. 16.

63. Ibid., p. 359.

64. Ibid., pp. 10–13; 22–27.

65. Ibid., p. 27; Polanyi, *The Tacit*, p. 64.

66. Polanyi, *Personal Knowledge*, pp. 63–64.

67. Shapin, *A Social*, p. 411.

68. Ibid., p. 375.

69. Ibid., pp. 376, 379.

70. Ibid., pp. 356, 381.

71. Ibid., p. 362.

72. Ibid., p. 379.

73. Ibid., pp. 362, 361.

74. It is surprising how deep the dialogical character is of those philosophical disciplines which claim to have their own subject matter, such as philosophy of science. What is dominant in the works of most philosophers of science is—in fact—dialogue with other philosophers of science.

75. The most extreme articulation of this idea is the Hegelian view that the self-reflectivity of cognition realizes itself entirely in philosophy. An important consequence is that the abyss between subject and object, existing in scientific cognition, is bridged and their identity is established. Hegel's belief that philosophy can achieve ultimate truth about itself does not seem plausible. Nevertheless, the very possibility of questioning itself indicates the ability of philosophical thought to surmount what has been achieved, and to elaborate a variety of viewpoints that have value in relation only to their perspective and only have validity from within their own points of view.

76. Kolakowski's picture of the priest/jester antagonism must not be treated uncritically. Philosophy is not done either by the solitary priest facing the absolute or the equally individual jester facing the theories of the priests. The work of most contemporary philosophers, in particular, seems to be a distillation of both approaches.

77. Kolakowski, "The Priest," p. 36; Znaniecki, *The Social Role*, pp. 95–106.

78. Kolakowski, "The Priest," p. 36.

79. It seems rather that—as Bernstein argues—there are no perennial problems in philosophy, "which are so deep that they escape historical contingencies" (R. J. Bernstein, "History, Philosophy, and the Question of Relativism," in *At the Nexus of Philosophy and History*, ed. B. P. Dauenhauer [Athens: University of Georgia Press, 1987], p. 17).

80. Gadamer, *Truth*, p. 452.

81. R. Tuomela, "Science, Protoscience, and Pseudo-Science," in *Rational Changes in Science: Essays in Scientific Reasoning*, ed. J. C. Pitt and M. Pera (Dordrecht: Reidel, 1987), p. 86.

82. Fleck, *Genesis*, p. 42; also Heelan, "Why a Hermeneutical," p. 277.

83. At first glance, to include science's technological involvement as part of the characteristics of scientific research as cognition may seem odd. It is, however, absolutely justified. Technologies and their products are instantiations and embodiments of scientific knowledge and, in this way, of scientific projections. Moreover, when understood as broadly as we employ them here, they are the means that connect science with other spheres of practice and give it a place of domination in relation to them.

84. Latour and Woolgar, *Laboratory Life*, pp. 36–37; Knorr-Cetina, "The Fabrication."

85. Latour and Woolgar, *Laboratory Life*, p. 179.

86. Woolgar, *Science*, pp. 72–77.

87. Ibid., p. 88; J. Golinski, "The Theory of Practice and the Practice of Theory: Sociological Approaches in the History of Science," *Isis* 81 (1990): 503.

88. Latour and Woolgar, *Laboratory Life*, p. 176; Woolgar, *Science*, pp. 72–77.

89. Latour and Woolgar, *Laboratory Life*, pp. 177, 176.

90. Ibid., p. 175.

91. Woolgar, *Science*, p. 69.

92. Latour and Woolgar, *Laboratory Life*, pp. 179–80.

93. Heelan, "Natural Science," p. 188.

94. Ibid., p. 189.

95. Ibid., p. 192.

96. D. Ihde, *Technics and Praxis* (Dordrecht: Reidel, 1979), p. 23.

97. Heelan, *Space-Perception*, pp. 5, 12.

98. D. Ihde, *Technics*, p. 22; Heelan, *Space-Perception*, p. 19.

99. P. Winch, *The Idea of a Social Science and Its Relation to Philosophy*, 11th ed. (London: Routledge and Kegan Paul, 1980), pp. 18–19.

100. R. Krohn, "On Gieryn on the 'Relativist/Constructivist' Programme in the Sociology of Science: Naivete and Reaction," *Social Studies of Science* 12 (1982): 325–26.

101. Krolick, *Recollective Resolve*, p. 56.

102. M. Eliade, *The Sacred and the Profane: The Nature of Religion*, trans. W. R. Trask (San Diego: Harcourt Brace & Co., 1987), p. 14.

103. The view that science and philosophy are two separate sub-practices is, of course, far too simplified and is applicable, at best, to their contemporary relationship. We cannot, however, engage this large issue at present.

104. Eliade, *The Sacred*, pp. 11, 155.

105. Krolick, *Recollective Resolve*, p. 67.

106. Ibid., p. 45.

107. Eliade, *The Sacred*, p. 105.

108. Ibid., pp. 85–90, 105.

109. Kolakowski, *The Presence*, p. 15.

110. Ibid., p. 19.

111. The difference between a mythical and a scientific view of our relation to the sacred or the objective, respectively, is closely connected to differences in how the part and whole are conceived and to the role ascribed to causality

(cf. E. Cassirer, *The Philosophy of Symbolic Forms*, vol. 2, *Mythical Thought*, trans. R. Manheim [New Haven: Yale University Press, 1955], pp. 49–50).

112. Gadamer, *Truth*, p. 476.

113. Weinsheimer, *Gadamer's Hermeneutics*, p. 4.

114. Ibid., p. 15.

115. Kockelmans, "Heidegger," pp. 159–60.

116. M. Heidegger, "The Age of World Picture," in *The Question Concerning Technology and Other Essays*, trans. W. Lovitt (New York: Harper and Row, 1977), p. 29.

117. Heidegger, *Being*, 363; Gadamer, *Truth*, p. 452; Caputo, "Heidegger's Philosophy," pp. 46–48; Richardson, *Existential Epistemology*, p. 49.

118. Dreyfus, "Holism," p. 81.

119. Heidegger, "The Age," p. 118.

120. Heidegger, *Being* (1962), 362.

121. Heidegger, *Being*, 362.

122. Rosner, *Hermeneutyka*, pp. 68–71.

123. Heidegger, *Being* (1962), 362.

124. Dreyfus, "Holism," pp. 198–204.

125. Caputo, "Heidegger's Philosophy," p. 46.

126. Heidegger, *Being*, 362–63.

127. G. E. M. Anscombe, *Causality and Determination: An Inaugural Lecture* (Cambridge: Cambridge University Press, 1971), pp. 1–7.

128. J. Faye, *The Reality of the Future* (Odense: Odense University Press, 1989), pp. 182–84; 200–206.

129. M. Friedman, "Remarks on the History of Science and the History of Philosophy," in *World Changes: Thomas Kuhn and the Nature of Science*, ed. P. Horwitch (Cambridge, Mass.: MIT Press, 1993), pp. 37–49.

130. I. Newton, "De Gravitatione et Aequipondio Fluidorum," in *Unpublished Scientific Papers of Isaac Newton: A Selection from the Portsmouth Collection in the University Library Cambridge*, ed. A. R. Hull and M. B. Hull (Cambridge: Cambridge University Press, 1962), pp. 121–52.

131. B. J. T. Dobbs, *The Janus Faces of Genius: The Role of Alchemy in Newton's Thought* (Cambridge: Cambridge University Press, 1991), pp. 122–68. She argues that *De Gravitatione* was probably written in 1684 and forms part of the sequence of writing that resulted in the composition of the *Principia*.

132. J. E. McGuire, "Predicates of Pure Existence: Newton on God's Space and Time," in *Philosophical Perspectives on Newtonian Science*, ed. P. Bricker and R. I. G. Hughes (Cambridge, Mass.: MIT Press, 1990), pp. 91–108; J. Carriero, "Newton on Space and Time: Comments on J. E. McGuire," in *Philosophical Perspectives on Newtonian Science*, ed. P. Bricker and R. I. G. Hughes (Cambridge, Mass.: MIT Press, 1990), pp. 109–32.

133. J. E. McGuire, "Phenomenalism, Relations, and Monadic Representation: Leibniz on Predicate Levels," in *How Things Are*, ed. J. Bogen and J. E. McGuire (Dordrecht: Reidel, 1981), pp. 205–33.

134. G. W. Leibniz, "Letter to Des Basses" [G II 435–37], in *Gottfried Wilhelm Leibniz: Philosophical Papers and Letters*, ed. L. E. Loemker (Dordrecht: Reidel, 1969), p. 600.

135. McGuire, "Phenomenalism," pp. 205-12.

136. D. Hume, A Treatise of Human Nature, trans. L. A. Selby-Bigge (Oxford: Clarendon Press, 1964), p. 222.

137. Ibid., bk. 1, pt. 3, sec. 2, 6.

138. G. E. M. Anscombe, *Causality and Determination,* p. 13.

139. J. E. McGuire, "Forces, Powers, Aethers, and Fields," *Boston Studies in the Philosophy of Science,* vol. 14 (Dordrecht: Reidel, 1974), p. 125.

140. J. Larmor, Aether and Matter (Cambridge: Cambridge University Press, 1900), p. 86.

141. McGuire, "Forces," p. 126.

142. Ibid., p. 125.

143. Ibid., p. 126.

144. Cf. B. Van Fraassen, *The Scientific Image* (Oxford: Clarendon Press, 1980), p. 122.

145. P. Teller, *An Interpretative Introduction to Quantum Field Theory* (Princeton, N.J.: Princeton University Press, 1995), pp. 16-35.

CHAPTER 5

1. Foucault, *The History,* p. 78.

2. Ibid., p. 82.

3. Arendt, *The Human,* p. 185.

4. Fackenheim, *Metaphysics,* p. 21.

5. J. Margolis, Historied Thought, *Constructed World: A Conceptual Primer for the Turn of the Millennium* (Berkeley: University of California Press, 1995), p. 105.

6. R. G. Collingwood, *The Idea of History* (Oxford: Oxford University Press 1956), p. 170; Ricoeur, "Narrative Time," p. 170.

7. Weinsheimer, *Gadamer's Hermeneutics,* p. 34.

8. E. Hobsbawm, *The Age of Extremes: A History of the World, 1914–1991* (New York: Pantheon Books, 1994), pp. 199-227, 255-86, 320-71.

9. J. Margolis, *The Flux of History and the Flux of Science* (Berkeley: University of California Press, 1993), pp. 48, 159.

10. Ruin, *Enigmatic Origins,* p. 134.

11. Dauenhauer, "History's Point," p. 167.

12. L. O. Mink, "History and Fiction as Modes of Comprehension," *New Literary History* 1 (1970): 557-58.

13. Fackenheim, *Metaphysics,* p. 24.

14. Ibid., pp. 24-25.

15. Ibid., p. 26.

16. Collingwood, *The Idea,* p. 169; Fackenheim, *Metaphysics,* pp. 16-24.

17. F. Ankersmit, "Historicism, Post-Modernism and Epistemology," in *Post-Modernism and Anthropology: Theory and Practice,* ed. K. Geuijen, D. Raven, and J. de Wolf (Assen: Van Gorcum, 1995), pp. 24-32; Collingwood, *The Idea,* pp. 169, 175-76.

18. Weinsheimer, *Gadamer's Hermeneutics,* p. 35.

19. P. Ricoeur, "The Narrative Function," in *Hermeneutics and the Human Sciences. Essays on Language, Action and Interpretation,* ed. and trans. J. B. Thompson (Cambridge: Cambridge University Press, 1981), p. 294.

20. Ricoeur, "Narrative Time," p. 165.

21. Carr, *Time,* pp. 45–116; MacIntyre, *After Virtue,* pp. 211–25; F. A. Olafson, *The Dialectic of Action* (Chicago: University of Chicago Press, 1979), pp. 102–3, 112–14; Ricoeur, "The Narrative Function," pp. 277–78.

22. Ricoeur, "Narrative Time," pp. 167, 277, and Dauenhauer, "History's Point," pp. 164–65.

23. Ricoeur, "The Narrative Function," p. 277.

24. Ricoeur, "Narrative Time," p. 180; Dauenhauer, "History's Point," p. 165.

25. Ricoeur, "The Narrative Function," p. 278.

26. Fackenheim, *Metaphysics,* p. 24.

27. Margolis, *The Flux,* p. 113.

28. Foucault, *The Archeology,* p. 127.

29. Weinsheimer, *Gadamer's Hermeneutics,* p. 14.

30. Ibid., p. 198.

31. Gadamer, *Truth,* p. 209.

32. Carr, *Time,* pp. 73–121.

33. Ricoeur, "The Narrative Function," p. 294.

34. Ruin, *Enigmatic Origins,* p. 64.

35. Wachterhauser, "Introduction," p. 7.

36. Weinsheimer, *Gadamer's Hermeneutics,* p. 198.

37. Ibid., p. 13.

38. Dauenhauer, "History's Point," p. 158.

39. Margolis, *Historied Thought,* p. 105.

40. Margolis, *The Flux,* p. 120.

41. N. Rotenstreich, "The Ontological Status of History," *American Philosophical Quarterly* 9 (1972): 56.

42. Ibid., p. 50.

43. Weinsheimer, *Gadamer's Hermeneutics,* p. 198.

44. Caputo, "Hermeneutics," p. 420.

45. Foucault, "Nietzsche," pp. 78, 79.

46. Heidegger, *Being,* 379.

47. L. Dupré, "Philosophy and Its History," in *At the Nexus of Philosophy and History,* ed. B. P. Dauenhauer (Athens: University of Georgia Press, 1987), p. 27; Heidegger, *Being,* 388.

48. Heidegger, *Being* (1962), 379.

49. Hoy, "History," p. 336.

50. Fackenheim, *Metaphysics,* pp. 31–32.

51. Ibid., pp. 35, 37.

52. Ibid., pp. 75–76.

53. Ibid., pp. 52–53, 89.

54. Campbell, *Truth,* pp. 405, 407.

55. Wachterhauser, "Must We," p. 225.

56. H.-G. Gadamer, "On the Problem of Self-Understanding," in *Philosophical Hermeneutics,* ed. and trans. D. E. Linge (Berkeley: University of California Press, 1976), p. 55.

57. Wachterhauser, "Must We," p. 225.

58. Jaspers, *Philosophy of Existence,* p. 72.

59. Krolick, *Recollective Resolve,* p. 85.

60. Hoffman, "Death," pp. 198–99.

61. Rotenstreich, "The Ontological Status," p. 54.

62. Foucault, "Truth and Power," p. 117.

63. Foucault, *The Archeology,* p. 12.

64. M. Foucault, "Discourse of History," in *Foucault Alive (Interviews, 1966–84),* ed. S. Lotringer and trans. J. Cascaito (New York: Semiotext(e), 1989), pp. 26–27; M. Foucault, "The Archeology of Knowledge," in *Foucault Alive (Interviews, 1966–84),* ed. S. Lotringer and trans. J. Johnston (New York: Semiotext(e), 1989), p. 51.

65. Foucault, "Truth and Power," p. 117; M. Foucault, "An Historian of Culture," in *Foucault Alive (Interviews, 1966–84),* ed. S. Lotringer and trans. J. Cascaito (New York: Semiotext(e), 1989), p. 78.

66. A. Megill, *Prophets of Extremity: Nietzsche, Heidegger, Foucault, Derrida* (Berkeley: University of California Press, 1985), pp. 211, 223.

67. Foucault, "Discourse," pp. 23–26; Foucault, "The Archeology of Knowledge," pp. 51–52.

68. Fackenheim, *Metaphysics,* p. 27.

69. Ricoeur, "Narrative Time," pp. 178, 184–85.

70. A. Schutz, *The Phenomenology of the Social World,* trans. G. Walsch and F. Lehnert (Evanston, Ill.: Northwestern University Press, 1967), pp. 142–43.

71. Fackenheim, *Metaphysics,* pp. 42, 49.

72. Ibid., p. 27.

73. Wachterhauser, "Introduction," p. 7.

74. Fackenheim, *Metaphysics,* p. 45.

75. Gadamer, "On the Problem," p. 49.

76. J. Stambaugh, *The Problem of Time in Nietzsche* (Canbury, N.J.: Associated University Presses, 1987), p. 44.

77. Bernstein, *Beyond,* p. 68.

78. Margolis, *The Flux,* pp. 161–62.

79. Ricoeur, *Time,* p. 113.

80. Ibid., pp. 114, 115, 116.

81. Stambaugh, *The Problem,* p. 41.

82. F. Nietzsche, *On the Advantage and Disadvantage of History for Life,* ed. and trans. P. Preuss (Indianapolis: Hackett, 1980), p. 9; translation modified.

83. Ibid.; translation modified.

84. Cf. Pannenberg, "Hermeneutics," p. 125.

85. Hoy, "Is Hermeneutics Ethnocentric?" p. 163.

86. Gadamer, *Truth,* p. 305.

87. Actually, Gadamer has changed his position. In the second edition of *Truth and Method* he talks about "a single horizon that embraces everything contained in historical consciousness" (H.-G. Gadamer, *Truth and Method,* trans. G. Barden and J. Cumming (trans. from 2d ed.) (New York: Seabury Press, 1975), p. 271; Margolis, *The Flux,* pp. 167–70, 227–28).

88. Gadamer, *Truth,* pp. 305–6.

89. Pannenberg, "Hermeneutics," p. 130.

90. Campbell, *Truth,* p. 400; also Fackenheim, *Metaphysics,* 40; Colling-wood, *The Idea,* p. 170.

91. Campbell, *Truth,* p. 400.

92. Gadamer, *Truth,* p. 397.

93. Nietzsche, *On the Advantage,* p. 9; also DiCenso, *Hermeneutics,* p. 49.

94. Nietzsche, *On the Advantage,* p. 11.

95. Ibid., p. 14; translation modified.

96. Ibid., p. 14.

97. Ibid., p. 13.

98. Ibid., pp. 21–22.

99. Ibid., pp. 22–32.

100. Stambaugh, *The Problem,* p. 50.

101. G. W. F. Hegel, *Lectures on the Philosophy of Religion,* ed. P. C. Hodgson and trans. R. F. Brown, P. C. Hodgson, J. M. Stewart (Berkeley: University of California Press, 1985), p. 212.

102. Campbell, *Truth,* pp. 404–8.

103. Gadamer, "On the Problem," p. 55.

104. H.-G. Gadamer, "The Philosophical Foundations of the Twentieth Century," in *Philosophical Hermeneutics,* ed. and trans. D. E. Linge (Berkeley: University of California Press, 1976), p. 123; Weinsheimer, *Gadamer's Hermeneutics,* p. 38.

105. Dreyfus, *Being-in-the-World,* p. 244.

106. Seibt, "Existence," p. 143.

107. Ibid., pp. 153–60; Rescher, *Process Metaphysics,* p. 65.

108. W. V. O. Quine, *Methods of Logic* (New York: Holt, 1950), pp. 210; Quine, *World,* p. 171.

109. Seibt, "Existence," p. 160.

110. P. T. Geach, *Some Problems About Time: Annual Philosophical Lecture,* Henrietta Hertz Trust (London: British Academy, 1965), p. 323.

111. Geach, *Some Problems,* pp. 323–29.

112. The distinction between mere variation and real change is made by Geach. He points out that according to the perdurantist scheme "man's growth would be regarded as the tapering of a four-dimensional body along its time-axis from earlier to later; but this again would no more be a change than is a poker's tapering along its length towards its point" (ibid., p. 323). References to different attributes, occurring at various positions along a time line, do not constitute an analysis of self-referential change with respect to one and the same thing across time.

113. Broad, *Examination,* pp. 155–56; Rescher, *Process Metaphysics,* p. 45.

114. Campbell, *Truth,* p. 400.

115. Nietzsche, *The Will,* pp. 636, 708.

116. Nehamas, *Nietzsche,* pp. 98–105.

117. Nietzsche, *The Will,* pp. 557, 635.

118. T. Merricks, "Endurance and Indiscernibility," *Journal of Philosophy* 91 (1994): 166.

119. Nietzsche, *The Will,* pp. 557, 689.

120. J. Richardson, *Nietzsche's System* (Oxford: Oxford University Press, 1996), p. 106.

121. F. Nietzsche, *The Gay Science,* trans. W. Kaufmann (New York: Vintage Books, 1974), p. 354; A. C. Danto, *Nietzsche as Philosopher* (New York: Macmillan 1965), pp. 116-22; Nehamas, *Nietzsche,* p. 85.

122. Nietzsche, *The Gay,* p. 354.

123. Within this contextualized, temporalized view, the puzzles of mind-body dualism are irrelevant: there is no question of two different equiprimordial substances—the mental and the physical—coming to relate in some mysterious way.

124. Nehamas, *Nietzsche,* p. 191.

125. O. Schutte, *Beyond Nihilism: Nietzsche without Masks* (Chicago: University of Chicago Press, 1986), pp. 76-104.

126. Nietzsche, *The Will,* pp. 538, 778.

127. Foucault, *The History,* p. 10.

128. Ibid., p. 6.

129. M. Foucault (writing as "Maurice Florence"), "Foucault, Michel, 1926-," in *The Cambridge Companion to Foucault,* ed. G. Gutting and trans. C. Porter (Cambridge: Cambridge University Press, 1994), p. 316.

130. MacIntyre, *After Virtue,* pp. 216-19; Nehamas, *Nietzsche,* p. 182.

131. See J. Haugeland, *Having Thought: Essays in the Metaphysics of Mind* (Cambridge, Mass.: Harvard University Press, 1998), pp. 127-70, for a perceptive account of three competing views of intentionality: that it resides in language-like internal representations, or in interactions between situated agents, or in the normatively sustained social practices of a community.

132. Strawson, *Individuals,* pp. 90, 102.

133. N. Elias, *The Civilizing Process: The History of Manners and State Formation and Civilization,* trans. E. Jephcott (Oxford: Basil Blackwell, 1994), pp. 134-56.

134. Ibid., pp. 155-56.

135. Ibid., p. 156.

136. R. Chartier, "Introduction" to chapter 2 in *A History of the Private Life: Passions of the Renaissance,* ed. R. Chartier and trans. A. Goldhammer (Cambridge: Belknap Press, 1989), 3:163-65; J. Revel, "The Uses of Civility," in *A History of the Private Life.*

137. Elias, *The Civilizing Process,* pp. 335-439.

138. Ibid., p. 347.

139. J. Gélis, "The Child: From Anonymity to Individuality," in *A History of the Private Life: Passions of the Renaissance,* ed. R. Chartier and trans. A. Goldhammer (Cambridge: Belknap Press, 1989), 3:309-26; Chartier, "Introduction" (to chap. 3), 3:399-401.

140. M. Hunter, *Science and Society in Restoration England* (Cambridge: Cambridge University Press, 1981), pp. 3-4.

141. Elias, *The Civilizing Process,* pp. 443-56.

142. P. Connerton, "Bakhtin and the Representation of the Body," *Journal of the Institute of Romance Studies* 1 (1992): 361.

143. Chartier, "Introduction" (to chap. 1), 3:16.

144. C. Taylor, *Sources of the Self: The Making of the Modern Identity* (Cambridge, Mass.: Harvard University Press, 1989), pp. 166–76.

145. Chartier, "Introduction," 3:18.

146. M. Heyd, "The Emergence of Modern Science as an Autonomous World of Knowledge in the Protestant Tradition of the Seventeenth Century," *Knowledge and Society: Studies in the Sociology of Culture Past and Present* 7: *Cultural Traditions and Worlds of Knowledge: Explorations in the Sociology of Knowledge*, S. N. Eisenstadt and Ilana Friedrich Silber, guest eds. (1988): 165–79.

147. Elias, "Scientific Establishments," p. 11.

148. Elias, *The Society*, p. 28.

149. Foucault, "An Historian," pp. 78–79; J. Mohanty, "Foucault as a Philosopher," in *Institutions, Normalization, and Power*, ed. J. Caputo and M. Yount (Philadelphia: Penn State University Press, 1993), pp. 36–39.

150. Taylor, *Sources*, pp. 131–36.

151. Elias, *The Society*, pp. 93–95, 103–5.

152. Wittgenstein, *Philosophical*, I, secs. 91–92.

Chapter 6

1. See Bernstein, *Beyond*, p. 74; and M. Wartofsky, "The Relation Between Philosophy of Science and History of Science," in *Models* (Dordrecht: Reidel, 1979), pp. 133–34, in reference to Kuhn and Lakatos; E. McMullin, "Discussion Review: Laudan's Progress and Its Problems," *Philosophy of Science* 46 (1979): 630 in reference to Laudan.

2. Ruin, *Enigmatic Origins*, p. 134.

3. D. Haraway, "Situated Knowledges: The Science Question in Feminism and the Privilege of Partial Perspective," *Feminist Studies* 14 (1988): 584.

4. Kuhn, *The Structure;* Toulmin, *Human Understanding;* D. Hull, *Science as a Process;* R. N. Giere, "Evolutionary Models of Science," in *Evolution, Cognition, and Realism: Studies in Evolutionary Epistemology*, ed. N. Rescher (Lanham-New York-London: University of America Press, 1990). Popper's claim that the development of science is a prolongation of biological evolution also belongs to the naturalist approach, but his model belongs to epistemological models.

5. Cf. R. Hilpinen, *On the Characterization of Cognitive Progress*, Reports from the Department of Theoretical Philosophy no. 16 (Turku: University of Turku, 1986), pp. 2–7; R. Wachbroit, "Progress: Metaphysical and Otherwise," *Philosophy of Science* 53 (1986): 356–57.

6. Bernstein, *Beyond*, pp. 82–83; Suppe, "The Search," p. 53.

7. K. R. Popper, *Objective Knowledge: An Evolutionary Approach* (Oxford: Clarendon Press, 1972), p. 47; I. Lakatos, "History of Science and Its Rational Reconstructions," in *I. Lakatos: Philosophical Papers*, vol. 1, ed. J. Worrall, G. Currie (Cambridge: Cambridge University Press, 1978), pp. 113–14.

8. Popper, *Objective Knowledge*, pp. 118–19, 243–44.

9. I. Lakatos, "Falsification and the Methodology of Scientific Research Programmes," in *Philosophical Papers*, vol. 1, ed. J. Worrall, G. Currie (Cam-

bridge: Cambridge University Press, 1978), pp. 69–72, 76, 87–90; E. Zahar, *Einstein's Revolution: A Study in Heuristic* (La Salle, Ill.: Open Court, 1989), p. 23.

10. Lakatos, "Falsification," pp. 31–32, 69.

11. Giere, *Explaining Science*, p. 34.

12. Kuhn, *The Structure*, p. 12.

13. Ibid., p. 185.

14. T. S. Kuhn, "Objectivity, Value Judgment, and Theory Choice," in *Introductory Readings in the Philosophy of Science*, rev. ed., ed. E. D. Klemke, R. Hollinger, A. D. Kline (Buffalo: Prometheus Books, 1988), p. 280; Mulkay and Gilbert, "Theory Choice," p. 133.

15. Kuhn, *The Structure*, p. 204. Objectively speaking, at the core of scientific revolutions is a change of taxonomic categories that enable describing and generalizing, because revolutions "alter the knowledge of nature that is intrinsic to the language itself." This change also has a psychological basis, namely, "a change in one's sense of what is similar to what, and of what is different" (T. S. Kuhn, "What Are Scientific Revolutions?" in *The Probabilistic Revolution*, vol. 1, *Ideas in History*, ed. L. Krüger, L. J. Daston, M. Heidelberger [Cambridge, Mass.: MIT Press, 1987], pp. 20–21).

16. P. K. Feyerabend, *Farewell to Reason* (London: Verso, 1987), p. 118.

17. Rouse, "Philosophy of Science," p. 141.

18. L. Laudan, *Science and Values: The Aims of Science and Their Role in Scientific Debate* (Los Angeles: University of California Press, 1984), pp. 50–62, 65.

19. D. Ginev, "Scientific Progress and the Hermeneutical Circle," *Studies in History and Philosophy of Science* 19 (1988): 393.

20. Heidegger, *Being*, 9.

21. Caputo, "Heidegger's Philosophy," p. 55.

22. Ibid., p. 54.

23. Ibid., p. 55.

24. Heidegger, "The Age," p. 117.

25. Gadamer, *Truth*, p. 283.

26. Ibid., p. 283.

27. Margolis, *The Flux*, p. 48.

28. Ibid., pp. 48–49.

29. Ibid., pp. 120, 129.

30. Feyerabend criticizes the idea that the creativity of the human mind is a source of novel theories in science because it presupposes some mysterious human ability (*Farewell*, pp. 128–43). However, as Millstone, following Sartre, emphasizes: "Our thoughts, beliefs, knowledge and actions are not reducible to the social and material conditions which occasion them" ("A Framework for the Sociology of Knowledge," *Social Studies of Science* 8 [1978]: 113). As we have pointed out, we are capable of going beyond the given, due to natality that enables not creation ex nihilo, but the beginning of something new that surpasses an existing situation.

31. Lakatos, "History," p. 119.

32. Ibid., p. 119.

33. Bernstein, *Beyond*, p. 68.

34. The access to the Truth would allow us to state simply that Bellarmine's arguments were deficient because they were false.

35. Ibid., p. 67.

36. Margolis, *The Flux*, pp. 161, 173.

37. Ibid., pp. 180–81.

38. Ibid., p. 176.

39. M. Hesse, *The Structure of Scientific Inference* (London: Macmillan Press, 1974), p. 290.

40. P. Horwich, *Truth* (Oxford: Basil Blackwell, 1990), pp. 54–70; M. Devitt, *Realism and Truth* (Princeton: Princeton University Press 1984), p. 227.

41. Tuchańska, B., "Can Relativism Be Reconciled with Realism and Causalism?" *International Studies in the Philosophy of Science* 4 (1990): 289.

42. Margolis, *The Flux*, pp. 189–93.

43. Ibid., pp. 196, 205.

44. Ibid., p. 197.

45. Ibid., p. 160.

46. Ibid., pp. 159–61, 188–93.

47. Heidegger, *Being*, 11, 357.

48. Heidegger, *The End*, p. 23.

49. Cf. Margolis, *The Flux*, pp. 140–41, 168; D. Shapere, *Reason and the Search for Knowledge* (Dordrecht: Reidel, 1984), pp. 211, 250–54.

50. Shapere, *Reason*, p. 212.

51. Knorr-Cetina, "The Ethnographic," p. 134.

52. Knorr-Cetina, *The Manufacture*, p. 19.

53. Latour and Woolgar, *Laboratory Life*, p. 36.

54. Gadamer, *Truth*, p. 357.

55. By "the world" we mean here not only nature as it exists for the natural sciences, but also society as it exists for the social sciences and history as it exists for the science of history.

56. Weinsheimer, *Gadamer's Hermeneutics*, p. 39.

57. Gadamer, *Truth*, p. 398.

58. Lawson, *Reflexivity*, p. 9.

59. Rouse, *Knowledge*, p. 117.

60. See Taylor, "Heidegger," pp. 114–20, and "Overcoming Epistemology," pp. 9–10, for an interesting analysis of the wider implications of this point.

61. Heidegger, *Being*, 205.

62. Ibid., 203.

63. Ibid., 155.

64. Ibid., 208.

65. Ibid., 201.

66. Ibid., 212.

67. Ibid., 228.

68. Ibid., 221.

69. Ibid., 219.

70. Taylor, "Overcoming," pp. 11–12.

71. Heidegger, *Being*, 218. See also 154–64.

72. Bernasconi and Schürmann give a particularly good account of the dialectical interplay of concealment and unconcealment. R. Bernasconi, *The Question of Language in Heidegger's History of Being* (New Jersey, N.J.: Humanities Press, 1985), pp. 15–23; R. Schürmann, *Heidegger on Being and Acting: From Principles to Anarchy* (Bloomington: Indiana University Press, 1987), pp. 1–74. These writers provide a compelling account of why the primary locus of truth lies with the totality of Dasein's engagements rather than with epistemic acts of abstract asserting. Untruth (falsity) is neither coordinated with truth nor its direct logical opposite, the view of those who locate both primarily with the logic of asserting. Thus, falsehood is not a sentence's failure to correspond with reality. It is more a matter of covering something up, of distorting it, and is done in ways other than sheer asserting—for example, by omission and by nonverbal actions.

73. Heidegger, *Being,* 218–19. Heidegger explicitly rejects representationalism as constitutive of the primordial notion of truth. In an act of disclosing something "Representations are not compared, neither among themselves nor in *relation* to the real thing" (218). In this context Heidegger parts company with representationalism as a theory of knowledge. It clearly fails to solve the problem it claims to solve. That is, it has to assume some sort of awareness of "real" things (which it holds to be impossible) in order to explain why being aware of them is possible only through awareness of their representations. Heidegger's point is that we directly encounter entities in their involvements with us and in their relations to other entities without the mediation of mentalistic entities (e.g., sense-data, intentional states) the ontic status of which is dubious.

74. Ibid., 220, 226, 227.

75. Ibid., 227.

76. Ibid.

77. Ibid., 222.

78. Misgeld, "On Gadamer's Hermeneutics," pp. 150–51.

79. Dreyfus, "Holism," p. 234.

80. Many writers do not realize that for Kuhn and Feyerabend the concept of incommensurability was mainly a critical idea. Kuhn introduced it to emphasize that the transition between certain normal-scientific traditions cannot be accomplished with the help of "semantically neutral techniques" and described as a logical relation (Kuhn, *The Structure,* pp. 150, 266–71; Kuhn, "Commensurability"; T. S. Kuhn, "Rationality and Theory Choice," *Journal of Philosophy* 80 [1983], pp. 669–70). Feyerabend, who presents his concept of incommensurability as opposed to that of Kuhn (P. K. Feyerabend, *Science in a Free Society* [London: New Left Books, 1978], p. 67; P. K. Feyerabend, *Against Method: Outline of an Anarchistic Theory of Knowledge,* rev. ed. [London: Verso, 1988], pp. 218–26), stresses the relative nature of incommensurability that appears only from a certain philosophical viewpoint. Incommensurable theories cannot be compared by reference to their "content or verisimilitude," though they can be compared on the basis of the criteria of theoretical preference: simplicity and coherence of a theory, the number of facts predicted, and the

theory's conformity with other basic theories and metaphysical principles (Feyerabend, *Science,* pp. 68–70).

81. Misgeld, "On Gadamer's Hermeneutics," pp. 151–52.

82. Kisiel, "The Happening," pp. 12–13.

83. Dupré, "Philosophy," p. 20.

84. Kuhn, *The Structure,* p. 140; Gadamer, *Truth,* p. 283.

85. Gadamer, *Truth,* p. 282.

86. Gadamer, "Text," p. 392.

87. J. Rouse, "Interpretation in Natural and Human Science," in The In*terpretive Turn: Philosophy, Science, Culture,* ed. D. R. Hiley, J. F. Bohman, R. Shusterman (Ithaca: Cornell University Press, 1991), p. 44.

88. Ibid., pp. 48–50.

89. Ibid., p. 55.

90. Wachterhauser, "Introduction," p. 13.

91. Gadamer, *Truth,* p. 285.

92. Pannenberg, "Hermeneutics," p. 122.

93. Ibid., p. 126.

94. Gadamer, *Truth,* pp. 387–88, 377.

95. Fleck, *Genesis,* p. 22.

96. Gadamer, *Truth,* p. 537.

97. Galison, *Image and Logic,* pp. 781–844.

98. Kockelmans, "Heidegger," pp. 156–57.

99. Knorr-Cetina, *The Manufacture,* p. 142; Rouse, *Knowledge,* pp. 76, 95.

100. Heidegger, "The Age," p. 127; Heidegger, *Being,* 65.

101. Ingram, "Hermeneutics," p. 35; Heidegger, *Being,* 73–74, 158.

102. Heidegger, *Being,* 61. Circumspective deliberation is a different attitude from circumspection, the "looking around" that belongs to our everyday practice and discovers entities as tools (69; Dreyfus, *Being-in-the-World,* p. 82; Richardson, *Existential Epistemology,* p. 18).

103. Heidegger, *Being,* 359; Caputo, "Heidegger's Philosophy," p. 80.

104. Dreyfus, *Being-in-the-World,* p. 81; Brandom, "Heidegger's Categories," p. 403–4.

105. Heidegger, *Being,* 70.

106. Ibid., 206.

107. Ibid., 62.

108. Ibid., 361–62.

109. Rouse, *Knowledge,* p. 77.

110. Dreyfus, *Being-in-the-World,* pp. 116, 121.

111. Ibid., pp. 206, 116, 121.

112. Heidegger, *Being,* (1962) 65.

113. Dreyfus, *Being-in-the-World,* p. 123.

114. Richardson, *Existential Epistemology,* p. 48.

115. Dreyfus, *Being-in-the-World,* p. 113; also Heidegger, *Being,* 69.

116. Heidegger, *Being,* 65.

117. Ibid., 357, 408.

118. Knorr-Cetina, *The Manufacture,* p. 142.

119. Richardson, *Existential Epistemology*, pp. 68–71.

120. In fact, the explanation of Dasein's concernful dealing with tools requires a parallel (though more basic) assumption that refers to the cultural community as an ontological foundation of practical meaning. And Heidegger admits that, together with equipment, Dasein encounters the public world and the surrounding world of nature that is "accessible to everyone" (Heidegger, *Being*, 70–71).

121. Heidegger's argument for the ontological priority of encountering beings as tools is the claim that "nothing is intelligible to us unless it first shows up as already integrated into our world, fitting into our coping practices" (Dreyfus, *Being-in-the-World*, p. 115). Let us add here that not only Husserl, but also von Uexküll has elaborated a view which is an inversion of Heidegger's conception. For Husserl, an encounter with equipment presupposes the synthesis of a passive experience, which gives us "something beforehand" as an object, not as a tool (E. Husserl, *Cartesian Meditations*, trans. D. Cairns [The Hague: Martinus Nijhoff, 1960], p. 78). According to Uexküll, entities become objects when they are placed within the principle of causality. When a subjective rule of acting is imposed on objects, they turn into meaningful and functional things (J. J. von Uexküll, *Theoretische Biologie* [Frankfurt/M 1973], p. 127). We notice a ladder first as an object and later as a tool (ibid., pp. 130, 199).

122. Elias, "Scientific Establishments," pp. 37–38.

123. It is well to recall, in this regard, that the Church never entirely rejected claims to truth based on reason. Indeed, there was no abatement in the traditional debate concerning the issue of basing Divine knowledge on reason, revelation, or on reason in relation to experience.

124. R. Feldhay, "Catholicism and the Emergence of Galilean Science: A Conflict Between Science and Religion," *Knowledge and Society: Studies in the Sociology of Culture Past and Present* 7: *Cultural Traditions in the Sociology of Knowledge: Explorations in the Sociology of Knowledge*, S. N. Eisenstadt and I. Friedrich Silber, guest eds. (1988): 141.

125. Heyd, "The Emergence," p. 166.

126. Ibid., p. 174.

127. A. Funkenstein, *Theology and the Scientific Imagination: From the Middle Ages to the Seventeenth Century* (Princeton, N.J.: Princeton University Press, 1986), pp. 298–99, 5–9, 23–31, 70–116, 192–201, 296–99.

128. J. E. McGuire, "Boyle's Conception of Nature," *Journal of the History of Ideas* 33 (1972): 523–42; Funkenstein, *Theology*, pp. 117–24.

129. Webster, *The Great Instauration;* Hunter, *Science";* Shapiro, *Probability;* Shapin and Schaffer, *Leviathan.*

130. The neopositivist understanding of the language that constitutes the foundation of science went through well-known phases of psychologism, phenomenalism, and physicalism. The ideals of the intersubjectivity of empirical data and the unity of science were the crucial criteria preserved during these changes.

131. C. Glymour, "Realism and the Nature of Theories," in *Introduction to the Philosophy of Science: A Text by Members of the Department of the History and Philosophy of Science of the University of Pittsburgh*, by M. H. Salomon, J. Earman, C.

Glymour, J. G. Lennox, P. Machamer, J. E. McGuire, J. D. Norton, W. C. Salomon, K. F. Schaffner (Englewood Cliffs, N.J.: Prentice-Hall, 1992), p. 114.

132. One of the presuppositions of logical empiricism is the idea of a symmetry between empirical and analytical truths, between proofs within logic or mathematics and the verification of empirical propositions by observations (G. Baker, *Wittgenstein, Frege and the Vienna Circle* [Oxford: Basil Blackwell 1988], p. 223). Even though they were aware that statements can be logically justified only by other statements, they still believed that perceptual experience provides a basic kind of justification for observational statements (K. R. Popper, *The Logic of Scientific Discovery* [London: Hutchinson of London, 1959], p. 43), that ideas are "derived" from experience (Baker, *Wittgenstein*, p. 169), that there is a "direct logical link" between evidence and theory (McMullin, "Values," p. 369).

133. D. Shapere, "Notes Toward a Post-Positivistic Interpretation of Science," in *The Legacy of Logical Positivism: Studies in the Philosophy of Science*, ed. P. Achinstein and S. F. Barker (Baltimore: Johns Hopkins Press, 1969), p. 117.

134. Dreyfus, "Holism," p. 228.

135. H. Putnam, "What Theories Are Not," in *Introductory Readings in the Philosophy of Science*, ed. E. D. Klemke, R. Hollinger, and A. D. Kline (Buffalo: Prometheus Books, 1988), pp. 179–82.

136. Cf. J. Bogan and J. Woodward, "Saving the Phenomena," *Philosophical Review* 97, no. 3 (1988): 303–6.

137. Polanyi, *Personal Knowledge*, p. 30.

138. M. Hesse, *Revolutions and Reconstructions in the Philosophy of Science* (Brighton: Harvester Press, 1980), pp. 172–73. In the light of the theoretical character of scientific experience (and its sociohistorical situatedness), one of Kuhn's methodological values, accuracy of predictions, is problematic. It requires comparison of theories in similar experimental contexts; but there may be no similar experimental contexts (J. Z. Buchwald, "Design for Experimenting," in *World Changes: Thomas Kuhn and the Nature of Science*, ed. P. Horwich [Cambridge, Mass.: MIT Press, 1993], p. 204).

139. Popper, *The Logic*, p. 106; van Fraassen, *The Scientific*, pp. 4, 18–19.

140. Lakatos, "Falsification," p. 44.

141. Ibid., pp. 52, 65–66.

142. P. K. Feyerabend, "Explanation, Reduction, and Empiricism," in *Minnesota Studies in Philosophy*, vol. 3, *Scientific Explanation, Space and Time*, ed. H. Feigl, and G. Maxwell (Minneapolis: University of Minnesota Press, 1962), pp. 50–51.

143. Feyerabend, *Against Method*, p. 265.

144. P. K. Feyerabend, "Classical Empiricism," in *Philosophical Papers*, vol. 2, *Problems of Empiricism* (Cambridge: Cambridge University Press, 1981), pp. 37–38.

145. Feyerabend, *Science*, p. 25.

146. W. V. Quine, "Three Indeterminacies," in *Perspectives on Quine: Proceedings of Perspectives on Quine, An International Conference*, April 9–13, 1988, at *Washington University in St. Louis*, ed. R. B. Barrett and R. F. Gibson (Oxford: Basil Blackwell, 1990); W. V. Quine, "Structure and Nature," *Journal of Philosophy*

89 (1992); Hesse, *The Structure,* pp. 11-12; Hesse, *Revolutions,* p. xvi; Kuhn, "Second Thoughts," p. 473; McDowell, *Mind,* pp. 24-33.

147. Gadamer, *Truth,* p. 352.

148. Ibid., p. 346.

149. Ibid., pp. 346-47.

150. Ibid., p..353.

151. Fleck, *Genesis,* pp. 96-97, 94.

152. P. Dear, "Narratives, Anecdotes, and Experiments: Turning Experience into Science in the Seventeenth Century," in *The Literary Structure of Scientific Argument: Historical Studies,* ed. P. Dear (Philadelphia: University of Pennsylvania Press, 1991), pp. 135-63.

153. Funkenstein, *Theology,* pp. 155-56.

154. This was more than what had been accomplished earlier in Greek philosophy, which deprived everyday experience of the right to explain reality, to know it as it ought to be known in virtue of revealing its essence.

155. D. Dubarle, "Galileo's Methodology of Natural Science," in *Galileo: Man of Science,* ed. and trans. E. McMullin (Princeton: Scholar's Bookshelf, 1988), p. 303.

156. Ibid., pp. 301, 311.

157. Gadamer, *Truth,* p. 449.

158. Gadamer, *Reason,* pp. 144-45.

159. Feyerabend, *Against Method,* p. 55.

160. Ibid., pp. 124-26.

161. P. K. Machamer, "Feyerabend and Galileo: The Interaction of Theories, and the Reinterpretation of Experience," *Studies in History and Philosophy of Science* 4 (1973): 30.

162. Dubarle, "Galileo's Methodology," p. 307.

163. Feyerabend, *Against Method,* p. 108.

164. T. S. Kuhn, "Mathematical versus Experimental Traditions in the Development of Physical Science," in *The Essential Tension: Selected Studies in Scientific Tradition and Change* (Chicago: University of Chicago Press, 1977), p. 44.

165. Shapin, *A Social,* p. 195.

166. Ibid., pp. 194-200; L. Daston, "Objectivity and the Escape from Perspective," *Social Studies of Science* 22 (1992): 597-618; P. Dear, "Totius in Verba: Rhetoric and Authority in the Early Royal Society," *Isis* 76 (1985).

167. R. Descartes, "Rules for the Direction of the Mind," in *The Essential Descartes,* ed. M. D. Wilson (New York: A Meridian Book, 1969), p. 77.

168. Ibid., pp. 82, 85.

169. Kant, *Critique,* Bxiii Bxi-Bxvi, A480/B508, A792/B820-A793/ B821; S. L. de C. Fernandes, *Foundations of Objective Knowledge: The Relations of Popper's Theory of Knowledge to That of Kant* (Dordrecht: Reidel, 1985), pp. 70-123.

170. I. Kant, *Prolegomena to Any Future Metaphysics that Will Be Able to Come Forward as Science,* trans. P. Carus and J. W. Ellington (Indianapolis: Hachett, 1977), pp. 49, 39.

171. Giere, *Explaining Science,* p. 93.

172. L. Daston and P. Galison, "The Image of Objectivity," *Representations* 40 (1992): 83–84.

173. Giere, *Explaining Science*, p. 128.

174. I. Hacking, *Representing and Intervening: Introductory Topics in the Philosophy of Natural Science* (Cambridge: Cambridge University Press, 1983), pp. 154–59.

175. Ibid., pp. 220–21, 230.

176. Giere, *Explaining Science*, p. 138.

177. Ibid., p. 138.

178. P. K. Feyerabend, "Realism and the Historicity of Knowledge," *Journal of Philosophy* 86 (1989): 404–5.

179. M. N. Wise, "Mediations: Enlightenment Balancing Acts, or the Technologies of Rationalism," in *World Changes: Thomas Kuhn and the Nature of Science*, ed. P. Horwich (Cambridge, Mass.: MIT Press, 1993), pp. 213–14, 244–45, 248–49.

180. Kuhn, "Afterwords," p. 332.

181. Hacking, *Representing*, pp. 21–31, 262–63.

182. Latour and Woolgar, *Laboratory Life*, p. 179.

183. Latour, "Give Me a Laboratory," p. 146.

184. For a full exchange of their views, see I. Hacking, "On the Stability of Laboratory Sciences," *Journal of Philosophy* 85 (1988): 507–14, and P. Heelan, "Experiment and Theory: Constitution and Reality," *Journal of Philosophy* 85 (1988): 515–24.

185. Heelan, *Space-Perception*, p. 193.

186. Heelan, "Natural Science," p. 192.

187. Heelan, *Space-Perception*, pp. 204–5.

Chapter 7

1. Richardson, *Existential Epistemology*, p. 49.

2. Daston, "Objectivity," p. 598.

3. Cf. ibid., p. 607.

4. P. K. Machamer, "The Concept of the Individual and the Idea(l) of Method in Seventeenth-Century Natural Philosophy," in *Scientific Controversies*, ed. A. Baltas, P. Machamer, and M. Pera (New York: Oxford University Press, 2000), pp. 82, 84, and 97; Westphal, "Hegel," pp. 71–72.

5. Fleck, *Genesis*, p. 144; Daston and Galison, "The Image," p. 81.

6. Fleck, *Genesis*, p. 144.

7. Daston, "Objectivity," p. 599.

8. Ibid., p. 600.

9. Ibid., p. 599.

10. Cf. C. G. Hempel, "Valuation and Objectivity in Science," in *Physics, Philosophy and Psychoanalysis: Essays in Honor of Adolf Grünbaum*, ed. R. S. Cohen, L. Laudan (Dordrecht: Reidel, 1983), p. 93.

11. K. R. Popper, *The Open Society*, p. 216.

12. Cf. Longino, "The Fate," pp. 144–45; Longino in Callebaut, *Taking*, pp. 25–28; M. Solomon, "A More Social Epistemology," in *Socializing Epistemology: The Social Dimensions of Knowledge*, ed. F. F. Schmitt (Lanham: Rowman and Littlefield, 1994), p. 226.

13. Bernstein, *Beyond*, p. 60.

14. Toulmin, *Human Understanding*, pp. 485, 495.

15. Ibid., p. 370.

16. L. Laudan, "Progress or Rationality? The Prospects for Normative Naturalism," *American Philosophical Quarterly* 24 (1987): 29.

17. Cf. Hempel, "Valuation," pp. 75–78, 81–85.

18. Popper, *Objective Knowledge*, pp. 52–60; cf. J. W. N. Watkins, "The Unity of Popper's Thought," in *The Philosophy of Karl Popper*, vol. 1, ed. P. A. Schilpp (La Salle, Ill.: Open Court, 1974), p. 400; R. Curtis, "Institutional Individualism and the Emergence of Scientific Rationality," *Studies in History and Philosophy of Science* 20 (1989): 78–98; W. H. Newton-Smith, *The Rationality of Science* (London: Routledge & Kegan Paul, 1981), pp. 46–76.

19. Bernstein, *Beyond*, pp. 56–57.

20. Newton-Smith, *The Rationality*, pp. 240–43; Toulmin, *Human Understanding*, pp. 134–35, 150, 364.

21. H. Siegel, "What Is the Question Concerning the Rationality of Science?" *Philosophy of Science* 52 (1985): 520–21.

22. Newton-Smith, *The Rationality*, pp. 4–8, 269–70; Siegel, "What Is," p. 529; Toulmin, *Human Understanding*, pp. 134, 150.

23. Newton-Smith, *The Rationality*, p. 269; Curtis, "Institutional," pp. 95–96; Toulmin, *Human Understanding*, p. 134.

24. Newton-Smith, *The Rationality*, p. 255; Toulmin, *Human Understanding*, p. 498; H. Meynell, "On the Limits of the Sociology of Knowledge," *Social Studies of Science* 7 (1977): 494–95.

25. Laudan, "Progress," pp. 21, 25, 29; Laudan, *Science*, pp. 39–41; Ginev, "Scientific," pp. 392–93.

26. Laudan, *Science*, pp. 39–41, 50–62.

27. Laudan, "Progress," p. 24.

28. G. Doppelt, "Relativism and the Reticulated Model of Scientific Rationality," *Synthese* 69 (1986); H. Siegel, "Laudan's Normative Naturalism," *Studies in History and Philosophy of Science* 21 (1990); D. Stump, "Fallibilism, Naturalism and the Traditional Requirements for Knowledge," *Studies in History and Philosophy of Science* 22 (1991).

29. Siegel, "Laudan's," pp. 308–10.

30. Hempel, "Valuation," p. 75.

31. Siegel, "Laudan's," p. 310.

32. Ibid., p. 311; Doppelt, "Relativism," p. 231.

33. L. Laudan, "Aim-less Epistemology?" *Studies in History and Philosophy of Science* 21 (1990): 320.

34. Feyerabend, *Against Method*, pp. 243–44.

35. Toulmin, *Human Understanding*, p. 484.

36. Feyerabend, *Farewell*, pp. 27–28.

37. Ibid., pp. 24, 26, 164–65.

38. Feyerabend, *Against Method*, pp. 243–44, 250.

39. Siegel, "What Is," p. 529; also W. W. Bartley III, "Theories of Rationality," in *Evolutionary Epistemology, Rationality, and the Sociology of Knowledge*, ed. G. Radnitzky and W. W. Bartley III (La Salle, Ill.: Open Court, 1987), pp. 209–11.

40. Daston, "Objectivity," p. 608.

41. Lynch, *Scientific Practice*, p. 317.

42. The distinction made by sociologists indicates that Daston fails to distinguish clearly between the anti-psychologistic view that "individual idiosyncrasy" may be eliminated if one discovers objective sense and truth (as in Frege or Peirce) and the social or hermeneutic view that the elimination of "individual idiosyncrasy" requires "the prolonged 'averaging' of viewpoints by communication" (Daston, "Objectivity," p. 607). The first view remains individualistic: no communication and the averaging of viewpoints, or the fusing of horizons, is necessary for reaching objective knowledge; the second view is truly collective.

43. Experimental continuity and stability are achieved over time by means of narrative histories constructed out of the systematic articulation of the data recorded: a life-world of research is brought to be discursively.

44. Rouse, *Knowledge*, p. 119.

45. Warnke, *Gadamer*, p. 80.

46. Bernstein, *Beyond*, pp. 77–78.

47. Knorr-Cetina, *The Manufacture*, p. 125.

48. Rouse, *Knowledge*, p. 121.

49. Ibid., p. 121.

50. H. M. Collins, *Changing Order: Replication and Induction in Scientific Practice* (London: SAGE Publ. 1985), pp. 18–19, 20.

51. Rouse, *Knowledge*, pp. 108, 109.

52. Lynch, *Scientific Practice*, pp. 85–86.

53. Fleck, *Genesis*, p. 85.

54. Ibid., p. 85.

55. Knorr-Cetina, *The Manufacture*, p. 52.

56. Collins, "Tacit Knowledge," p. 54.

57. Rouse, *Knowledge*, p. 117.

58. J. R. Ravetz, *Scientific Knowledge and Its Social Problems* (Oxford: Clarendon Press, 1971), p. 194.

59. Rouse, *Knowledge*, pp. 113, 114.

60. Ibid., p. 110.

61. Ibid., p. 96.

62. Ibid., p. 113.

63. Bernstein, *Beyond*, pp. 77–78.

64. Misgeld, "On Gadamer's Hermeneutics," p. 154.

65. The notion of cognitive values is also central to pragmatism (cf. C. I. Lewis, *An Analysis of Knowledge and Valuation* [La Salle, Ill.: Open Court, 1946]).

66. Kuhn, *The Structure*, p. 199; Kuhn, "Rationality"; Kuhn, "Objectivity"; Laudan, *Science;* E. McMullin, "Values in Science," in *Introductory Readings in the Philosophy of Science*, rev. ed., ed. D. Klemke, R. Hollinger, and A. D. Kline (Buffalo: Prometheus Books, 1988); E. McMullin, "Rationality and Paradigm

Change in Science," *World Changes: Thomas Kuhn and the Nature of Science,* ed. P. Horwich (Cambridge: MIT Press, 1993); R. Rudner, "The Scientist *Qua* Scientist Makes Value Judgments," in *Introductory Readings in the Philosophy of Science,* rev. ed., ed. D. Klemke, R. Hollinger, and A. D. Kline (Buffalo: Prometheus Books, 1988); also Bernstein, *Beyond,* pp. 54–57.

67. McMullin, "Values," pp. 351–52, 364.

68. Kuhn, "Objectivity," p. 287; Kuhn, "Afterwords," p. 338.

69. Kuhn, "Objectivity," pp. 284–86; McMullin, "Values," p. 354.

70. Kuhn, "Objectivity," p. 287.

71. McMullin, "Values," pp. 360–61.

72. McMullin, "Rationality," pp. 67–69.

73. Bartley, "Theories," pp. 211–14; G. Radnitzky, "In Defense of Self-Applicable Critical Rationalism," in *Evolutionary Epistemology, Rationality, and the Sociology of Knowledge,* ed. G. Radnitzky and W. W. Bartley III (La Salle, Ill.: Open Court, 1987), p. 303.

74. A. Musgrave, "The Objectivism of Popper's Epistemology," in *The Philosophy of Karl Popper,* ed. P. A. Schilpp (La Salle, Ill.: Open Court, 1974), p. 565.

75. Newton-Smith notes that Popper has failed to establish a link between falsification and the end of science (*The Rationality,* p. 75); Radnitzky, that cognitive progress cannot be achieved by criticism that only eliminates error ("In Defense," p. 288); and Rescher, that a better based estimate of the truth of a picture of nature does not mean it is a better picture (*The Limits of Science* [Berkeley: University of California Press, 1984], pp. 73–74).

76. L. Kolakowski, *Religion: If There Is No God . . . On God, the Devil, Sin and Other Worries of the So-Called Philosophy of Religion* (London: Fontana Paperbacks, 1982), p. 185.

77. N. Rescher, *A System of Pragmatic Idealism,* vol. 1, *The Validity of Values: A Normative Theory of Evaluative Rationality* (Princeton, N.J.: Princeton University Press, 1993), pp. 81–92.

78. Lewis, *An Analysis,* pp. 23, 365–78.

79. Kolakowski, *Religion,* pp. 193–97.

80. Ibid., pp. 193, 175.

81. Polanyi, *Personal Knowledge,* p. xiii.

82. Ibid., pp. 134–36. This immersion of (scientific) cognitive acts into the realm of meanings and values is notoriously disregarded by naturalist epistemologies (pp. 262–63).

83. Ibid., pp. 64–65, 300.

84. Ibid., p. 308.

85. Siegel, "What Is," pp. 528–29.

86. Polanyi, *Personal Knowledge,* p. 268.

87. Ibid., pp. 183–84.

88. Ibid., p. 309.

89. Ibid., p. 164.

90. Polanyi, *The Tacit,* p. 81.

91. Ibid., p. 80.

92. Shapin, *A Social,* p. 26.

93. Ibid., pp. 36, 350.

94. Daston, Galison, "The Image," p. 103.

95. Ibid., p. 83.

96. Rouse, *Knowledge,* p. 120.

97. S. Woolgar, "Irony in the Social Study of Science," in *Science Observed: Perspectives on the Social Study of Science,* ed. K. D. Knorr-Cetina and M. Mulkay (London: SAGE Publ. 1983), p. 240.

98. Ibid., pp. 243–45.

99. B. Barnes, *Interests and the Growth of Knowledge* (London: Routledge and Kegan Paul, 1977), pp. 10, 25–26; D. Bloor, "Reply to Steven Lukes," *Studies in History and Philosophy of Science* 13 (1982): 320.

100. Woolgar, "Irony," p. 244.

101. It suffices to mention that realism stretches from the view that the world consists of mind-independent, nonsocial objects (H. Putnam, *Reason, Truth and History* [Cambridge: Cambridge University Press, 1981], p. 49) to the view that "there is good reason to believe in the existence of entities *substantially like* those postulated by theories that have been successful over the long term" (E. McMullin, "Explanatory Success and the Truth of Theory," in *Scientific Inquiry in Philosophical Perspective,* ed. N. Rescher [Lanham: University Press of America, 1987], p. 57), to an even more radical view that "there is nothing except the postulated entities of science," so that sensed objects, as well as thoughts, perceptions, and feelings, "are in fact aggregates of postulated theoretical entities" (G. Gutting, "Scientific Realism," in *The Philosophy of Wilfrid Sellars: Queries and Extensions,* ed. J. C. Pitt [Dordrecht: Reidel, 1978], pp. 105, 111, 117). This last view entails that there are no features of reality independent of scientific cognition. For example, witches or Alice's wonderland do not exist, since science sees no need to postulate them.

102. McMullin, "Explanatory Success," pp. 61–62. Hacking's realism is, thus, in direct opposition to Gutting's characterization of realism.

103. M. Devitt, "Aberrations of the Realism Debate," *Philosophical Studies* 61 (1991): 45.

104. R. Tuomela, "The Myth of the Given and Realism," *Erkenntnis* 29 (1988): 192; Newton-Smith, *The Rationality,* p. 38; Popper, *Objective Knowledge,* p. 35; also Putnam, *Reason,* p. 49.

105. R. Bhaskar, *A Realist Theory of Science* (Brighton: Harvester Press 1978), p. 250; Tuomela, "The Myth," pp. 197–99.

106. A. Fine, "Piecemeal Realism," *Philosophical Studies* 61 (1991): 83; Devitt, "Aberrations," p. 49.

107. Murphy, "Scientific Realism," p. 296.

108. A. Fine, "Unnatural Attitudes: Realist and Instrumentalist Attachments to Science," *Mind* 95 (1986): 156; A. C. Genova, "Ambiguities About Realism and Utterly Distinct Objects," *Erkenntnis* 28 (1988): 87–90.

109. H. I. Brown, "Prospective Realism," *Studies in History and Philosophy of Science* 21 (1990): 226.

110. M. C. Banner, *The Justification of Science and the Rationality of Religious Beliefs* (Oxford: Clarendon Press 1990), p. 34; Newton-Smith, *The Rationality,* p. 29; Rouse, "Interpretation," p. 44.

111. Fine, "Unnatural Attitudes," p. 150.

112. Banner, *The Justification*, p. 35.

113. Newton-Smith, *The Rationality*, p. 39; McMullin, "Explanatory Success," pp. 59–60.

114. Y. Elkana, "Two-Tier Thinking: Philosophical Realism and Historical Relativism," *Social Studies of Science* 8 (1978): 311.

115. Putnam, *Reason*, p. 54; Fine, "Unnatural Attitudes," p. 151.

116. Putnam, *Reason*, pp. 49–52.

117. T. S. Kuhn, "Second Thoughts on Paradigms" and "Discussion," in *The Structure of Scientific Theories*, ed. F. Suppe (Urbana-Chicago: University of Illinois Press, 1977), p. 473.

118. Putnam, *Reason*, "pp. 52–53.

119. C. L. Elder, "The Case Against Irrealism," *American Philosophical Quarterly* 20 (1983): 239; Genova, "Ambiguities," pp. 88–89; Kuhn, "The Road," p. 6; Quine, "Structure and Nature," pp. 8–9.

120. Genova, "Ambiguities," p. 87.

121. Devitt, *Realism*, p. 15.

122. Brown, "Prospective Realism," p. 21.

123. Putnam, *Reason*, "p. 49.

124. F. L. Will, "Reason, Social Practice, and Scientific Realism," *Philosophy of Science* 48 (1981): 1.

125. This difficulty is one part of the problem of *reciprocity;* its other part appears when realists accept that we causally interact with things (with whatever we observe). Contrary to what they presuppose, the information we retrieve from such interactions refers to "interacted-with things" (Fine, "Unnatural Attitudes," p. 151).

126. Even thinking of the possibility that such objects exist is excluded, since that possibility only mediates the fact that the world is its correlative: the existence of the world that we cannot think of depends on the possibility of its existence that we think of. It would be like thinking of the possibility of the existence of something without specifying the *what* that it is, which seems to be an absurd case of thinking. It is difficult to imagine how an idea of thinking about the possibility of the existence of something that cannot be thought of could be articulated.

127. Putnam, *Reason*, p. 50; Kuhn, "The Road," p. 6.

128. Brown, "Prospective Realism," p. 215.

129. Ibid., p. 218.

130. Ibid., p. 219; Kuhn, "The Road," pp. 3–6; Newton-Smith, *The Rationality*, pp. 40–43, 158, 179–82.

131. McMullin, "Explanatory Success," p. 66.

132. Murphy, "Scientific Realism," p. 296.

133. McMullin, "Explanatory Success," p. 66.

134. Ibid., p. 66.

135. Elkana, "Two-Tier Thinking," p. 317.

136. Putnam, *Reason*, p. 50.

137. Ibid., pp. 55–56.

138. Genova, "Ambiguities," pp. 89–90.

139. Fine, "Unnatural Attitudes," p. 157.

140. Van Fraassen, *The Scientific,* pp. 3–4, 5, 12.

141. Putnam's expression. See *Realism and Reason,* vol. 3 of *Philosophical Papers* (Cambridge: Cambridge University Press, 1983), pp. viii–xii and 205–28.

142. Collins and Yearley, "Epistemological Chicken," p. 308; Elkana, "Two-Tier Thinking," p. 317; Latour and Woolgar, *Laboratory Life,* p. 179.

143. Latour and Woolgar, *Laboratory Life,* p. 179.

144. Weinsheimer, *Gadamer's Hermeneutics,* p. 29.

145. Heelan, *Space-Perception,* pp. 210–12.

146. Ibid., p. 203.

147. Ibid., p. 19.

148. Ibid., p. 175; Heelan, "Natural Science," p. 202.

149. Heelan, *Space-Perception,* pp. 173–74.

150. Ibid., p. 177.

151. Ibid., p. 178.

152. Crease, "The Play of Nature," p. 69.

153. Ibid., p. 71.

154. Ibid., p. 76.

155. Ibid., p. 78; Crease, *The Play,* pp. 123–26.

156. Ihde, *Technics,* p. 49.

157. Tiryakian, *Sociologism,* p. 161.

158. Ingram, "Hermeneutics," pp. 36–37.

159. Cf. Grondin, "Hermeneutics," p. 46.

160. Feyerabend, *Science,* p. 34.

161. Ibid., p. 70.

162. Kuhn, *The Structure,* p. 111.

163. Ibid., p. 112.

164. Ibid., pp. 120, 121. Elaborating details of his position, Kuhn refers to the distinction between immutable stimula and changeable sensations in order to avoid both individual and social solipsism (ibid., p. 193).

165. Gadamer, *Truth,* p. 447.

166. Ibid., pp. 451, 452.

167. Ibid., p. 447.

168. Ibid., p. 447.

169. Ibid., p. 247.

170. Heidegger, History, p. 217; Kockelmans, "Heidegger," p. 149; Mulhall, *Heidegger,* p. 103.

171. Heidegger, *Being,* 226.

172. Bernstein, *Beyond,* pp. 23, 68.

173. Caputo, "Heidegger's Philosophy," p. 52.

174. For a recent, perceptive study of such epistemic issues written with an eye to their historical development, see A. Plantinga, *Warrant: The Current Debate* (Oxford: Oxford University Press, 1993).

175. Dreyfus, *Being-in-the-World,* p. 252.

176. Heidegger, *Being,* 363.

177. Dreyfus, *Being-in-the-World,* p. 206.

178. Ibid., p. 254.

179. Ibid., pp. 251–52.

180. Ibid., p. 257.

181. Heidegger, *Being,* 212.

182. Ibid., 207.

183. Collins, *Changing Order,* p. 9.

184. Woolgar, *Science,* p. 56.

185. Ibid., pp. 72, 67; Knorr-Cetina, "The Ethnographic Study," p. 136.

186. Woolgar, *Science,* p. 55.

187. Fleck, *Genesis,* p. 95.

188. Latour, *The Pasteurization,* p. 159.

189. Fleck, *Genesis,* p. 101.

190. Ibid., pp. 101–2.

191. Ibid., p. 102.

192. F. Nietzsche, *On the Genealogy of Morality,* ed. K. Ansell-Pearson and trans. C. Diethe (Cambridge: Cambridge University Press, 1994), p. 92.

193. Grondin, "Hermeneutics," p. 52.

194. Latour, *We Have Never,* pp. 112, 113.

195. Ibid., p. 127.

196. Ibid., pp. 127–28.

197. Fleck, *Genesis,* p. 100.

198. Woolgar, *Science,* pp. 55–56.

199. P. A. Heelan, "After Experiment: Realism and Research," *American Philosophical Quarterly* 26 (1989): 297.

200. The (arbitrary) ontological presupposition that they derive from the everyday life-world makes them necessarily secondary in the ontic sense and also prohibits the historical reversal of that order, because in every moment of their being they are ordered in the same way. Thus, it is not only a historical fact that the world of science grew out of the everyday life-world in the seventeenth century; it is and will remain derivative from it. The everyday life-world cannot transform into one that grows out of the world of science. However, this seems to us, for good or ill, the real situation nowadays of the everyday life-world.

201. As Rouse argues, the view that language works representationally and that (scientific) knowledge in itself represents is a nonargued presumption of philosophy and sociology of science ("Against Representation: Davidsonian Semantics and Cultural Studies of Science," in J. Rouse, *Engaging Science: How to Understand Its Practices Philosophically* [Ithaca: Cornell University Press, 1996], pp. 206–7).

202. Rouse, *Knowledge,* p. 101. In reference to scientific objects transferred outside science, Heelan uses the phrase "naturalized parts of the furniture of the lifeworld." Heelan, "Why a Hermeneutical Philosophy," pp. 289–90.

203. Weinsheimer, *Gadamer's Hermeneutics,* p. 248.

204. Kolakowski, *Religion,* p. 164.

205. Shapin, *A Social,* p. 350.

206. McMullin, "Values," p. 364.

207. M. Mulkay, "Knowledge and Utility," in *Sociology of Science: A Sociological Pilgrimage* (Philadelphia: Open University Press, 1991), p. 93.

208. Mulkay and Gilbert, "Theory Choice," p. 152.

209. Feyerabend, *Science,* p. 34.

210. Rouse, *Knowledge,* p. 121.

211. Husserl, *The Crisis,* pp. 50–51.

212. J. Rouse, "Husserlian Phenomenology and Scientific Realism," *Philosophy of Science* 54 (1987): 227.

213. Ibid., p. 229.

Epilogue

1. Kolakowski, *The Presence,* p. 17.

2. Kolakowski, *Religion,* p. 63.

3. Kolakowski, *The Presence,* p. 15.

REFERENCES

Ankersmit, F. "Historicism, Post-Modernism and Epistemology." In *Post-Modernism and Anthropology: Theory and Practice,* ed. K. Geuijen, D. Raven, J. de Wolf, pp. 21–51. Assen: Van Gorcum, 1995.

Anscombe, G. E. M. *Causality and Determination: An Inaugural Lecture.* Cambridge: Cambridge University Press, 1971.

Arendt, H. *The Human Condition.* Chicago: University of Chicago Press, 1989 [1958].

Aristotle. "Ethica Nicomachea." In *The Basic Works of Aristotle,* ed. R. McKeon, trans. W. D. Ross, pp. 927–1112. New York: Random House, 1941.

———. "Metaphysica." In *The Works of Aristotle,* ed. W. D. Ross. Vol. 8. Oxford: Clarendon Press, 1960 [1908].

Aronowitz, S. *Science as Power: Discourse and Ideology in Modern Society.* Basingstoke: Macmillan Press, 1988.

Baker, G. *Wittgenstein, Frege and the Vienna Circle.* Oxford: Basil Blackwell, 1988.

Ballard, B. W. *The Role of Mood in Heidegger's Ontology.* Lanham: University of America Press, 1991.

Banner, M. C. *The Justification of Science and the Rationality of Religious Beliefs.* Oxford: Clarendon Press, 1990.

Barnes, B. *Interests and the Growth of Knowledge.* London: Routledge and Kegan Paul, 1977.

———. *About Science.* Oxford: Basil Blackwell, 1985.

Barnes, J. "Metaphysics." In *Cambridge Companion to Aristotle,* ed. J. Barnes, pp. 66–108. Cambridge: Cambridge University Press, 1995.

Bartley, W. W., III. "Theories of Rationality." In *Evolutionary Epistemology, Rationality, and the Sociology of Knowledge,* ed. G. Radnitzky and W. W. Bartley III, pp. 205–14. La Salle, Ill.: Open Court, 1987.

Bauman, Z. *Hermeneutics and Social Science: Approaches to Understanding.* Aldershot: Gregg Revivals, 1992 [1978].

Berger, P., and T. Luckmann. *The Social Construction of Reality: A Treatise in the Sociology of Knowledge.* Garden City, N.Y.: Doubleday, 1966.

Bernasconi, R. *The Question of Language in Heidegger's History of Being.* New Jersey: Humanities Press, 1985.

———. *Heidegger in Question: The Art of Existing.* New Jersey: Humanities Press, 1993.

Bernstein, R. J. *Praxis and Action: Contemporary Philosophies of Human Activity.* Philadelphia: University of Pennsylvania Press, 1971.

———. *Beyond Objectivism and Relativism: Science, Hermeneutics and Praxis.* Philadelphia: University of Pennsylvania Press, 1983.

———. "History, Philosophy, and the Question of Relativism." In *At the Nexus of Philosophy and History,* ed. B. P. Dauenhauer, pp. 3-19. Athens: University of Georgia Press, 1987.

———. "Heidegger's Silence: *Ethos* and Technology. In *The New Constellation: The Ethical-Political Horizons of Modernity/Postmodernity,* pp. 79-141. Cambridge, Mass.: MIT Press, 1993.

Bhaskar, R. *A Realist Theory of Science.* Brighton: Harvester Press, 1978.

Biagioli, M. *Galileo Courtier: The Practice of Science in the Culture of Absolutism.* Chicago: University of Chicago Press, 1993.

Bloor, D. *Knowledge and Social Imagery.* London: Routledge and Kegan Paul, 1976.

———. "Reply to Steven Lukes." *Studies in History and Philosophy of Science* 13 (1982): 319-22.

Bogan, J., and J. Woodward. "Saving the Phenomena." *Philosophical Review* 97, no. 3 (1988): 303-52.

Bourdieu, P. "Intellectual Field and Creative Project." *Social Science Information* 8 (1969): 89-119.

———. "The Specificity of the Scientific Field and the Social Conditions of the Progress of Reason." Trans. R. Nice. *Social Science Information* 14 no. 6 (1975): 19-47.

———. "The Genesis of the Concepts of *Habitus* and of *Field*." *Sociocriticism* 2 (1985): 11-24.

———. *The Logic of Practice.* Trans. R. Nice. Stanford: Stanford University Press, 1990.

Bowen-Moore, P. *Hannah Arendt's Philosophy of Natality.* New York: St. Martin's Press, 1989.

Brandom, R. B. "Heidegger's Categories in *Being and Time*." *Monist* 66 (1983): 387-409.

———. *Making It Explicit: Reasoning, Representing, and Discursive Commitment.* Cambridge, Mass.: Harvard University Press, 1994.

Broad, C. D. *Examination of McTaggart's Philosophy.* Vol. 1. Cambridge: Cambridge University Press, 1933.

Brock, W. "An Account of 'Being and Time.'" In *Existence and Being,* ed. M. Heidegger, pp. 11-116. Washington, D.C.: Gateway Editions, 1988 [1949].

Brown, H. I. "Prospective Realism." *Studies in History and Philosophy of Science* 21 (1990): 211-42.

Buchwald, J. Z. "Design for Experimenting." In *World Changes: Thomas Kuhn and the Nature of Science,* ed. P. Horwich, pp. 169-206. Cambridge, Mass.: MIT Press, 1993.

Cahoone, L. E. *The Dilemma of Modernity. Philosophy, Culture, and AntiCulture.* Albany: SUNY Press, 1988.

Callebaut, W., ed. *Taking the Naturalistic Turn, or How Real Philosophy of Science Is Done.* Chicago: University of Chicago Press, 1993.

Callon, M., and B. Latour. "Don't Throw the Baby Out with the Bath School! A Reply to Collins and Yearly." In *Science as Practice and Culture,* ed. A. Pickering, pp. 343-68. Chicago: University of Chicago Press, 1992.

Campbell, R. *Truth and Historicity.* Oxford: Oxford University Press, 1992.

Caputo, J. D. "The Thought of Being and the Conversation of Mankind: The

Case of Heidegger and Rorty." In *Hermeneutics and Praxis,* ed. R. Hollinger, pp. 248-71. Notre Dame: University of Notre Dame Press, 1985.

———. "Heidegger's Philosophy of Science: The Two Essences of Science." In *Rationality, Relativism and the Human Sciences,* ed. J. Margolis, M. Krausz, R. M. Burian, pp. 43-60. Dordrecht: Martinus Nijhoff Publ., 1986.

———. "Hermeneutics as the Recovery of Man." In *Hermeneutics and Modern Philosophy,* ed. B. R. Wachterhauser, pp. 416-45. Albany: SUNY Press, 1986.

Carr, D. *Time, Narrative, and History.* Bloomington and Minneapolis: Indiana University Press, 1991.

Carriero, J. "Newton on Space and Time: Comments on J. E. McGuire." In *Philosophical Perspectives on Newtonian Science,* ed. P. Bricker and R. I. G. Hughes, pp. 109-33. Cambridge, Mass.: MIT Press, 1990.

Cassirer, E. *The Philosophy of Symbolic Forms.* Vol. 2, *Mythical Thought.* Trans. R. Manheim. New Haven: Yale University Press, 1955 [1925].

Chartier, R. Introduction to chapter 1. In *A History of the Private Life: Passions of the Renaissance,* ed. R. Chartier and trans. A. Goldhammer, 3:15-19. Cambridge: Belknap Press, 1989.

———. Introduction to chapter 2. In *A History of the Private Life: Passions of the Renaissance,* ed. R. Chartier and trans. A. Goldhammer, 3:163-65. Cambridge: Belknap Press, 1989.

———. Introduction to chapter 3. In *A History of the Private Life: Passions of the Renaissance,* ed. R. Chartier and trans. A. Goldhammer, 3:399-401. Cambridge: Belknap Press, 1989.

Chubin, D. E., and S. Restivo. "The 'Mooting' of Science Studies: Research Programmes and Science Policy." In *Science Observed: Perspectives on the Social Study of Science,* pp. 53-83. London: SAGE Publ., 1983.

Coffa, J. A. *The Semantic Tradition from Kant to Carnap: To the Vienna Station,* ed. L. Wessels. Cambridge: Cambridge University Press, 1991.

Collingwood, R. G. *The Idea of History.* Oxford: Oxford University Press, 1956.

Collins, H. M. "Tacit Knowledge and Scientific Networks." In *Science in Context: Readings in the Sociology of Science,* ed. B. Barnes and D. Edge, pp. 44-46. Cambridge, Mass.: MIT Press, 1982.

———. "An Empirical Relativist Programme in the Sociology of Scientific Knowledge." *Science Observed: Perspectives on the Social Studies of Science,* ed. K. D. Knorr-Cetina and M. Mulkay, pp. 85-113. London: SAGE Publ., 1983.

———. *Changing Order: Replication and Induction in Scientific Practice.* London: SAGE Publ., 1985.

Collins, H. M., and S. Restivo. "Development, Diversity, and Conflict in the Sociology of Knowledge." *Sociological Quarterly* 24 (1983): 185-200.

Collins, H. M., and S. Yearley. "Epistemological Chicken." In *Science as Practice and Culture,* ed. A. Pickering, pp. 301-26. Chicago: University of Chicago Press, 1992.

Connerton, P. "Gadamer's Hermeneutics." *Comparative Criticism* 5 (1983): 107-28.

———. "Bakhtin and the Representation of the Body." *Journal of the Institute of Romance Studies* 1 (1992): 349-62.

Crease, R. P. *The Play of Nature: Experimentation as Performance.* Bloomington: Indiana University Press, 1993.

———. "The Play of Nature: Experimentation as Performance." In *Continental and Postmodern Perspectives in the Philosophy of Science,* ed. B. E. Babich, D. B. Bergoffen, S. V. Glynn, pp. 69–88. Avebury: Aldershot, 1995.

———. "Hermeneutics and the Natural Sciences: Introduction." *Man and World* 30 (1997): 259–70.

Crombie, A. C. *Styles of Thinking in the European Tradition.* London: Duckworth, 1994.

Curtis, R. "Institutional Individualism and the Emergence of Scientific Rationality." *Studies in History and Philosophy of Science* 20 (1989): 77–113.

Danto, A. C. *Nietzsche as Philosopher.* New York: Macmillan, 1965.

Daston, L. "Objectivity and the Escape from Perspective." *Social Studies of Science* 22 (1992): 597–618.

Daston, L., and P. Galison. "The Image of Objectivity." *Representations* 40 (1992): 81–128.

Dauenhauer, B. P. "History's Point and Subject Matter: A Proposal." In *At the Nexus of Philosophy and History,* ed. B. P. Dauenhauer, pp. 157–77. Athens: University of Georgia Press, 1987.

Dear, P. "Totius in Verba: Rhetoric and Authority in the Early Royal Society." *Isis* 76 (1985): 145–61.

———. "Narratives, Anecdotes, and Experiments: Turning Experience into Science in the Seventeenth Century." In *The Literary Structure of Scientific Argument: Historical Studies,* ed. P. Dear, pp. 135–63. Philadelphia: University of Pennsylvania Press, 1991.

Deleuze, G. *Foucault.* Trans. S. Hand. Minneapolis: University of Minnesota Press, 1988.

Dennett, D. C. *Kinds of Minds: Toward an Understanding of Consciousness.* New York: Basic Books, 1996.

Descartes, R. "Rules for the Direction of the Mind." In *The Essential Descartes,* ed. M. D. Wilson, pp. 35–105. New York: A Meridian Book, 1969 [1701].

Devitt, M. *Realism and Truth.* Princeton: Princeton University Press, 1984.

———. "Aberrations of the Realism Debate." *Philosophical Studies* 61 (1991): 43–63.

DiCenso, J. *Hermeneutics and the Disclosure of Truth: A Study in the Work of Heidegger, Gadamer, and Ricoeur.* Charlottesville: University Press of Virginia, 1990.

Dobbs, B. J. T. *The Janus Faces of Genius: The Role of Alchemy in Newton's Thought.* Cambridge: Cambridge University Press, 1991.

Doppelt, G. "Relativism and the Reticulated Model of Scientific Rationality." *Synthese* 69 (1986): 225–52.

Dreyfus, H. L. "Holism and Hermeneutics." In *Hermeneutics and Praxis,* ed. R. Hollinger, pp. 227–47. Notre Dame: University of Notre Dame Press, 1985.

———. *Being-in-the-World: A Commentary on Heidegger's Being and Time. Division I.* Cambridge, Mass.: MIT Press, 1991.

Dreyfus, H. L., and J. Haugeland. "Husserl and Heidegger: Philosophy's Last Stand." In *Heidegger and Modern Philosophy: Critical Essays,* ed. M. Murray, pp. 222–38. New Haven: Yale University Press, 1978.

Dubarle, D. "Galileo's Methodology of Natural Science." In *Galileo: Man of Science*, ed. and trans. E. McMullin, pp. 295–314. Princeton: Scholar's Bookshelf, 1988 [1967].

Dupré, J. *The Disorder of Things: Metaphysical Foundations of the Disunity of Science*. Cambridge, Mass.: Harvard University Press, 1993.

Dupré, L. "Philosophy and Its History." In *At the Nexus of Philosophy and History*, ed. B. P. Dauenhauer, pp. 20–41. Athens: University of Georgia Press, 1987.

Elder, C. L. "The Case against Irrealism." *American Philosophical Quarterly* 20 (1983): 239–53.

Eliade, M. *The Myth of the Eternal Return*. New York: Pantheon Books, 1954.

———. *The Sacred and the Profane: The Nature of Religion*, trans. W. R. Trask. San Diego: Harcourt Brace & Co., 1987 [1957].

Elias, N. "Scientific Establishments." In *Scientific Establishments and Hierarchies*, ed. N. Elias, H. Martins, and R. Whitley, pp. 3–69. Dordrecht: Reidel, 1982.

———. *The Society of Individuals*. Ed. M. Schröter and trans. E. Jephcott. Oxford: Basil Blackwell, 1991.

———. *The Civilizing Process: The History of Manners and State Formation and Civilization*. Trans. E. Jephcott. Oxford: Basil Blackwell, 1994 [1939].

Elkana, Y. "Two-Tier-Thinking: Philosophical Realism and Historical Relativism." *Social Studies of Science* 8 (1978): 309–26.

Emmet, D. *The Passage of Nature*. Philadelphia: Temple University Press, 1992.

Fackenheim, E. L. *Metaphysics and Historicity*. Milwaukee: Marquette University Press, 1961.

———. *To Mend the World: Foundations of Future Jewish Thought*. New York: Schocken Books, 1982.

Faye, J. *The Reality of the Future*. Odense: Odense University Press, 1989.

Feldhay, R. "Catholicism and the Emergence of Galilean Science: A Conflict between Science and Religion?" *Knowledge and Society: Studies in the Sociology of Culture Past and Present* 7 (1988): 139–63.

Fernandes, S. L., de C. *Foundations of Objective Knowledge: The Relations of Popper's Theory of Knowledge to That of Kant*. Dordrecht: Reidel, 1985.

Feyerabend, P. K. "Explanation, Reduction, and Empiricism." In *Minnesota Studies in Philosophy*. Vol. 3, *Scientific Explanation, Space and Time*, ed. H. Feigl and G. Maxwell, pp. 28–97. Minneapolis: University of Minnesota Press, 1962.

———. *Science in a Free Society*. London: New Left Books, 1978.

———. "Classical Empiricism." In *Philosophical Papers*. Vol. 2, *Problems of Empiricism*, pp. 34–51. Cambridge: Cambridge University Press, 1981.

———. "Wittgenstein's *Philosophical Investigations*." In *Philosophical Papers*. Vol. 2, *Problems of Empiricism*, pp. 99–130. Cambridge: Cambridge University Press, 1981 [1955].

———. *Farewell to Reason*. London: Verso, 1987.

———. *Against Method: Outline of an Anarchistic Theory of Knowledge*. Rev. ed. London: Verso, 1988.

———. "Realism and the Historicity of Knowledge." *Journal of Philosophy* 86 (1989): 393–406.

Fine, A. "Unnatural Attitudes: Realist and Instrumentalist Attachments to Science." *Mind* 95 (1986): 149–79.

———. "Piecemeal Realism." *Philosophical Studies* 61 (1991): 79–96.

Fleck, L. *Genesis and Development of a Scientific Fact.* Chicago: University of Chicago Press, 1979 [1936].

Ford, K. M., C. Glymour, and P. J. Hayes. "Introduction." In *Android Epistemology,* ed. K. M. Ford, C. Glymour, and P. J. Hayes, pp. xi–xvii. Menlo Park and Cambridge, Mass.: AAAI Press and MIT Press, 1995.

Foucault, M. *The Archeology of Knowledge and the Discourse on Language.* Trans. A. M. S. Smith. New York: Pantheon Books, 1972.

———. *The Order of Things. An Archeology of the Human Sciences.* New York: Random House, 1973.

———. "Theatrum Philosophicum." In M. Foucault, *Language, Counter-Memory, Practice. Selected Essays and Interviews,* ed. D. F. Bouchard and trans. D. F. Bouchard and S. Simon, pp. 165–96. Ithaca: Cornell University Press, 1977.

———. "Truth and Power." In M. Foucault, *Power/Knowledge: Selected Interviews and Other Writings 1972–1977,* ed. C. Gordon and trans. C. Gordon, L. Marshall, J. Mepham, K. Soper, pp. 109–33. New York: Pantheon Books, 1980.

———. "The Eye of Power." In M. Foucault, *Power/Knowledge: Selected Interviews and Other Writings 1972–1977,* ed. C. Gordon and trans. C. Gordon, L. Marshall, J. Mepham, K. Soper, pp. 146–65. New York: Pantheon Books, 1980.

———. "Two Lectures." In M. Foucault, *Power/Knowledge: Selected Interviews and Other Writings 1972–1977,* ed. C. Gordon and trans. C. Gordon, L. Marshall, J. Mepham, K. Soper, pp. 78–108. New York: Pantheon Books, 1980.

———. "Afterward: The Subject and Power." In *Michel Foucault: Beyond Structuralism and Hermeneutics. With an Afterword by Michel Foucault,* ed. H. L. Dreyfus and P. Rabinow, pp. 208–26. Chicago: University of Chicago Press, 1982.

———. "What Is Enlightenment?" In *The Foucault Reader,* ed. P. Rabinow, pp. 32–50. New York: Pantheon, 1984.

———. "Nietzsche, Genealogy, History." In *The Foucault Reader,* ed. P. Rabinow, pp. 76–100. New York: Pantheon, 1984 [1971].

———. "Preface to the Second Volume of *The History of Sexuality.*" In *The Foucault Reader,* ed. P. Rabinow, pp. 333–39. New York: Pantheon, 1984.

———. *The History of Sexuality.* Vol. 2, *The Uses of Pleasure.* Trans. R. Hurley. New York: Vintage Books, 1986.

———. "Questions of Method: An Interview with Michel Foucault." In *After Philosophy: End or Transformation?* Ed. K. Baynes, J. Bohman, T. McCarthy, pp. 100–117. Cambridge, Mass.: MIT Press, 1987 [1980].

———. "The Archeology of Knowledge." In *Foucault Alive (Interviews, 1966–84),* ed. S. Lotringer and trans. J. Johnston, pp. 45–56. New York: Semiotext(e), 1989 [1969].

———. "Clarifications on the Question of Power." In *Foucault Alive (Interviews, 1966–84),* ed. S. Lotringer and trans. J. Cascaito, pp. 179–92. New York: Semiotext(e), 1989 [1978].

———. "Discourse of History." In *Foucault Alive (Interviews, 1966–84)*, ed. S. Lotringer and trans. J. Cascaito, pp. 11–33. New York: Semiotext(e), 1989 [1967].

———. "An Historian of Culture." In *Foucault Alive (Interviews, 1966–84)*, ed. S. Lotringer and trans. J. Cascaito, pp. 73–88. New York: Semiotext(e), 1989 [1973].

———. [As "Maurice Florence"]. "Foucault, Michel, 1926–." In *The Cambridge Companion to Foucault*, ed. G. Gutting and trans. C. Porter, pp. 314–19. Cambridge: Cambridge University Press, 1994.

———. *Discipline and Punish: The Birth of the Prison*. Trans. A. Sheridan. New York: Vintage Books (Random House), 1995.

Frede, D. "The Question of Being: Heidegger's Project." In *The Cambridge Companion to Heidegger*, ed. C. B. Guignon, pp. 42–69. Cambridge: Cambridge University Press, 1993.

Frede, M. "Categories in Aristotle." In M. Frede, *Essays in Ancient Philosophy*, pp. 29–48. Minneapolis: University of Minnesota Press, 1987.

———. "The Unity of General and Special Metaphysics: Aristotle's Conception of Metaphysics." In M. Frede, *Essays in Ancient Philosophy*, pp. 81–95. Minneapolis: University of Minnesota Press, 1987.

Freudenthal, G. "The Role of Shared Knowledge in Science: The Failure of the Constructivist Programme in the Sociology of Science." *Social Studies of Science* 14 (1984): 285–95.

Friedman, M. "Remarks on the History of Science and the History of Philosophy." In *World Changes: Thomas Kuhn about the Nature of Science*, ed. P. Horwitch, pp. 37–49. Cambridge, Mass.: MIT Press, 1993.

———. "Overcoming Metaphysics: Carnap and Heidegger." In *Origins of Logical Empiricism*, ed. R. Giere and A. Richardson. *Minnesota Studies*, 17:45–79. Minneapolis: University of Minnesota Press, 1996.

Funkenstein, A. *Theology and the Scientific Imagination: From the Middle Ages to the Seventeenth Century*. Princeton, N.J.: Princeton University Press, 1986.

Fynsk, C. *Heidegger, Thought and Historicity*. Ithaca: Cornell University Press, 1986.

Gadamer, H.-G. *Truth and Method*. 2d ed. Trans. G. Barden and J. Cumming. New York: Seabury Press, 1975.

———. "On the Problem of Self-Understanding." In H.-G. Gadamer, *Philosophical Hermeneutics*. Ed and trans. D. E. Linge, pp. 44–58. Berkeley: University of California Press, 1976 [1962].

———. "The Philosophical Foundations of the Twentieth Century." In H.-G. Gadamer, *Philosophical Hermeneutics*. Ed and trans. D. E. Linge, pp. 107–29. Berkeley: University of California Press, 1976 [1962].

———. *Reason in the Age of Science*. Trans. F. G. Lawrance. Cambridge, Mass.: MIT Press, 1981.

———. "Rhetoric, Hermeneutics, and the Critique of Ideology: Metacritical Comments on *Truth and Method*." In *The Hermeneutics Reader: Texts of the German Tradition from the Enlightenment to the Present*. Ed. K. Mueller-Vollmer, pp. 274–92. New York: Continuum, 1985.

———. "On the Scope and Function of Hermeneutical Reflection." Trans. G. B. Hess and R. E. Palmer. In *Hermeneutics and Modern Philosophy*, ed. B. R. Wachterhauser, pp. 277–99. Albany: SUNY Press, 1986 [1967].

———. "Text and Interpretation." In *Hermeneutics and Modern Philosophy*, ed. B. R. Wachterhauser, pp. 377–96. Albany: SUNY Press, 1986.

———. *Truth and Method*. Trans. J. Weinsheimer and D. G. Marshall. New York: Crossroad, 1989 [1960].

———. "The Universality of Hermeneutical Problem." In *The Hermeneutical Tradition: From Ast to Ricoeur*, ed. G. L. Ormiston and A. D. Schrift, pp. 147–58. Albany: SUNY Press, 1990.

———. "Reply to My Critics." In *The Hermeneutical Tradition: From Ast to Ricoeur*, ed. G. L. Ormiston and A. D. Schrift, pp. 273–97. Albany: SUNY Press, 1990.

Galison, P. *Image and Logic: A Material Culture of Microphysics*. Chicago: University of Chicago Press, 1997.

Gatens-Robinson, E. "Why Falsification Is the Wrong Paradigm for Evolutionary Epistemology: An Analysis of Hull's Selection Theory." *Philosophy of Science* 60 (1993): 535–57.

Geach, P. T. *Some Problems About Time: Annual Philosophical Lecture*. Henrietta Hertz Trust. London: British Academy, 1965.

Gélis, J. "The Child: From Anonymity to Individuality." In *A History of Private Life*. Vol. 3, *Passions of the Renaissance*, ed. R. Chartier and trans. A. Goldhammer, pp. 309–26. Cambridge: Belknap Press, 1989.

Genova, A. C. "Ambiguities about Realism and Utterly Distinct Objects." *Erkenntnis* 28 (1988): 87–95.

Giddens, A. *Central Problems in Social Theory: Action, Structure and Contradiction in Social Analysis*. Berkeley: University of California Press, 1979.

Giere, R. N. *Explaining Science: A Cognitive Approach*. Chicago: University of Chicago Press, 1988.

———. "Evolutionary Models of Science." In *Evolution, Cognition, and Realism: Studies in Evolutionary Epistemology*, ed. N. Rescher, pp. 21–32. Lanham, New York, and London: University of America Press, 1990.

Gilson, E. *Being and Some Philosophers*. 2d ed. Toronto: Pontifical Institute of Mediaeval Studies, 1961 [1949].

Ginev, D. "Scientific Progress and the Hermeneutical Circle." *Studies in History and Philosophy of Science* 19 (1988): 391–95.

Glymour, C. "Realism and the Nature of Theories." In *Introduction to the Philosophy of Science: A Text by Members of the Department of the History and Philosophy of Science of the University of Pittsburgh*, ed. M. H. Salomon, J. Earman, C. Glymour, J. G. Lennox, P. Machamer, J. E. McGuire, J. D. Norton, W. C. Salomon, K. F. Schaffner, pp. 104–31. Englewood Cliffs, N.J.: Prentice Hall, 1992.

Golinski, J. "The Theory of Practice and the Practice of Theory: Sociological Approaches in the History of Science." *Isis* 81 (1990): 492–505.

Grene, M. *Dreadful Freedom: A Critique of Existentialism*. Chicago: University of Chicago Press, 1948.

———. "Evolution, 'Typology' and 'Population Thinking.'" *American Philosophical Quarterly* 27, no. 3 (1990): 237-44.

Grice, P. *Studies in the Way of Words.* Cambridge, Mass.: Harvard University Press, 1989.

Grondin, J. "Hermeneutical Truth and Its Historical Presuppositions: A Possible Bridge between Analysis and Hermeneutics." In *Anti-Foundationalism and Practical Reasoning,* ed. E. Simpson, pp. 45-58. Edmonton: Academic Printing & Publishing, 1987.

———. "Hermeneutics and Relativism." In *Festivals of Interpretation. Essays on Hans-Georg Gadamer's Work,* ed. K. Wright, pp. 42-62. Albany: SUNY Press, 1990.

Guignon, C. B. *Heidegger and the Problem of Knowledge.* Indianapolis: Hackett Publising Co., 1983.

———. "Pragmatism or Hermeneutics? Epistemology After Foundationalism." In *The Interpretative Turn: Philosophy, Science, Culture,* ed. D. R. Hiley, J. F. Bohman, R. Shusterman, pp. 81-101. Cornell: Cornell University Press, 1991.

Gutting, G. "Conceptual Structures and Scientific Change." *Studies in History and Philosophy of Science* 4 (1973): 209-30.

———. "Scientific Realism. In *The Philosophy of Wilfrid Sellars: Queries and Extensions,* ed. J. C. Pitt, pp. 105-28. Dordrecht: Reidel, 1978.

Habermas, J. *The Philosophical Discourse of Modernity: Twelve Lectures.* Trans. F. G. Lawrence. Cambridge, Mass.: MIT Press, 1993.

Hacking, I. *Representing and Intervening: Introductory Topics in the Philosophy of Natural Science.* Cambridge: Cambridge University Press, 1983.

———. "On the Stability of Laboratory Sciences." *Journal of Philosophy* 85 (1988): 507-14.

———. "Statistical Language, Statistical Truth, and Statistical Reason: The Self-Authentification of a Style of Scientific Reasoning." In *The Social Dimensions of Science,* ed. E. McMullin, pp. 130-57. Notre Dame: University of Notre Dame Press, 1992.

Haraway, D. "Situated Knowledges: The Science Question in Feminism and the Privilege of Partial Perspective." *Feminist Studies* 14 (1988): 575-99.

Haugeland, J. "Heidegger on Being a Person." *Noûs* 16 (1982): 15-25.

———. *Having Thought. Essays in the Metaphysics of Mind.* Cambridge, Mass.: Harvard University Press, 1998.

Heelan, P. "Natural Science as Hermeneutic of Instrumentation." *Philosophy of Science* 50 (1983): 181-204.

———. *Space-Perception and the Philosophy of Science.* Berkeley: University of California Press, 1983.

———. "Experiment and Theory: Constitution and Reality." *Journal of Philosophy* 85: (1988): 515-24.

———. "After Experiment: Realism and Research." *American Philosophical Quarterly* 26 (1989): 297-308.

———. "Hermeneutical Phenomenology and the Philosophy of Science." In *Gadamer and Hermeneutics: Science, Culture, Literature. Plato, Heidegger, Barthes, Ricoeur, Habermas, Derrida.* Ed. H. J. Silverman, pp. 213-28. New York: Routledge, 1991.

———. "Why a Hermeneutical Philosophy of the Natural Sciences." *Man and World* 30 (1997): 271-98.

Hegel, G. W. F. *Lectures on the Philosophy of Religion*. Ed. P. C. Hodgson, R. F. Brown, P. C. Hodgson, and trans. J. M. Stewart . Berkeley: University of California Press, 1985 [1832].

Heidegger, M. *Being and Time*. Trans. J. MacQuarrie, E. Robinson. New York: Harper and Row, 1962 [1927].

———. *The End of Philosophy*. Trans. J. Stambaugh. New York: Harper and Row, 1973 [1954].

———. "The Age of World Picture." In *The Question Concerning Technology and Other Essays*, trans. W. Lovitt, pp. 115-54. New York: Harper and Row, 1977 [1952].

———. "Science and Reflection." In *The Question Concerning Technology and Other Essays*, trans. W. Lovitt, pp. 155-82. New York: Harper and Row, 1977 [1954].

———. *Basic Problems of Phenomenology*. Trans. A. Hofstadter. Bloomington: Indiana University Press, 1982.

———. *History of the Concept of Time. Prolegomena*. Trans. T. Kisiel. Bloomington: Indiana University Press, 1985.

———. "Hölderlin and the Essence of Poetry." In M. Heidegger, *Existence and Being*, trans. D. Scott, pp. 270-91. Washington, D.C.: Gateway Editions, 1988 [1936].

———. *Being and Time*. Trans. J. Stambaugh. Albany: SUNY Press, 1996 [1927].

Hekman, S. J. *Hermeneutics and the Sociology of Knowledge*. Cambridge: Polity Press, 1986.

Hempel, C. G. "Valuation and Objectivity in Science." In *Physics, Philosophy and Psychoanalysis. Essays in Honor of Adolf Grünbaum*, ed. R. S. Cohen and L. Laudan, pp. 73-100. Dordrecht: Reidel, 1983.

Hesse, M. *The Structure of Scientific Inference*. London: Macmillan Press, 1974.

———. *Revolutions and Reconstructions in the Philosophy of Science*. Brighton: Harvester Press, 1980.

Heyd, M. "The Emergence of Modern Science as an Autonomous World of Knowledge in the Protestant Tradition of the Seventeenth Century." *Knowledge and Society: Studies in the Sociology of Culture Past and Present* 7 (1988): 165-79.

Hilpinen, R. *On the Characterization of Cognitive Progress*. Reports from the Department of Theoretical Philosophy no. 16. Turku: University of Turku, 1986.

Hobsbawm, E. *The Age of Extremes: A History of the World, 1914-1991*. New York: Pantheon Books, 1994.

Hoffman, P. "Death, Time, History: Division II of *Being and Time.*" In *The Cambridge Companion to Heidegger*, ed. C. B. Guignon, pp. 195-214. Cambridge: Cambridge University Press, 1993.

Horwitch, P. *Truth*. Oxford: Basil Blackwell, 1990.

Howard, R. J. *Three Faces of Hermeneutics: An Introduction to Current Theories of Understanding*. Berkeley: University of California Press, 1982.

Hoy, D. C. "History, Historicity, and Historiography in *Being and Time.*" In *Heidegger and Modern Philosophy: Critical Essays*, ed. M. Murray, pp. 329-53. New Haven: Yale University Press, 1978.

———. "Is Hermeneutics Ethnocentric?" In *The Interpretative Turn: Philosophy,*

Science, Culture, ed. D. R. Hiley, J. F. Bohman, R. Shusterman, pp. 155–75. Ithaca: Cornell University Press, 1991.

Hull, D. *Science as a Process: An Evolutionary Account of the Social and Conceptual Development of Science.* Chicago: University of Chicago Press, 1988.

Hume, D. *A Treatise of Human Nature.* Trans. L. A. Selby-Bigge. Oxford: Clarendon Press, 1964.

Hunter, M. *Science and Society in Restoration England.* Cambridge: Cambridge University Press, 1981.

Husserl, E. *Ideas: General Introduction to Pure Phenomenology.* Trans. W. R. B. Gibson. London: George Allen and Unwin; New York: Macmillan, 1952 [1913].

———. *Cartesian Meditations.* Trans. D. Cairns. The Hague: Martinus Nijhoff, 1960 [1931].

———. *The Crisis of European Sciences and Transcendental Phenomenology: An Introduction to Phenomenological Philosophy.* Trans. D. Carr. Evanston: Northwestern University Press, 1970 [1954].

Ihde, D. *Technics and Praxis.* Dordrecht: Reidel, 1979.

Ingold, T. "The Evolution of Society." In *Evolution: Society, Science and the Universe,* ed. A. C. Fabian. Cambridge: Cambridge University Press, 1998.

Ingram, D. "Hermeneutics and Truth." In *Hermeneutics and Praxis,* ed. R. Hollinger, pp. 32–53. Notre Dame: University of Notre Dame Press, 1985.

Irzik, G., and T. Grunberg. "Whorfian Variations on Kantian Themes: Kuhn's Linguistic Turn and Its Relation to Carnap's Philosophy." *Studies in History and Philosophy of Science* 29A (1998): 207–21.

Jaspers, K. *Philosophy of Existence.* Trans. R. F. Grabau. Philadelphia: University of Pennsylvania Press, 1971 [1937].

Johnson, W. E. *Logic: In Three Parts.* Pt. 1, *Propositions and Relations.* New York: Dover Publications, 1964 [1921].

Johnson-Laird, P. N. *Mental Models: Towards a Cognitive Science of Language, Inference, and Consciousness.* Cambridge, Mass.: Harvard University Press, 1983.

Kant, I. *Critique of Pure Reason.* Trans. N. Kemp Smith. New York: St. Martin's Press, 1965 [1781].

———. *Prolegomena to Any Future Metaphysics that Will Be Able to Come Forward as Science.* Trans. P. Carus, J. W. Ellington. Indianapolis: Hachett, 1977 [1783].

Kisiel, T. "The Happening of Tradition: The Hermeneutics of Gadamer and Heidegger." In *Hermeneutics and Praxis,* ed. R. Hollinger, pp. 3–31. Notre Dame: University of Notre Dame Press, 1985.

———. *The Genesis of Heidegger's* Being and Time. Berkeley: University of California Press, 1993.

———. "A Hermeneutics of the Natural Science? The Debate Updated." *Man and World* 30 (1997): 329–41.

Kitcher, P. *The Advancement of Science: Science without Legend, Objectivity without Illusions.* New York: Oxford University Press, 1993.

———. "Contrasting Conceptions of Social Epistemology." In *Socializing Epistemology: The Social Dimensions of Knowledge,* ed. F. F. Schmitt, pp. 111–34. Lanham: Rowman and Littlefield, 1994.

Kmita, J. *Problems in Historical Epistemology*. Dordrecht: Reidel; Warsaw: PWN, 1988 [1980].

———. "The Controversy About the Determinants of the Growth of Science." In J. Kmita, *Essays on the Theory of Scientific Cognition*, trans. J. Chołówka, pp. 78–115. Dordrecht: Kluwer; Warsaw: PWN, 1991 [1990].

Knorr-Cetina, K. D. *The Manufacture of Knowledge: An Essay on the Constructivist and Contextual Nature of Science*. Oxford: Pergamon Press, 1981.

———. "The Ethnographic Study of Scientific Work: Towards a Constructivist Interpretation of Science." In *Science Observed: Perspectives on the Social Study of Science*, pp. 115–40. London: SAGE Publ., 1983.

———. "The Fabrication of Facts: Toward a Microsociology of Scientific Knowledge." In *Society and Knowledge: Contemporary Perspectives in the Sociology of Knowledge*, ed. N. Stehr and V. Meja, pp. 223–44. New Brunswick: Transaction Books, 1984.

Knorr-Cetina, K. D., and M. Mulkay. "Emerging Principles in Social Studies of Science." In *Science Observed: Perspectives on the Social Study of Science*, ed. K. D. Knorr-Cetina and M. Mulkay, pp. 1–17. London: SAGE Publ., 1983.

Kockelmans, J. J. "Heidegger on the Essential Difference and Necessary Relationship between Philosophy and Science." In *Phenomenology and the Natural Sciences*, ed. J. J. Kockelmans and T. J. Kisiel, pp. 147–66. Evanston: Northwestern University Press, 1970.

———. "Beyond Realism and Idealism: A Response to Patrick A. Heelan." In *Gadamer and Hermeneutics. Science, Culture, Literature. Plato, Heidegger, Barthes, Ricoeur, Habermas, Derrida*, ed. H. J. Silverman, pp. 229–44. New York: Routledge, 1991.

Kolakowski, L. "The Priest and the Jester." In *Toward a Marxist Humanism: Essays on the Left Today*, trans. J. Zielonko Peel, pp. 9–37. New York: Grove Press, 1968.

———. "Karl Marx and the Classical Definition of Truth." In *Toward a Marxist Humanism. Essays on the Left Today*. Trans. J. Zielonko Peel, pp. 38–66. New York: Grove Press, 1968.

———. "In Praise of Inconsistency." In *Toward a Marxist Humanism: Essays on the Left Today*, trans. J. Zielonko Peel, pp. 211–20. New York: Grove Press, 1968.

———. *Religion. If There Is no God . . . On God, the Devil, Sin and Other Worries of the So-called Philosophy of Religion*. London: Fontana Paperbacks, 1982.

———. *Main Current of Marxism*. Vol. 1, *The Founders*. Oxford: Oxford University Press, 1982.

———. *The Presence of Myth*. Chicago: University of Chicago Press, 1989.

———. "Looking for the Barbarians: The Illusions of Cultural Universalism." In L. Kolakowski, *Modernity on Endless Trial*, trans. A. Kolakowska. Chicago: University of Chicago Press, 1990.

Krohn, R. "On Gieryn on the 'Relativist/Constructivist' Programme in the Sociology of Science: Naivete and Reaction." *Social Studies of Science* 12 (1982): 325–28.

Krolick, S. *Recollective Resolve: A Phenomenological Understanding of Time and Myth*. Macon, Ga.: Mercer University Press, 1987.

Kuhn, T. S. *The Structure of Scientific Revolutions.* 2d ed. Chicago: University of Chicago Press, 1970.

———. "Mathematical versus Experimental Traditions in the Development of Physical Science." In T. S. Kuhn, *The Essential Tension: Selected Studies in Scientific Tradition and Change,* pp. 31-65. Chicago: University of Chicago Press, 1977.

———. "Second Thoughts on Paradigms" and "Discussion." In *The Structure of Scientific Theories,* ed. F. Suppe, pp. 459-82, 500-517. Urbana-Chicago: University of Illinois Press, 1977.

———. "Commensurability, Comparability, Communicability." In *PSA 1982: Proceedings of the 1982 Biennial Meeting of the Philosophy of Science Association.* Vol. 2, *Symposia.* Ed. P. D. Asquith and T. Nickles, pp. 669-88. East Lansing, Mich.: PSA, 1983.

———. "Rationality and Theory Choice." *Journal of Philosophy* 80 (1983): 563-70.

———. "What Are Scientific Revolutions?" In *The Probabilistic Revolution.* Vol. 1, *Ideas in History,* ed. L. Krüger, L. J. Daston, M. Heidelberger, pp. 7-22. Cambridge, Mass.: MIT Press, 1987.

———. "Objectivity, Value Judgment, and Theory Choice." In *Introductory Readings in the Philosophy of Science,* ed. E. D. Klemke, R. Hollinger, A. D. Kline, pp. 277-91. Buffalo: Prometheus Books, 1988.

———. "The Road Since Structure." In *PSA 1990: Proceedings of the 1990 Biennial Meeting of the Philosophy of Science Association.* Vol. 2. Ed. A. Fine, M. Forbes, L. Wessels, pp. 3-13. East Lansing, Mich.: PSA, 1991.

———. "Afterwords." In *World Changes: Thomas Kuhn and the Nature of Science,* ed. P. Horwitch, pp. 311-41. Cambridge, Mass.: MIT Press, 1993.

Lakatos, I. "Falsification and the Methodology of Scientific Research Programmes." In I. Lakatos, *Philosophical Papers.* Vol. 1, ed. J. Worrall and G. Currie, pp. 8-101. Cambridge: Cambridge University Press, 1978.

———. "History of Science and Its Rational Reconstructions." In I. Lakatos, *Philosophical Papers.* Vol. 1, ed. J. Worrall and G. Currie, pp. 102-38. Cambridge: Cambridge University Press, 1978.

Larmor, J. *Aether and Matter.* Cambridge: Cambridge University Press, 1900.

Latour, B. "Give Me a Laboratory and I Will Raise the World." In *Science Observed: Perspectives on the Social Study of Science,* ed. K. D. Knorr-Cetina and M. Mulkay, pp. 141-70. London: SAGE Publ., 1983.

———. *Science in Action: How to Follow Scientists and Engineers through Society.* Cambridge, Mass.: Harvard University Press, 1987.

———. *The Pasteurization of France.* Trans. A. Sheridan and J. Law. Cambridge, Mass.: Harvard University Press, 1988.

———. "Postmodern? No, Simply Amodern! Steps Towards an Anthropology of Science." *Studies in History and Philosophy of Science* 21 (1990): 145-71.

———. "One More Turn After the Social Turn." In *The Social Dimensions of Science,* ed. E. McMullin, pp. 272-94. Notre Dame: University of Notre Dame Press, 1992.

———. *We Have Never Been Modern.* Cambridge, Mass.: Harvard University Press, 1993.

Latour, B., and S. Woolgar. *Laboratory Life: The Construction of Scientific Facts.* 2d ed. Princeton, N.J.: Princeton University Press, 1986.

Laudan, L. *Science and Values: The Aims of Science and Their Role in Scientific Debate.* Los Angeles: University of California Press, 1984.

———. "Progress or Rationality? The Prospects for Normative Naturalism." *American Philosophical Quarterly* 24 (1987): 19–31.

———. "Aim-less Epistemology?" *Studies in History and Philosophy of Science* 21 (1990): 315–22.

Lawson, H. *Reflexivity: The Post-Modern Predicament.* London: Hutchinson, 1985.

Lehrer, K., and C. Wagner. *Rational Consensus in Science and Society: A Philosophical and Mathematical Study.* Dordrecht: Reidel, 1981.

Leibniz, G. W. "Letter to Des Basses." [G II 435-37]. In *Gottfried Wilhelm Leibniz: Philosophical Papers and Letters,* ed. L. E. Loemker, pp. 600–601. Dordrecht: Reidel, 1969 [1875].

Lenoir, T. "Practice, Reason, Context: The Dialogue between Theory and Experiment." *Science in Context* 2, no. 1 (1988): 3–22.

Levinas, E. *Totality and Infinity: An Essay on Exteriority.* Trans. A. Lingis. Pittsburgh: Duquesne University Press, 1969.

Lewis, C. I. *An Analysis of Knowledge and Valuation.* La Salle, Ill.: Open Court, 1946.

Longino, H. "Multiplying Subjects and the Diffusion of Power." *Journal of Philosophy* 88, no. 11 (1991): 666–74.

———. "Essential Tensions—Phase Two: Feminist, Philosophical, and Social Studies of Science." In *The Social Dimensions of Science,* ed. E. McMullin, pp. 198–216. Notre Dame: University of Notre Dame Press, 1992.

———. "The Fate of Knowledge in Social Theories of Science." In *Socializing Epistemology: The Social Dimensions of Knowledge,* ed. F. F. Schmitt, pp. 135–57. Lanham: Rowman and Littlefield, 1994.

Lukács, G. "Existentialism or Marxism?" In *Existentialism versus Marxism: Conflicting Views on Humanism,* ed. G. Novack, pp. 134–53. New York: A Delta Book, 1966 [1949].

Lynch, M. *Scientific Practice and Ordinary Action: Ethnomethodology and Social Studies of Science.* Cambridge: Cambridge University Press, 1993.

Machamer, P. K. "Feyerabend and Galileo: The Interaction of Theories, and the Reinterpretation of Experience." *Studies in History and Philosophy of Science* 4 (1973): 1–46.

———. "Selection, System and Historiography." In *Trends in the Historiography of Science,* ed. K. Gavroglu et al., pp. 149–60 Dordrecht: Reidel Publ. Co., 1994.

———. "Kitcher and the Achievement of Science." *Philosophy and Phenomenological Research* 55, no. 3 (1995): 629–36.

———. "The Concept of Individual and the Idea (l) of Method in Seventeenth-Century Natural Philosophy." In "Scientific Controversies: Philosophical and Historical Perspectives," ed. A. Baltas, P. Machamer, M. Pera, pp. 81–99. New York: Oxford University Press, 2000.

MacIntyre, A. *After Virtue: A Study in Moral Theory.* Notre Dame: University of Notre Dame Press, 1984.

Margolis, J. *Texts without Referents: Reconciling Science and Narrative.* Oxford: Basil Blackwell, 1989.

———. "Praxis and Meaning: Marx's Species Being and Aristotle's Political Animal." In *Marx and Aristotle: Nineteenth-Century German Social Theory and Classical Antiquity,* ed. G. E. McCarthy, pp. 329-55. Lanham: Rowman & Littlefield, 1992.

———. *The Flux of History and the Flux of Science.* Berkeley: University of California Press, 1993.

———. *Historied Thought, Constructed World: A Conceptual Primer for the Turn of the Millennium.* Berkeley: University of California Press, 1995.

McDowell, J. *Mind and the World.* Cambridge, Mass.: Harvard University Press, 1996.

McGuire, J. E. "Boyle's Conception of Nature." *Journal of the History of Ideas* 33 (1972): 523-42.

———. "Forces, Powers, Aethers, and Fields." *Boston Studies in the Philosophy of Science,* 14:119-59. Dordrecht: Reidel, 1974.

———. "Phenomenalism, Relations, and Monadic Representation: Leibniz on Predicate Levels." In *How Things Are,* ed. J. Bogen and J. E. McGuire, pp. 205-33. Dordrecht: Reidel, 1981.

———. "Predicates of Pure Existence: Newton on God's Space and Time." In *Philosophical Perspectives on Newtonian Science,* ed. P. Bricker and R. I. G. Hughes, pp. 91-108. Cambridge, Mass.: MIT Press, 1990.

———. "Scientific Change: Perspectives and Proposals." In M. H. Salmon, J. Earman, C. Glymour, J. G. Lennox, P. Machamer, J. E. McGuire, J. D. Norton, W. C. Salmon, K. F. Schaffner, *Introduction to the Philosophy of Science. A Text by Members of the Department of the History and Philosophy of Science of the University of Pittsburgh,* pp. 132-78. Englewood Cliffs, N.J.: Prentice Hall, 1992.

McGuire, J. E., and S. K. Strange. "An Annotated Translation of Plotinus *Ennead* iii 7: *On Eternity and Time.*" *Ancient Philosophy* 8 (1988): 251-71.

McMullin, E. "Discussion Review: Laudan's Progress and Its Problems." *Philosophy of Science* 46 (1979): 623-44.

———. "Explanatory Success and the Truth of Theory." In *Scientific Inquiry in Philosophical Perspective,* ed. N. Rescher, pp. 51-73. Lanham: University Press of America, 1987.

———. "Values in Science." In *Introductory Readings in the Philosophy of Science,* ed. D. Klemke, R. Hollinger, A. D. Kline, pp. 349-71. Buffalo: Prometheus Books, 1988.

———. "Rationality and Paradigm Change in Science." *World Changes: Thomas Kuhn and the Nature of Science,* ed. P. Horwich, pp. 55-78. Cambridge: MIT Press, 1993.

Megill, A. *Prophets of Extremity. Nietzsche, Heidegger, Foucault, Derrida.* Berkeley: University of California Press, 1985.

Mellor, H. "In Defense of Dispositions." *Philosophical Review* 83 (1974): 157-81.

Mennell, S. *Norbert Elias. Introduction.* Oxford: Basil Blackwell, 1992.

Merricks, T. "Endurance and Indiscernibility." *Journal of Philosophy* 91 (1994): 165-84.

Merton, R. K. *Science, Technology and Society in Seventeenth Century England*. New York: Harper and Row, 1970 [1938].

———. "The Perspectives of Insider and Outsider." In R. K. Merton, *The Sociology of Science: Theoretical and Empirical Investigations*, ed. N. W. Storer, pp. 99-136. Chicago: University of Chicago Press, 1973.

Meynell, H. "On the Limits of the Sociology of Knowledge." *Social Studies of Science* 7 (1977): 489-500.

Millstone, E. "A Framework for the Sociology of Knowledge." *Social Studies of Science* 8 (1978): 117-25.

Mink, L. O. "History and Fiction as Modes of Comprehension." *New Literary History* 1 (1970): 541-58.

Misgeld, D. "On Gadamer's Hermeneutics." In *Hermeneutics and Praxis*, ed. R. Hollinger, pp. 143-70. Notre Dame: University of Notre Dame Press, 1985.

Mohanty, J. "Foucault as a Philosopher." In *Institutions, Normalization, and Power*, ed. J. Caputo and M. Yount, pp. 27-40. Philadelphia: Penn State University Press, 1993.

Mulhall, S. *Heidegger and Being and Time*. New York: Routledge, 1996.

Mulkay, M. "Knowledge and Utility." In M. Mulkay, *Sociology of Science: A Sociological Pilgrimage*, pp. 90-105. Philadelphia: Open University Press, 1991.

Mulkay, M., and N. Gilbert. "Theory Choice." In M. Mulkay, *Sociology of Science: A Sociological Pilgrimage*, pp. 131-53. Philadelphia: Open University Press, 1991.

Murphy, N. "Scientific Realism and Postmodern Philosophy." *British Journal for the Philosophy of Science* 41 (1990): 291-303.

Musgrave, A. "The Objectivism of Popper's Epistemology." In *The Philosophy of Karl Popper*, ed. P. A. Schilpp, pp. 560-99. La Salle, Ill.: Open Court, 1974.

Nancy, J.-L. "Introduction." In *Who Comes After the Subject?* ed. E. Cadava, P. Connor, J.-L. Nancy, pp. 1-8. New York: Routledge, 1991.

Nehamas, A. *Nietzsche: Life as Literature*. Cambridge, Mass.: Harvard University Press 1985.

Newton, I. "De Gravitatione et Aequipondio Fluidorum." In *Unpublished Scientific Papers of Isaac Newton. A Selection from the Portsmouth Collection in the University Library Cambridge*, ed. A. R. Hull and M. B. Hull, pp. 121-52. Cambridge: Cambridge University Press, 1962.

Newton-Smith, W. H. *The Rationality of Science*. London: Routledge & Kegan Paul, 1981.

Nickles, T. "Good Science as Bad History: From Order of Knowing to Order of Being." In *The Social Dimensions of Science*, ed. E. McMullin, pp. 85-129. Notre Dame: University of Notre Dame Press, 1992.

Nietzsche, F. *The Will to Power*. Ed. W. Kaufmann, trans. W. Kaufmann, R. J. Hollingdale. New York: Vintage Books, 1968 [1901].

———. *The Gay Science*. Trans. W. Kaufmann. New York: Vintage Books, 1974 [1882].

———. *On the Advantage and Disadvantage of History for Life*. Ed. and trans. P. Preuss. Indianapolis: Hackett Publ. Co. 1980 [1874].

———. *On the Genealogy of Morality*. Ed. K. Ansell-Pearson, trans. C. Diethe. Cambridge: Cambridge University Press, 1994.

Nowak, L. "Essence, Idealization, Praxis. An Attempt at a Certain Interpretation of the Marxist Concept of Science." *Poznań Studies in the Philosophy of the Sciences and the Humanities* 2, no. 3 (1976): 1-28.

———. *The Structure of Idealization. Towards a Systematic Interpretation of the Marxian Idea of Science*. Dordrecht: Reidel, 1980.

Okrent, M. *Heidegger's Pragmatism. Understanding, Being, and the Critique of Metaphysics*. Ithaca: Cornell University Press, 1988.

Olafson, F. A. *The Dialectic of Action*. Chicago: University of Chicago Press, 1979.

Owen, G. E. L. "Logic and Metaphysics in Some Earlier Works of Aristotle." In *Logic, Science and Dialectic*, ed. M. Nussbaum, pp. 180-99. Ithaca: Cornell University Press, 1986.

Pannenberg, W. "Hermeneutics and Universal History." In *Hermeneutics and Modern Philosophy*, ed. B. R. Wachterhauser, pp. 111-46. Albany: SUNY Press, 1986.

Pickering, A. "From Science as Knowledge to Science as Practice." In *Science as Practice and Culture*, ed. A. Pickering, pp. 1-26. Chicago: University of Chicago Press, 1992.

———. *The Mangle of Practice: Time, Agency and Science*. Chicago: University of Chicago Press, 1995.

Pinker, S. *How the Mind Works*. New York: W. W. Norton & Co., 1997.

———. "Down to Darwin," an interview by H. Blume. *The Boston Book Review* 4, no. 9 (1997/9): 4-5.

Plantinga, A. *Warrant: The Current Debate*. Oxford: Oxford University Press, 1993.

Polanyi, M. *Personal Knowledge: Towards a Post-Critical Philosophy*. New York: Harper & Row, 1964.

———. *The Tacit Dimension*. New York: Anchor Books, 1967.

Popper, K. R. *The Open Society and Its Enemies*. Vol. 2, *The High Tide of Prophecy: Hegel, Marx, and the Aftermath*. London: Routledge and Kegan Paul, 1952.

———. *The Logic of Scientific Discovery*. London: Hutchinson of London, 1959.

———. *Objective Knowledge: An Evolutionary Approach*. Oxford: Clarendon Press, 1972.

Putnam, Hilary. *Realism and Reason*. Vol. 3 of *Philosophical Papers*. Cambridge: Cambridge University Press, 1983.

———. *Reason, Truth and History*. Cambridge: Cambridge University Press, 1981.

———. "What Theories Are Not." In *Introductory Readings in the Philosophy of Science*, ed. E. D. Klemkc, R. Hollinger, A. D. Kline, pp. 178-83. Buffalo: Prometheus Books, 1988 [1962].

Quine, W. V. O. *Methods of Logic*. New York: Holt, 1950.

———. "On What There Is." In *From the Logical Point of View*, pp. 1-19. Cambridge, Mass.: Harvard University Press, 1953.

———. *World and Object*. Cambridge, Mass.: MIT Press; New York: Jon Wiley & Sons, 1960.

———. "Three Indeterminacies." In *Perspectives on Quine. Proceedings of Perspectives on Quine: An International Conference, April 9–13, 1988, at Washington University in St. Louis,* ed. R. B. Barrett and R. F. Gibson, pp. 1–16. Oxford: Basil Blackwell, 1990.

———. "Structure and Nature." *Journal of Philosophy* 89 (1992): 5–9.

Radnitzky, G. "In Defense of Self-Applicable Critical Rationalism." In *Evolutionary Epistemology, Rationality, and the Sociology of Knowledge,* ed. G. Radnitzky, W. W. Bartley III, pp. 279–312. La Salle, Ill.: Open Court, 1987.

Ravetz, J. R. *Scientific Knowledge and Its Social Problems.* Oxford: Clarendon Press, 1971.

Rescher, N. *The Limits of Science.* Berkeley: University of California Press, 1984.

———. *A System of Pragmatic Idealism.* Vol. 1, *The Validity of Values. A Normative Theory of Evaluative Rationality.* Princeton, N.J.: Princeton University Press, 1993.

———. *Process Metaphysics: An Introduction to Process Philosophy.* Albany: SUNY Press, 1996.

Revel, J. "The Uses of Civility." In *A History of the Private Life.* Vol. 3, *Passions of the Renaissance,* ed. R. Chartier, trans. A. Goldhammer, pp. 167–205. Cambridge: Belknap Press, 1989.

Rheinberger, H.-J. *Toward a History of Epistemic Things.* Stanford: Stanford University Press, 1997.

Richardson, J. *Existential Epistemology: A Heideggerian Critique of Cartesian Project.* Oxford: Clarendon Press, 1986.

———. *Nietzsche's System.* Oxford: Oxford University Press, 1996.

Ricoeur, P. "Narrative Time." In *On Narrative,* ed. W. J. T. Mitchell, pp. 165–86. Chicago: University of Chicago Press, 1981.

———. "The Narrative Function." In *Hermeneutics and the Human Sciences: Essays on Language, Action and Interpretation,* ed. and trans. J. B. Thompson, pp. 274–96. Cambridge: Cambridge University Press, 1981.

———. *The Reality of the Historical Past.* Milwaukee: Marquette University Press, 1984.

———. *Time and Narrative.* Vol. 3. K. Blamey and D. Pellauer. Chicago: University of Chicago Press, 1988.

Ringer, F. "The Intellectual Field, Intellectual History, and the Sociology of Knowledge." *Theory and Society* 19 (1990): 269–94.

Rockmore, T. "Epistemology as Hermeneutics." *Monist* 73 (1990): 115–33.

Rorty, R. *Philosophy and the Mirror of Nature.* Princeton, N.J.: Princeton University Press, 1979.

———. "Pragmatism, Relativism, and Irrationalism." In *Consequences of Pragmatism: Essays, 1972–1980,* pp. 160–75. Minneapolis: University of Minnesota Press., 1982.

———. "Solidarity or Objectivity?" In *Post-Analytic Philosophy,* ed. J. Rajchman and C. West, pp. 3–19. New York: Columbia University Press, 1985.

———. "Inquiry as Recontextualization: An anti-Dualist Account of Interpretation." In *Philosophical Papers.* Vol. 1, *Objectivity, Relativism, and Truth,* pp. 93–110. Cambridge: Cambridge University Press, 1991.

——. "Cosmopolitanism without Emancipation: A Response to Jean-François Lyotard." In *Philosophical Papers.* Vol. 1, *Objectivity, Relativism, and Truth,* pp. 211–22. Cambridge: Cambridge University Press, 1991.

Rosner, K. *Hermeneutyka jako krytyka kultury: Heidegger, Gadamer, Ricoeur* [Hermeneutics as a criticism of culture: Heidegger, Gadamer, Ricoeur]. Warszawa: PIW, 1991.

Rotenstreich, N. "The Ontological Status of History," *American Philosophical Quarterly* 9 (1972): 49–58.

——. *Theory and Practice: An Essay in Human Intentionalities.* The Hague: Martinus Nijhoff, 1977.

Rouse, J. *Knowledge and Power: Toward a Political Philosophy of Science.* Ithaca: Cornell University Press, 1987.

——. "Husserlian Phenomenology and Scientific Realism." *Philosophy of Science* 54 (1987): 222–32.

——. "Philosophy of Science and the Persistent Narratives of Modernity." *Studies in History and Philosophy of Science* 22 (1991): 141–62.

——. "The Dynamics of Power and Knowledge in Science." *Journal of Philosophy* 88, no. 11 (1991): 658–65.

——. "Interpretation in Natural and Human Science." In *The Interpretive Turn. Philosophy, Science, Culture,* ed. D. R. Hiley, J. F. Bohman, R. Shusterman, pp. 42–56. Ithaca: Cornell University Press, 1991.

——. "The Significance of Scientific Practices." In J. Rouse, *Engaging Science: How to Understand Its Practices Philosophically,* pp. 125–57. Ithaca: Cornell University Press, 1996.

——. "Against Representation: Davidsonian Semantics and Cultural Studies of Science." In J. Rouse, *Engaging Science: How to Understand Its Practices Philosophically,* pp. 205–63. Ithaca: Cornell University Press, 1996.

Rudner, R. "The Scientist *Qua* Scientist Makes Value Judgments." In *Introductory Readings in the Philosophy of Science,* ed. D. Klemke, R. Hollinger, A. D. Kline, pp. 327–33. Buffalo: Prometheus Books, 1988 [1953].

Ruin, H. *Enigmatic Origins. Tracing the Theme of Historicity through Heidegger's Works.* Acta Universitatis Stockholmiensis, Stockholm Studies in Philosophy 15. Stockholm: Alquist and Wiksell International, 1994.

Sartre, J.-P. *Being and Nothingness.* New York: Gramercy Books, 1994.

Schatzki, T. R. *Social Practices. A Wittgensteinian Approach to Human Activity and the Social.* Cambridge: Cambridge University Press, 1996.

Schrag, C. O. "Heidegger on Repetition and Historical Understanding." *Philosophy East and West* 20 (1970): 287–95.

Schürmann, R. *Heidegger on Being and Acting: From Principles to Anarchy.* Bloomington: Indiana University Press, 1987.

Schutte, O. *Beyond Nihilism: Nietzsche without Masks.* Chicago: University of Chicago Press, 1986.

Schutz, A. *The Phenomenology of the Social World.* Trans. G. Walsch, F. Lehnert. Evanston, Ill.: Northwestern University Press, 1967 [1932].

——. "Common-Sense and Scientific Interpretation of Human Action." In *Collected Papers.* Vol. 1, *The Problem of Social Reality,* ed. M. Natanson. The Hague: Martinus Nijhoff, 1971 [1953], pp. 3–27.

Searle, J. *Intentionality: An Essay in the Philosophy of Mind*. Cambridge: Cambridge University Press, 1983.

———. *The Construction of Social Reality*. New York: Free Press, 1995.

Seibt, J. "Existence in Time: From Substance to Process." In *Perspectives on Time*, ed. J. Faye, U. Scheffler, M. Urchs, pp. 143–83. Dordrecht: Kluwer, 1996.

Sellars, W. *Science, Perception and Reality*. London: Routledge and Kegan Paul, 1963.

Shapere, D. "Notes Toward a Post-Positivistic Interpretation of Science." In *The Legacy of Logical Positivism: Studies in the Philosophy of Science*, ed. P. Achinstein, S. F. Barker, pp. 115–60. Baltimore: Johns Hopkins Press, 1969.

———. *Reason and the Search for Knowledge*. Dordrecht: Reidel, 1984.

Shapin, S. *A Social History of Truth: Civility and Science in Seventeenth-Century England*. Chicago: University of Chicago Press, 1994.

Shapin, S., and S. Schaffer. *Leviathan and the Air-Pump: Hobbes, Boyle, and the Experimental Life*. Princeton: Princeton University Press, 1985.

Shapiro, B. *Probability and Certainty in Seventeenth-Century England: The Relationship between Natural Science, Religion, History, Law and Literature*. Princeton: Princeton University Press, 1983.

Shiner, L. "Reading Foucault: Anti-Method and the Genealogy of Power-Knowledge." *History and Theory* 21 (1982): 382–98.

Siegel, H. "What Is the Question Concerning the Rationality of Science." *Philosophy of Science* 52 (1985): 517–37.

———. "Laudan's Normative Naturalism." *Studies in History and Philosophy of Science* 21 (1990): 295–313.

Simons, P. *Parts: A Study in Ontology*. Oxford: Clarendon Press; New York: Oxford University Press, 1987.

Solomon, M. "A More Social Epistemology." In *Socializing Epistemology: The Social Dimensions of Knowledge*, ed. F. F. Schmitt, pp. 217–33. Lanham: Rowman and Littlefield, 1994.

Sperber, D. *Explaining Culture: A Naturalistic Approach*. Cambridge: Blackwell Publ., 1966.

Sperber, D., and D. Wilson. *Relevance, Communication and Cognition*. Oxford: Blackwell, 1986.

Stambaugh, J. *The Problem of Time in Nietzsche*. Canbury, N.J.: Associated University Presses, 1987 [1959].

———. "Translator's Preface." In M. Heidegger, *Being and Time*, trans. J. Stambaugh, pp. xiii–xvi. Albany: SUNY Press, 1996.

Steiner, G. *Martin Heidegger*. Chicago: University of Chicago Press, 1991 [1978].

Strawson, P. *Individuals: An Essay in Descriptive Metaphysics*. London: Methuen, 1959.

Stump, D. "Fallibilism, Naturalism and the Traditional Requirements for Knowledge." *Studies in History and Philosophy of Science* 22 (1991): 451–69.

Suppe, F. "The Search for Philosophic Understanding of Scientific Theories." In *The Structure of Scientific Theories*, ed. F. Suppe, pp. 3–232. Chicago: University of Illinois Press, 1977.

Taminiaux, J. "Finitude and the Absolute: Remarks on Hegel and Heidegger as Interpreters of Kant." In *Dialectic and Difference: Finitude in Modern Thought,* ed. J. Decker, R. Crease, trans. T. Sheehan, pp. 55-77. Atlantic Highlands, N.J.: Humanities Press, 1985.

———. "Marx, Art, and Truth." In *Dialectic and Difference: Finitude in Modern Thought,* ed. J. Decker, R. Crease, trans. J. Decker. Atlantic Highlands, N. J.: Humanities Press, 1985 [1974], pp. 39-54.

———. *Heidegger and the Project of Fundamental Ontology.* Albany: SUNY Press, 1991.

Taylor, C. "Interpretation and the Sciences of Man." In *Interpretive Social Science: A Reader,* ed. P. Rabinow, W. M. Sullivan, pp. 25-71. Berkeley: University of California Press, 1979.

———. "Overcoming Epistemology." In *Philosophical Arguments,* pp. 1-19. Cambridge, Mass.: Harvard University Press, 1995.

———. "Heidegger, Language, and Ecology." In *Philosophical Arguments,* pp. 100-126. Cambridge, Mass.: Harvard University Press, 1995.

———. *Sources of the Self: The Making of the Modern Identity.* Cambridge, Mass.: Harvard University Press, 1989.

Teller, P. *An Interpretative Introduction to Quantum Field Theory.* Princeton, N.J.: Princeton University Press, 1995.

Theunissen, M., *The Other: Studies in the Social Ontology of Husserl, Heidegger, Sartre and Buber.* Cambridge, Mass.: MIT Press, 1984 [1965].

Tiryakian, E. A. *Sociologism and Existentialism: Two Perspectives on the Individual and Society.* Englewood Cliffs, N.J.: Prentice-Hall, 1962.

Toulmin, S. E. *Human Understanding.* Vol. 1, *General Introduction* and *Part I.* Princeton, N.J.: Princeton University Press, 1972.

Touraine, A. *The Self-Production of Society.* Trans. D. Coltman. Chicago: University of Chicago Press, 1977.

Tuchańska, B. "The Methodological Problem of the Development of Science versus the Historical Problem of How Science Performs Its Social Functions." *Polish Sociological Bulletin* 3 (1980): 5-24.

———. "The Marxian Conception of Man." *Dialectics and Humanism,* 1983/3, 127-139.

———. "The Problem of Cognition as an Ontological Question." *Acta Universitatis Lodziensis, Folia Philosophica* 6 (1988): 31-43. A revised version in Polish, In *Racjonalność, nauka, społeczeństwo,* ed. H. Kozakiewicz, E. Mokrzycki, M. J. Siemek, pp. 241-63. Warsaw: PWN, 1989.

———. "Can Relativism Be Reconciled with Realism and Causalism?" *International Studies in the Philosophy of Science* 4 (1990): 285-94.

Tuomela, R. "Science, Protoscience, and Pseudo-science." In *Rational Changes in Science: Essays in Scientific Reasoning,* ed. J. C. Pitt, M. Pera. Dordrecht: Reidel, 1987.

———. "The Myth of the Given and Realism." *Erkenntnis* 29 (1988): 181-200.

Turner, S. *The Social Theory of Practices: Tradition, Tacit Knowledge, Presuppositions.* Chicago: University of Chicago Press, 1994.

Uexküll, J. J., von. *Theoretische Biologie.* Frankfurt/M 1973.

Van Fraassen, B. *The Scientific Image.* Oxford: Clarendon Press, 1980.

Visker, R. *Michel Foucault. Genealogy as Critique*. Trans. C. Turner. London: Verso, 1995.

Vogel, L. *The Fragile "We." Ethical Implications of Heidegger's "Being and Time."* Evanston, Ill.: Northwestern University Press, 1994.

Volpi, F. "Dasein as *Praxis:* The Heideggerian Assimilation and Radicalization of the Practical Philosophy of Aristotle." In *Critical Heidegger*, ed. C. Macann, pp. 27–66. London: Routledge, 1996.

Wachbroit, R. "Progress: Metaphysical and Otherwise." *Philosophy of Science* 53 (1986): 354–71.

Wachterhauser, B. R. "Introduction: History and Language in Understanding." In *Hermeneutics and Modern Philosophy*, ed. B. R. Wachterhauser, pp. 5–61. Albany: SUNY Press, 1986.

———. "Must We Be What We Say? Gadamer on Truth in the Human Sciences." In *Hermeneutics and Modern Philosophy*, ed. B. R. Wachterhauser, pp. 219–40. Albany: SUNY Press, 1986.

Warnke, G. *Gadamer. Hermeneutics, Tradition and Reason*. Cambridge: Polity Press, 1987.

Wartofsky, M. "The Relation between Philosophy of Science and History of Science." In M. Wartofsky, *Models*. Dordrecht: Reidel, 1979.

Watkins, J. W. N. "The Unity of Popper's Thought." In *The Philosophy of Karl Popper*. Vol. 1, ed. P. A. Schilpp, pp. 371–412. La Salle, Ill.: Open Court, 1974.

Webster, C. *The Great Instauration: Science, Medicine, and Reform, 1626–1660*. New York: Holmes and Meir, 1975.

Weinsheimer, J. *Gadamer's Hermeneutics: A Reading of Truth and Method*. New Haven: Yale University Press, 1985.

Westphal, M. "Hegel and Gadamer." In *Hermeneutics and Modern Philosophy*, ed. B. R. Wachterhauser, pp. 65–86. Albany: SUNY Press, 1986.

Wiehl, R. "Heidegger, Hermeneutics, and Ontology." In *Hermeneutics and Modern Philosophy*, ed. B. R. Wachterhauser, pp. 460–84. Albany: SUNY Press, 1986.

Will, F. L. "Reason, Social Practice, and Scientific Realism." *Philosophy of Science* 48 (1981): 1–18.

Winch, P. *The Idea of a Social Science and Its Relation to Philosophy*. 11th ed. London: Routledge and Kegan Paul, 1980 [1958].

Wise, M. N. "Mediations: Enlightenment Balancing Acts, or the Technologies of Rationalism." In *World Changes: Thomas Kuhn and the Nature of Science*, ed. P. Horwich, pp. 207–56. Cambridge, Mass.: MIT Press, 1993.

Wittgenstein, L. *Philosophical Investigations*. Trans. G. E. M. Anscombe. Oxford: Basil Blackwell, 1974 [1953].

Woolgar, S. "Irony in the Social Study of Science." In *Science Observed: Perspectives on the Social Study of Science*, ed. K. D. Knorr-Cetina, M. Mulkay, pp. 239–66. London: SAGE Publ. 1983.

———. "On the Alleged Distinction between Discourse and *Praxis*." *Social Studies of Science* 16 (1986): 309–16.

———. *Science: The Very Idea*. Chichester: Ellis Harwood Ltd. Publ.; London: Tavistock Publ., 1988.

————. Ashmore, M. "The Next Step: An Introduction to the Reflexive Project." In *Knowledge and Reflexivity: New Frontiers in the Sociology of Knowledge*, pp. 1–11. London: SAGE Publ., 1988.

Wright, K. "Gadamer: The Speculative Structure of Language." In *Hermeneutics and Modern Philosophy*, ed. B. R. Wachterhauser, pp. 193–218. Albany: SUNY Press, 1986.

Zahar, E. *Einstein's Revolution: A Study in Heuristic*. La Salle, Ill.: Open Court, 1989.

Znaniecki, F. *The Social Role of the Man of Knowledge*. New York: Octagonal Books, 1965 [1940].

Zybertowicz, A. *Przemoc i poznanie: Studium z nieklasycznej socjologii wiedzy. [Violence and cognition: A Study in Non-Classical Sociology of Science]. Toruń: UMK, 1995.*

INDEX